Sieve Methods

HEINI HALBERSTAM

and

HANS-EGON RICHERT

DOVER PUBLICATIONS, INC.
Mineola, New York

Bibliographical Note

This Dover edition, first published in 2011, is an unabridged republication of the work originally published in 1974 by Academic Press, Inc., London. Heini Halberstam has provided a new Preface and a new Errata List for this edition.

Library of Congress Cataloging-in-Publication Data

Halberstam, H. (Heini)
Sieve methods / Heini Halberstam and Hans-Egon Richert.
p. cm.
Originally published: London ; New York : Academic Press, 1974.
Includes bibliographical references and index.
ISBN-13: 978-0-486-47939-2
ISBN-10: 0-486-47939-0
1. Sieves (Mathematics) I. Richert, H.-E. (Hans-Egon), 1924– II. Title.

QA246.H35 2011
512.7'.3—dc22

2010039623

Printed in Canada
47939004 2025
www.doverpublications.com

Preface to the Dover Edition

There is never a right time to publish a systematic account of a subject that is in a state of healthy growth; even as the late Ted Richert and I were embarking on the writing of *Sieve Methods* forty years ago, inspired by the 1965 seminal of Jurkat and Richert, the young Henryk Iwaniec was discovering what has become known as the Rosser-Iwaniec method, which has since lead to many advances and applications, by himself and others. There is now in the literature an "enveloping" sieve (Hooley), a "vector" sieve (Brudern and Fourvy), a "prime detecting" sieve (Harman); and probably there will be others.

There are now also other books about sieves that the interested reader might wish to consult: there is the influential discussion of sieves by one of the founders of the subject, Atle Selberg, in Volume II of his *Collected Papers* (1991), there is *Sieves in Number Theory* (2001) by the late George Greaves, there is Glyn Harman's monograph on "Prime Detecting Sieves" (2007), and recently Harold Diamond and I have published *A Higher Dimensional Sieve Method* (2009), building on Iwaniec's approach. Several current books on Number Theory also devote sections or chapters to sieves.

Our book has by now been long out of print, but demand for it continues and second-hand copies, when they can be found, command exorbitant prices. I am very grateful therefore to Dover Publications for reprinting it. I apologize for the long list of Errata; that they were necessary in the original was entirely my fault.

Heini Halberstam
March, 2010

Preface

We conceived the idea of writing an introduction to applicable sieve theory during a brief encounter at Syracuse airport in the spring of 1966; we had both been lecturing on sieves, and it seemed to us that the time for a systematic exposition was long overdue. Our original plan was modest: to to supply, in lecture notes form, the theoretical background for the method of Jurkat–Richert [1] and to illustrate it by means of significant applications; also, to give a first connected account of the large sieve. As we burrowed in the literature and came to realize its extent and wealth—the notes we prepared on some two hundred papers in the summer of 1968 alone would have filled a monograph—the inadequacy of our original conception became increasingly clear. Not only did we find that we should have to assimilate much classical material into a revised scheme but, as we progressed, new results began to appear which affected it profoundly; so much so that, viewing the matter in retrospect, if we had called a halt at any earlier stage and published then, the book would have been much the poorer. In the end we abandoned the chapters on the large sieve altogether—happily the monographs of Montgomery [2] and Huxley [2] have appeared in the meantime and provide admirable introductions to that subject and much else beside—and concentrated on the "small" sieves of Brun and Selberg; even so, critics should have no trouble in pointing to omissions, and we accept any such criticisms in advance. Indeed, we have been at pains to point out important extensions that time and space forced us to exclude, and we hope only that what we have done here will make it easier for others to cover these too in due course.

The intermittent opportunities we have had of working together on the book have often shared the nature of that first encounter; even so, they would have been much rarer still but for the generous support of the National Science Foundation and of the U.S. Army Research Office, and for the equally generous hospitality of Syracuse University during several happy summers of uninterrupted work. We express our due gratitude to these institutions for all their help. We owe a special word of thanks also to Dr. R. C. Vaughan of Imperial College, London, for having read with great care several chapters of the manuscript and for having removed some blemishes that had escaped our notice; also for having pointed out to us a notable simplification in Chapter 11. The final version of the manuscript was typed by Mrs. Shoshana Mandel of Tel Aviv University and Mrs. Angela Fullerton of Nottingham, and we are indebted to them both for undertaking a task that must often have been extremely trying. We thank the London Mathematical Society for accepting the book into their series of research monographs. Finally, we are grateful to Academic Press for all their help and patience during the course of publication; Mr. Hitchings, of Roystan Printers Ltd., told us that our manuscript had been typographically the most difficult he had met in twenty years of printing mathematics.

The companionship of a shared endeavour has been a memorable experience for us both. Now that we have completed, with a mixture of sadness and relief, the task we set ourselves eight years ago, we hope that there will be some readers for whom our book will prove to be a helpful introduction to a beautiful topic that has been shrouded for too long in unnecessary mystery.

July, 1974 H. Halberstam

H. -E. Richert

Contents

Notation

Standard Nomenclature

$[x]$ always denotes the largest integer not exceeding x.

Where no confusion is possible with the notation for intervals, (a, b) denotes the highest common factor of integers a, b and $[a, b]$ their lowest common multiple.

We denote the Möbius function by $\mu(n)$ (for a definition see p. 34). We write $\nu(n)$ for the number of distinct prime divisors of n, $\Omega(n)$ for the number of prime divisors of n counted according to multiplicity, $\tau(n)$ for the number of positive integer divisors of n, and $\phi(n)$ for Euler's function (the order of the multiplicative group of reduced residue classes modulo n).

We denote the number of primes not exceeding x by $\pi(x)$ and the number of primes not exceeding x that are congruent to l modulo k by $\pi(x; k, l)$.

Sieve Notation

Throughout, $\mathscr{A}$ stands for the general integer sequence to be "sifted" and $\mathfrak{P}$ for a "sifting" set of primes. The special sets $\mathfrak{P}_K$, $\mathfrak{P}_{l,k}$ are defined on pages 24, 73 respectively.

The following table indicates where the various principal functions featuring in our account of sieves are first introduced:

the multiplicative functions	ω_0 on page	15
	ω	28
	g	29
the remainders	r_d	15
	R_d	28
the summatory functions	$G(z)$	29
	$G_k(z)$	98
	$G(x, z)$	29
the products	$P(z)$	25
	$P_{y,z}$	38
	$W(z)$	28
	$V(z)$	29

the transcendental functions	$\sigma_\kappa(u)$	194
	$\eta_\kappa(u)$	211
	$F(u)$	226
	$f(u)$	226
	Λ_r	253
the sifting functions	$S(\mathscr{A}; \mathfrak{P}, z)$	25
	$W(\mathscr{A}; \mathfrak{P}, v, u, \lambda)$	243
	$W_\vartheta(\mathscr{A}; \mathfrak{P}, v, u, \lambda)$	292
	$W_\vartheta{}^*(\mathscr{A}; \mathfrak{P}, z_1, z, y, \lambda)$	294

The unique real root of the equation $\eta_\kappa(u) = 1$, written ν_κ, is introduced on page 212.

All our general theorems have been formulated subject to certain basic conditions. The symbolic representations of these conditions, and the pages on which they are first described, are as follows:

(Ω_0)	on page 30
(Ω_1)	29
$(\Omega_2(\kappa))$	52
$(\Omega_2(\kappa, L))$	142
$(\Omega_2{}^*(\kappa))$	252
(Ω_3)	253
(R)	30
$(R(\kappa, \alpha))$	219
(R_0)	64
$(R_1(\kappa, \alpha))$	64

Constants

The symbols π and e have their usual meaning throughout, and γ always denotes Euler's constant.

The letters A, B, C with or without suffices or superscripts denote constants $\geqslant 1$ throughout. The A-constants, together with the constants κ and α, feature in the basic conditions and we therefore refer to them sometimes as *structure constants*. The dependence of the B- and C-constants is first described at the end of Chapter 1, Section 4 (pp. 29, 30), and there are also reminders at appropriate stages throughout the text.

The O- and o-notations of Landau, and the $\ll$-notation of I. M. Vinogradov, are now so well established as not to require definition; for the dependence of the constants implied by the use of these notations see top of p. 30.

Conventions

We often use $:=$ as the symbol for equality by definition; see e.g. p. 15.

Theorem 5.4 is the fourth theorem in Chapter 5; Corollary 5.4.2 is the second corollary of Theorem 5.4. Lemmas are numbered in the same way.

In any one chapter we attach to a formula the notation (3.6) if it is the sixth numbered formula in Section 3 of that chapter. If we wish in Chapter 7 to refer to formula (3.6) of Chapter 5 we do so by referring to it as (5.3.6).

To avoid excessive length in the statement of a general theorem (some statements are very long as it is!) we use a short-hand way of indicating which of the basic conditions the theorem rests on; for a description see footnotes on pp. 31, 57.

Abbreviations

RH denotes Riemann's conjecture concerning the location of zeros of the Riemann ζ-function; *GRH* denotes the generalized Riemann conjecture, concerning the location of zeros of L-functions.

H–W stands for Hardy and Wright, "An Introduction to the theory of numbers", 4th edition, Oxford 1960. *D* stands for Davenport's "Multiplicaative Number Theory", Markham, Chicago 1967. We refer to these texts for various classical results.

ERRATA

The following corrections should be made in the text:

p.1, *l*.6: for from some point onward read greater than 2

p.9, NOTES *l*.1: derives from Schinzel

p.10, *l*.14: for H^* read H_N^*

p.14, *l*.–11: for half a century ago read at the beginning of the 20th century

p.15, *l*.–7: after we then introduce add the remainders

p.20, *l*.–6: omit full stop

p.21, *l*.8: in $\pi(x; \ldots.)$ write l' for l

p.23, *l*.11: replace = by – and insert } before $+ \vartheta\rho(d)$

l.15: for $\rho_1{}^*$ read $\rho_1{}^*(d)$

p.26, (4.7): insert comma at end of line

p.28, *l*.(4.13) – 1: correct misprint 'numbers'

p.29, *l*.–8: comma at end of (4.18)

l.–4: for be write by

p.34, *l*.2: Staatsexamensarbeit, Marburg 1970

p.34, *l*.–2: The reference for (1.3.17) to Nagel [1] is not helpful, because there he derives this formula from Landau's deep Prime Ideal Theorem whereas Landau gave a much simpler proof elsewhere. For an account of this, see the note by Diamond and Halberstam, 'Congruences and Ideals' in 'Analytic Number Theory, Essays in Honor of Klaus Roth, Cambridge 2009, 164–169 .

p.36, *l*.3: on right of inequality read $\geq \sum_{1 \leq n < z} 1/n \geq$

p.37, *l*.(1.2) + 1: for σ(n) read $u(n)$

47, *l*.1: for v(d) read ν(d)

p.48, *l*.–3: for any positive integer read any positive odd integer

p.50, *l*.4: for (as $x \rightarrow \infty$) read (as $X \rightarrow \infty$)

p.52, *l*.(3.2)–1: for $\Omega_2(k)$ read $\Omega_2(\kappa)$

p.57, *l*.(4.3): before } on the right insert $\exp((2b+2)c_1 / (\lambda \log z))\} + \cdots$

p.61, *l*.(4.18) & (4.18) – 2: for $e\Lambda - 1$ read $e^{\Lambda} - 1$

p.62, *l*.5: for ε^{Λ} read $\varepsilon\Lambda$

p.65, *l*.6: for $(e2\lambda/\kappa - 1)$ read $(e^{2\lambda/\kappa} - 1)$
p.67, *l*.19: for *a* read *A*
p.68, *l*.(5.2): for $\prod_{p<X}(-\omega(p)/p)$ read $\prod_{p<X}(1-\omega(p)/p))$
p.79, *l*.(7.14): at end, move comma from exponent to main line
p.83, *l*.(8.2): in O-term, for $(\varepsilon\kappa//2)$ read $(\varepsilon\kappa/(2\lambda)$
l.−8: for $\log(x/\alpha)$ read $\log(\kappa/\alpha)$
p.87, *l*.−1: for $\max_{\substack{1\le l\le d\\(l,d)=1}}|....|$ read $\max_{\substack{1\le l\le d\\(l,d)=1}}|.....|$
p.90, (9.6): in last line, for $j_1 = [....]$ read $j_0 = [....]$
(9.7): the last line, *l*.−9, for p^{β}_{2ji+2}...... read $p^{\beta}_{2j_1+2}$......
p.92, *l*.6: for probed read proved
l.7: for if read of
p.95, *l*.7: for $n =$ mod Dk read $n = l$ mod Dk
p.99, *l*.2: on right, for $\prod_{l/d} 1/g(l)$ read $\sum_{l/d} 1/g(l)$
p.102, *l*.(2.1): on right, for $\sum_{p/k}$ read $\prod_{p/k}\cdots$
p.106, *l*.−6: on extreme right, for $((z-1)/2)$ read $((z-1)/2)^2$
p.111, *l*.9: on left, for $1/\phi(l)$ read $1/\phi(n)$
p.112, *l*.−5: on right, first sum should read $\sum_{\substack{d\le x^{1/2}\\(d,l)=1}} 1/\phi(d)$
p. 118, *l*.(7.3): full stop at end
p.121, *l*.−4: for $d_1 = p^{\alpha_1}$ . . . read $d_1 = p^{\alpha_1}_1$.....
p.122, *l*.−2: R.R. Hall, Halving an estimate obtained from Selberg's upper bound method, Acta Arith. *25*(1973/74), 347–351
p.124, *l*.−4: for full stop at end read comma
p.128, *l*.8: on right, insert + O(log *k*)
l.9: the right side should read $\cdots \le \sum_{\substack{h=1\\(h,k)=1}}^{k} E^2(x^{1/2};\ k,h) + O(\log k)$

l.10: the right side should read $\cdots\sum_{k\le Q}\left(\sum_{\substack{h=1\\(h,k)=1}}^{k} E^2(x^{1/2};k,h)+O(\log k)\right)$

p.131, *l*.(1.2), +3: at the lower limit of integration, for 2 read 1
p.143, *l*.(1.1): on right, for log $|z/w|$ read log (z/w)
p.149, *l*.2: for (2.2.8) read (2.3.8)
p.151, *l*.4: at the lower limit of integration on right, for 0 read 1
p.154, *l*.−3: for c.f read cf.
p.179, *l*.−8: Omit full stop after 5.8.2
p.183, *l*.−3: On right, read $8\prod_{2<p\nmid k}(1-....)$

p.189, *l.*−5: this should read

$$\frac{\omega(D)}{D} = \frac{\omega(d_1)\,\omega(d_2)}{d_1 d_2} \bullet \frac{(d_1, d_2)}{\omega((d_1, d_2))}$$

pp.198–201: go to p.3

The proof of Lemma 6.1 (pp. 198–201) requires correction although the result as stated is correct. In the argument leading up to (4.14), and again in the last steps on p. 201, c_4 cannot be absorbed into the O-constant because it depends implicitly on L. However, this difficulty can be avoided. First of all, the error term in (4.5) is $O(L/W(z))$ by (4.1.1) and this leads to the generally better error term

$$O\left(\frac{L}{W(z)\log^{\kappa+1} w}\right) \tag{*}$$

in both (4.7) and (4.8). Now modify the definition of R(ξ, z) by writing (4.9) in the new form

$$T(\xi, z) = \frac{1}{2}\bar{\sigma}_\kappa(2\tau)\frac{\log z}{W(z)} + R(\xi, z), \qquad \tau = \frac{\log \xi}{\log z}. \tag{4.9$'$}$$

Then (4.11) is still true, but with error term (*) and subject to $z \leq w \leq \xi$. From this one deduces (4.12) in the new form

$$\frac{|R(\xi, z)|}{\log^{\kappa+1}\xi} \leq \frac{B_{12}L}{W(z)\log^{\kappa+1} z}(v-1)^{\kappa+1}, \quad v-1 < \tau \leq v. \tag{4.12$'$}$$

The equation on *l.* − 5 is correct subject to (5.3.3) (i.e. if $t \geq \exp(B_8 L)$); from this it follows, in the light of (4.9′), by (4.6), (4.1.1) and the properties of $\bar{\sigma}_\kappa$, that (4.14) holds in the weaker but adequate form

$$|R(t, z)| \leq \frac{B_{10}L}{W(z)} \text{ if } 1 < t \leq z. \tag{4.14$'$}$$

From here on the argument follows, subject to the obvious changes, as on p. 200 and leads to

$$T(\xi, z) = \frac{1}{2}\bar{\sigma}_\kappa(2\tau)\frac{\log z}{W(z)} + O\left(\frac{L}{W(z)}\tau^{2\kappa+2}\right) \text{ if } \xi > z \tag{4.16$'$}$$

in place of (4.16); the proof of the lemma via the revised (4.5) is then straightforward.

p.203, Note on 6.5, replace last full stop by ...., Onishi[1].

p.203, Note on 6.5, replace last full stop by ...., Onishi[1].
p.215, *l*.3: lower limit of integration is $\log \xi^2/\log z_2$
p.219, *l*.2: should read $0 < \alpha \le 1$
(6.2): for $\le$ read $\ge$
p.220, *l*.10: in parentheses, for ; read : $A = \{p+2 : p \le x\}$
p.222, *l*.5: Write for earlier lower sieve estimates
p.224, *l*.5: should read

$$+(-1)^b \sum_{\substack{z_1 \le p_b < \cdots p_1 < z \\ p_j < \xi_j,\, p_j \in \mathfrak{P} (j=1,\ldots b)}} S(A_{qp_1 \cdots p_b}; \mathfrak{P}, p_b)$$

p.225, *l*.1: omit second 'that'
p.228, *l*.5: should read $0 \le f(u_2) - f(u_1) \le \ldots\ldots$
p.230, *l*.−1: for lemma read theorem
p.232, *l*.8: for $\left(\frac{\log \xi_b{}^2}{\log p_1}\right)$ read $\left(\frac{\log \xi_b{}^2}{\log p_b}\right)$
p.236, *l*.4: write the right side as $W(z)e^{-\log \xi^2/\log z}\varepsilon^b$
p.253, *l*.(3.2): read $\Lambda_r = r + 1 + \log \frac{\{4(1+3^{-r})\}}{\log 3}$
l.(3.2) + 1: read $\Lambda_1 = 1$ for $\Lambda_1 = r$
p.257, *l*.(4.3) – 1: (cf.(1.3.8)) we now choose
p.267, *l*.11: and $[(r+1)/2]$-free for r > 5
p.269, *l*.20: if better results were even at this moment
p.283, *l*.−3: at end a comma not a full stop
p.287, *l*.(3.9): no full stop at end
p.308, *l*.(4.25): or l.–6: read $-C_{45}(1/\lambda + 1)\,(\log\log 3X)^2/\log X$
p.290, *l*.1: by $g \log g + (2 + \log 4.88\ldots)g$
p.317, *ll*.−2,−1: ....$\kappa < 1$; our remark here is misguided in the light of Iwaniec [2], who applies . . .
p.323, *l*.12: omit < 0.0165
l.14: on right side read $\ge 8(3 \log 3 - 5 \log 2 + 1/2) > 2.64081$
p.326, *l*.12: at end write 0.544 for 0.538
F/N: at end read ...estimate $0.490095\,(1+\varepsilon)$, $|\varepsilon| < 10^{-5}$
p.327, *l*.4: the second summation condition is $(N - rn, P(z)) = 1$
p.328, *l*.(3.3): Omit printing blemish
the summand is $\frac{\lambda_{d_1}\lambda_{d_2}}{\phi(D)}$
p.345, *l*. Chen;3: for larger read large
in reference add MR55, 7959

"Far and few, far and few,
Are the lands where the Jumblies live;
Their heads are green and their hands are blue,
And they went to sea in a Sieve."

E. Lear, *The Jumblies*

Introduction

1. Hypotheses H and H_N

Two of the oldest problems in the theory of numbers, and indeed in the whole of mathematics, are the so-called prime twins and binary Goldbach conjectures. The first of these asserts that

"*There exist infinitely many primes p such that* $p + 2$ *is a prime*",

and the second that

"*every even natural number N from some point onward can be expressed as the sum of two primes*".

To G. H. Hardy, addressing the Mathematical Society of Copenhagen in 1921, the binary Goldbach conjecture appeared "..... probably as difficult as any of the unsolved problems of mathematics"; yet at the time Hardy pronounced this judgement, Brun's now famous researches on the Eratosthenian sieve were already well under way, and in the previous year† Brun himself had established the following remarkable approximations to the two problems: there exist infinitely many integers n such that both n and $n + 2$ have at most *nine* prime factors; also, every sufficiently large even N is the sum of two numbers each having at most *nine* prime factors (cf. Chapter 2).

There are many who believe that Hardy's view is as valid now as it was then. Nevertheless, many dramatic advances have taken place in the intervening years, and in 1966 we set out to chart this progress from Brun's pioneering work to the present day. As we progressed, we found, however, that we were driven increasingly to view both questions as parts of a much grander design.

Mere difficulty has never been a bar to further speculation in mathematics; and over the years the two notorious conjectures had given rise to a host of related questions, each a generalization, in one sense or another, of the prime twins problem or of the Goldbach problem. We owe to Schinzel the classification of these and many other questions from classical prime number theory in terms of two grandly conceived conjectures, the hypotheses H and H_N below, which we shall now formulate.

† Actually, these results were announced already in 1919 (see Brun [4]).

Let $F_1, \ldots, F_g$ be distinct irreducible polynomials in $\mathbb{Z}[x]$ (with positive leading coefficients) and suppose that the product polynomial $F = F_1 \ldots F_g$ has no fixed prime divisor. Then

Hypothesis H: There exist infinitely many integers n such that each $F_i(n)$ $(i = 1, \ldots, g)$ is prime.

Furthermore, let N be a natural number and G another polynomial of $\mathbb{Z}[x]$ (with positive leading coefficient) such that $N - G$ is irreducible, and also such that $F(N - G)$ has no fixed prime divisor. Then

Hypothesis H_N: If N is large enough, there exists a positive integer n such that $N - G(n) > 0$ and such that each of $F_i(n)$ $(i = 1, \ldots, g)$ and $N - G(n)$ is prime.

It was against the background of these two comprehensive conjectures that the principal results of our book were finally conceived.

When $g = 1$ and $F_1(n) = an + b$ with $(a, b) = 1$, then, according to Hypothesis H, the arithmetic progression $an + b$ $(n = 0, 1, 2 \ldots)$ contains infinitely many primes. By the famous theorem of Dirichlet (1837) this is indeed true, and so provides the only instance of Hypothesis H that has been settled up to the present time. When $g = 2$, $F_1(n) = n$ and $F_2(n) = n+2$, then Hypothesis H coincides with the prime twins conjecture. Similarly, taking $g = 1$, $F_1(n) = G(n) = n$, Hypothesis H_N coincides with Goldbach's conjecture; and Goldbach's conjecture is evidently the simplest special case of H_N. There is no end to the fascinating illustrations of H or H_N that one might list; but we shall content ourselves for the moment with mentioning only two more: taking $g = 1$ and $F_1(n) = n^2 + 1$ in H, the conjecture claims that $n^2 + 1$ represents primes infinitely often; and the case of g linear polynomials in H (with suitable conditions on the coefficients) corresponds to the well-known prime g-tuplets problem.

In the absence of even a single case of H or H_N in which Dirichlet's achievement has been matched, a natural way to approach these seemingly intractable problems is to show that there exist positive integers n (in the case of H, infinitely many such n's) for which the polynomials concerned have very few prime factors. Let us make this more precise in the context of H (for technical algebraic reasons we shall pay less attention in this book to H_N, apart chiefly from approximations to the Goldbach case itself). We denote by P_r any integer having at most r prime factors, equal or distinct, and refer to such a number as an *almost-prime* (thus Brun's results assert that $P_9 + 2 = P_9'$ for infinitely many almost-primes P_9, and, if N is even and large enough, N is representable in the form $N = P_9 + P_9'$). Then Hypothesis H is equivalent, under the given conditions, to the statement that

$$F_1(n) \ldots F_g(n) = P_g \text{ for infinitely many positive integers } n; \tag{1}$$

and a reasonable way of approximating to H is therefore to show that there exists a relatively small integer r, depending at most on g and on the degree of F, such that

$$F_1(n) \ldots F_g(n) = P_r \textit{ for infinitely many positive integers } n. \tag{2}$$

One of our main objectives will be to establish, by means of sieve theory, such approximations to Hypothesis H. Thus, for example, we shall prove in Chapter 9 that

$$n^2 + 1 = P_3 \textit{ for infinitely many integers } n; \tag{3}$$

and, in Chapter 10, we shall show, as a special case of a quite general theorem, that when g is large (but fixed) and the polynomials $F_1, \ldots, F_g$ are linear, then (2) holds, subject only to necessary conditions on the coefficients of $F_1, \ldots, F_g$, with an r equal to about $g \log g$. Also in Chapter 10 we shall give Selberg's classical approximation (first announced in 1950) to the prime twins conjecture, namely that

$$n(n+2) = P_5 \textit{ for infinitely many integers } n; \tag{4}$$

and the same argument yields the corresponding approximation to the Goldbach problem, namely, that if N is even and sufficiently large, then

$$n(N-n) = P_5 \textit{ for some positive integer } n < N. \tag{5}$$

In each of the results (4) and (5) the two factors on the left may both be composite; but already in 1948 Rényi had shown, using appreciably deeper methods than Selberg's, that n could be taken as prime in each of (4) and (5) at the cost of having on the right, in place of P_5, a P_r with r some large but fixed integer (cf. Chapter 2). Thus Rényi was the first to prove that there exists an r such that

$$p + 2 = P_r \textit{ for infinitely many primes } p, \tag{6}$$

and that every sufficiently large even integer N is representable in the form

$$N = p + P_r. \tag{7}$$

Let us view the result (6) against the background of H. If there we take $g + 1$ in place of g and take $F_{g+1}(n) = n$, then, according to H and provided that $nF_1(n)\ldots F_g(n)$ has no fixed prime divisor,† there exist infinitely many primes p such that each $F_i(p)$ $(i = 1, \ldots, g)$ is simultaneously a prime; or, in our alternative formulation (cf. (1)),

$$F_1(p)\ldots F_g(p) = P_g \textit{ for infinitely many primes } p. \tag{8}$$

We shall find that we are able to approximate to (8) by results of the type

$$F_1(p)\ldots F_g(p) = P_r \textit{ for infinitely many primes } p, \tag{9}$$

where r is some relatively small positive integer depending only on g and the degree of F. Since (6) corresponds to $g = 1, F_1(n) = n + 2$, we shall therefore establish a result of type (6); indeed, we shall derive from our general theory (see Chapter 9) that (6) is true, in a rather strong quantitative form, with $r = 3$, and we shall do the same for (7). Moreover, we shall prove in Chapter 11 Chen's recent theorem that (7) is true even with $r = 2$; it will be clear there from our account that the same improvement holds also in (6). Without a doubt results (6) and (7) with $r = 2$ are the most dramatic achievements of the modern sieve method—an optimist might well say that we are now within a stone's throw of both, the Goldbach and the prime twins conjectures; we have certainly come very close.

Returning for a moment to (8) and (9), it is clear that while results of type (8) are just special cases of H, approximations such as (9) represent, compared with (2), sometimes different and sometimes better approaches to instances of H. To cite but two illustrations of the many results we shall prove in Chapters 9 and 10, it will be shown there that

$$p^2 + p + 1 = P_5 \textit{ for infinitely many primes } p, \tag{10}$$

and that

$$(p + 2)(p + 6)(p + 8) = P_{14} \textit{ for infinitely many primes } p. \tag{11}$$

The quality of the theorems in Chapters 9, 10 and 11, of which we have listed

† Let $\rho(p)$, $\rho'(p)$ denote the numbers of solutions respectively of

$$F(n) \equiv 0 \bmod p \quad \text{and} \quad nF(n) \equiv 0 \bmod p,\ F = F_1\ldots F_g.$$

The condition that F has no fixed prime divisor is expressible as

$$\rho(p) < p \quad \textit{for all primes } p.$$

It is easily seen (cf. Section 5.7) that

$$\rho'(p) = \begin{cases} \rho(p) + 1, & p \nmid F(0) \\ \rho(p), & p \mid F(0); \end{cases}$$

so that the requirement of $nF(n)$ having no fixed prime divisor is equivalent to

$$\rho'(p) < p \quad \textit{for all primes } p,$$

and this means, in terms of ρ, that

$$\rho(p) < p \textit{ for all primes } p \textit{ and } \rho(p) < p - 1 \textit{ for all primes } p \nmid F(0).$$

several samples above, has convinced us that the time is ripe for a systematic account of attacks on hypothesis H and hypothesis H_N in terms of "approximate" results of type (2) and (9), within the framework of a single theory. In the event, we have concentrated in this book chiefly on approximations to H and to Goldbach's conjecture; in principle our theory applies to approximations to H_N also, but would need to be augmented by a considerable amount of algebraic machinery for which we had, in the end, no room.

2. Sieve Methods

Any arithmetical sieve (in our sense of the word) is based on the following idea: given a finite integer sequence $\mathscr{A}$, a sequence $\mathfrak{P}$ of primes and a number z ($\geqslant 2$), if we remove (sift out) from $\mathscr{A}$ all those elements that are divisible by primes of $\mathfrak{P}$ less than z, then the residual (unsifted) members of $\mathscr{A}$ can have prime divisors from $\mathfrak{P}$ only if these are large (in the sense of being $\geqslant z$); *a fortiori*, each unsifted element of $\mathscr{A}$ has only few such prime divisors (provided that z is not too small compared with $\max_{a \in \mathscr{A}} |a|$). The object of any sieve theory is to estimate the number $S(\mathscr{A}; \mathfrak{P}, z)$ of unsifted elements of $\mathscr{A}$, or of some weighted form of this number.

There are many arithmetical investigations in which it suffices to have a good upper bound for this number—the famous Brun–Titchmarsh inequality is a classic example; for an account of this see Chapter 3—and therefore we establish in Chapters 2–5 several general upper bounds of $S(\mathscr{A}; \mathfrak{P}, z)$ under various weak sets of conditions; we also work out numerous applications.

Other investigations require an asymptotic formula for $S(\mathscr{A}; \mathfrak{P}, z)$, but in general this is an extremely difficult if not hopeless problem; nevertheless, we do obtain such formulae in Chapters 2 and 7, again subject only to rather weak conditions, provided that, essentially

$$\frac{\log|\mathscr{A}|}{\log z} \to \infty$$

(i.e. provided that z is not too large—we shall make everything precise in the text). Such asymptotic formulae are known in the literature as fundamental lemmas.

Finally, problems of type (2) and (9) require (positive) lower bounds for $S(\mathscr{A}; \mathfrak{P}, z)$ for appropriate sequences $\mathscr{A}$ and with z large. We shall show how to obtain such lower bounds in Chapters 7 and 8; and we shall show in the last three chapters how, in a wide variety of applications, the effectiveness of these lower bounds can be enhanced quite dramatically by means of combinatorial weighting procedures. Although, in the end, our results fall well short of (1) (and (8)), sieve methods of one kind or another do seem, at the time of writing, to offer the only approach to these profoundly difficult problems.

The first to devise an effective sieve method that goes substantially beyond the sieve of Eratosthenes was Viggo Brun, and we hope that our exposition of Brun's sieve (in Chapter 2) demonstrates how powerful and fruitful a method it is even today. Brun's sieve is heavily combinatorial in character (most sieves share this property, at any rate to some degree); its complicated structure and Brun's own early accounts tended to discourage closer study and it is fair to say that only a small number of experts ever mastered its technique or realized its full potential. Hardy and Littlewood used Brun's sieve only once, to derive an early version of the Brun–Titchmarsh inequality (see Chapter 3); and Hardy expressed the opinion that Brun's sieve did not seem "... sufficiently powerful or sufficiently profound to lead to a solution [of Goldbach's Problem]". Landau left Brun's manuscript untouched in his drawer for six years, until he saw Schnirelmann's skilful use of it in his proof that the primes form a basis for the natural numbers > 2. When he wrote up an account of both, Brun's upper bound sieve and Schnirelmann's theorems, he confessed: "Ich publiziere diese Darstellung; denn selbst ich hatte mich bisher nie der Mühe unterzogen, Herrn Brun's Originalarbeit genau durchzuarbeiten". It is probably true to say of number theory as of other mathematical disciplines that it was slow to accept the power of combinatorial methods; it is certain that Brun's ideas did not receive the attention they deserved. By a strange and, under the circumstances, perverse chance when, in the late forties, Berkeley Rosser discovered an important refinement of the Brun sieve his discovery remained unpublished for more than twenty years,† and even now there is no account of his theory at the level of generality of which it is capable. We give a brief indication of Rosser's method at the end of Chapter 2; but we have concentrated instead, from Chapter 3 onward, on constructing as comprehensive an account of what we shall call Selberg theory as the present state of knowledge permits. Nevertheless, we believe that the Brun approach is not exhausted (even when bearing in mind Rosser's sieve), and for that reason we have dwelt at some length in Chapter 2 on an exposition of his combinatorial ideas, without embroidering these too much with technical subtleties, in the hope that some readers will examine these ideas afresh.

Selberg published an account of his upper bound sieve in a short paper in 1947; and subsequently he indicated its place in the construction of lower bound sieves in several important expository articles, without actually ever publishing any details. The first attempts to supply these were made in the fifties, in Russia (notably by A. I. Vinogradov, Levin and Barban) but it is

† cf. Selberg [5] and Iwaniec [1, 2].

probably fair to say that Ankeny and Onishi in 1964, Jurkat and Richert in 1965, and Buchstab in the same year, were the first to exploit fully the power inherent in Selberg's method; moreover, each of these seminal papers introduced (different) combinatorial embellishments which have vastly improved the power of the Selberg sieve,† and probably contain the seeds of further progress. These combinatorial ideas are of two kinds: there is the use and iteration of combinatorial identities, a method that was pioneered in this context by Buchstab (already in association with Brun's sieve) and is also implicit in Rosser's work as well as in Selberg's survey lectures; and there is the introduction of weighted sieves for which the pioneering credit must go to Kuhn. The design of weighted sieves is still at the stage of trial and error; a systematic theory should be the objective of future research.

The earliest sieves were of a purely combinatorial character, and this gave them, as we have already said, a superficial appearance; while the intrinsic dependence of modern theories on substantial analytic apparatus from the field of differential-difference equations might cause some reappraisal of this impression, by far the more important factor in assessing their true arithmetical potential has been the fact that deep information from prime number theory has been incorporated successfully into the structure of sieves, with a notable improvement in the quality of results (for example, all the best results of type (8) derive from an interaction of this kind). Rényi was the first to demonstrate this (in his proof of (6) and (7)) and Selberg drew attention to the possibility of such interaction in several of his early lectures. We shall be more specific about the nature of the connection in the text (and in the Notes accompanying the text); sufficient to say here that the important Bombieri–Vinogradov theorem on the average distribution of primes in arithmetic progressions plays, as we shall show, a critical part in many of the finest sieve results. Indeed, we confidently expect that significant new information about the distribution of primes in arithmetic progressions will lead to substantial improvements in the quality of sieve results; but, having said this, it is important to add now a word of caution.

Sieves tend to suffer from the defects of their virtues. When properly formulated, they constitute a general method with quite an extraordinary range of applicability; yet it is just this broad range of relevance that militates against any particular sieve being the optimal method of attack in any one specific problem. We shall explain this more clearly in the Notes for Chapter 8. Selberg himself has investigated the natural limitations inherent in sieves, both in his early lectures and in a recent important memoir; but we shall not occupy ourselves with this question. Our point of view, we

† See Buchstab [10], Halberstam, Jurkat and Richert [1], Richert [1], Halberstam and Richert [2], and Chen [3].

emphasise, is highly practical: namely, to formulate the most effective sieve mechanisms known to us and to illustrate these in terms of numerous applications; and, in particular, to place on record the sharpest form of Selberg sieve theory available at the present time, with a full account of the contribution of this theory to the classical problem area covered by Hypothesis H and the most famous instance of H_N.

It is far too early to attempt a definitive judgement on the role of sieves in the theory of numbers. Anyone making an interim assessment should remember, however, that a sieve idea played a fundamental part in Vinogradov's three primes theorem, and that any elementary proof of the prime number theorem rests on what Selberg has called an instance of a 'local' sieve;† and we hope too that he might feel further encouraged as to the future of sieves by the principal results of this book.

3. Scope and Presentation

We conclude with some remarks on the scope of the book and also on some basic rules of procedure that we have followed in writing it.

First of all, we have restricted attention to sifting by primes only; we have not dealt with sieves consisting of several residue classes modulo each prime (cf. Halberstam–Roth [1], Chapter 4) and, in particular, we have not given an exposition of the large sieve or of Montgomery's sieve. There is an excellent account of the latter two, and of many attendent developments, in Montgomery [2] (see also Huxley [2]) and it is the subject of a lively on-going literature (see Notes for Chapter 3). However, we have made full use of the large sieve in Chapter 11.

Neither have we dealt with sieves consisting of numbers other than primes (see Notes for Chapter 1), nor with sieves in algebraic number fields (see Bibliography); it should be possible, at least in principle, to model methods appropriate to these situations on the theories described by us. Still less have we attempted to formulate sieves in pure combinatory analysis; for a first step in this direction the reader is referred to Wilson [1].

However, within the limits set by us, we have formulated our sieves in as general a manner as possible and with the utmost care so far as basic assumptions and dependences are concerned. We have worked out many applications, mostly within the orbit of approximations to Hypothesis H, and for this purpose we have analysed in Section 3 of Chapter 1 several basic arithmetical situations. In each application of a basic theorem we have gone to great lengths to verify the conditions of the theorem; from time to

† We learn from correspondence with Professor Bombieri that he has some new results which appear to be far reaching generalisations of this idea and have bearing on the subject matter of this book. These are to appear in *Acta Arithmetica*.

time this procedure might irritate an expert reader, but we hope the newcomer will find it helpful, and we have to add that the literature contains several erroneous papers where authors failed in just this respect. There are many applications which, to our regret, we have not been able to deal with. The reason in most cases has been, not that our theories did not apply, but that in order to apply them we should have had to devote too much space to developing the mathematical background to the problem and to verifying the basic conditions; as examples we might mention approximations to H_N, also the very interesting application of lower bound sieves to latin squares (see Chapter 8, Notes) and to the extension of the results of Chapters 9 and 10 to forms in several variables (see Notes attached to Chapters 9, 10).

In our exposition we have followed the practice of leaving the text untrammelled with attributions and historical comments; and the reader should look for these at the ends of the Chapters, in the Notes (one set of Notes for each chapter). The Notes also contain technical comments, a certain amount of discussion and brief mention of refinements, related results, principal applications and of work in progress. The Bibliography is partially classified and the fullest ever to have been published in this field; nevertheless, it has not been our intention to list every paper in which an application of the sieve has ever appeared.

Our account is self-contained with respect to sieve results, but we have felt free to refer to existing literature for basic results from elementary and analytic number theory. We have quoted also several results about the solutions of certain differential-difference equations from the work of de Bruijn and Ankeny–Onishi. Finally, although we have managed to avoid numerical integration in all our theoretical results, we have not hesitated to employ it in the context of illustrations and applications. These calculations and various other numerical computations mentioned in the text have been checked and rechecked; but our readers will readily understand that machines are notoriously prone to human error!

NOTES

1. Hypothesis H as stated here derives Schinzel–Sierpiński [1]. Case $g = 1$ of H is equivalent to an older conjecture of Bouniakowski (*Mem. Acad. Sci. St. Petersburg* 6 (1857), 305–329), and the case of g linear polynomials corresponds to a conjecture of Dickson [1].

Bateman–Horn [1, 2] have given H in the following quantitative form:

(H^*): If $Q(F_1, \ldots, F_g; N)$ denotes the number of integers n, $1 \leqslant n \leqslant N$, such that each of $F_i(n)$ is a positive prime, then

$$Q(F_1, \ldots, F_g; N) = \frac{C(F_1, \ldots, F_g)}{h_1 \ldots h_g} \int_2^N \frac{du}{\log^g u} \{1 + o(1)\}, \quad N \to \infty,$$

where $h_i = \deg F_i$ and, if $\rho(p)$ is the number of solutions of

$$F_1(n) \dots F_g(n) \equiv 0 \bmod p,$$

then

$$C(F_1, \dots, F_g) = \prod_p \left\{ \left(1 - \frac{1}{p}\right)^{-g} \left(1 - \frac{\rho(p)}{p}\right) \right\}.$$

They show that this product converges, and give heuristic reasons in support of H^*.

The famous conjectures B, D, E, F, K, P and Theorem X1 of Hardy–Littlewood [2], are all special cases of H^* (see Schinzel [2]).

Theorems 9.2 (inequality (2.1)), 9.7 and 9.8, as well as Theorems 10.4, 10.6, 10.11, 10.12 and 11.1 (together with all the various special cases derived from them) are all in quantitative form and therefore represent approximations to H^* (as well as H).

Schinzel [2] was led by comparison of H^* with the conjectures C, G, L of Hardy–Littlewood [2] to formulate H^* in the following quantitative form:

($H_N{}^*$): Denote by $X = X(N)$ the number of positive integers n such that $N - G(n) > 0$, and by $Q_1(F_1, \dots, F_g; G; N)$ the numbers of these n such that each of $F_1(n), \dots, F_g(n), N - G(n)$ is prime. Then

$$Q_1(F_1, \dots, F_g; G; N) = \frac{X}{\log^{g+1} X} (h_o h_1 \dots h_g)^{-1}$$

$$\times \prod_p \left(1 - \frac{\omega(p)}{p}\right) \left(1 - \frac{1}{p}\right)^{-g-1} \{1 + o(1)\}, \quad N \to \infty,$$

where $h_o = \deg G$ and $\omega(p)$ denotes the number of solutions of

$$F_1(n) \dots F_g(n) (N - G(n)) \equiv 0 \bmod p.$$

He remarked that conjecture A of Hardy–Littlewood 2 is included in C, and that conjectures H, I are included in G; he showed that C is included in $H_N{}^*$ and stated the same for conjectures G and L.

We repeat that H^* is a far-reaching generalization of the prime twin conjecture and $H_N{}^*$ of Goldbach's binary conjecture. $H_N{}^*$ lies deeper than H^* because the discriminant of $N - G$ depends in an important way on N, and this dependence makes for additional difficulties. In fact, the only contribution we describe towards $H_N{}^*$ is an approximation to Goldbach itself (Theorem 9.2, (2.2), including the more general (9.6.11); and Theorem 11.1) and the upper estimate Cor. 4.2.3. See Russell [1] for a good account of the algebraic apparatus associated with H_N.

Davenport and Schinzel [1] studied the constant $C(F_1, \dots, F_g)$ and computed it in two special cases associated with the Hardy–Littlewood conjecture K.

Much numerical evidence has been compiled in support of various special cases of H^*: see Hardy and Littlewood [2], D. H. Lehmer (reviewed in "Mathematical Tables and other aids to Computation", **13** (1959), 56–57), A. E. Western (*Proc. Cambridge Phil. Soc.* **21** (1922), 108–109), Shanks (*Math. Comp.* **14** (1960), 201–203; 321–332, *ibid* **15** (1961), 186–189, *ibid* **17** (1963), 188–193), Bateman–Stemmler [1], Bateman–Horn [1, 2], B. H. Mayoh (*Nordisk. Tidskr. Inf. Beh.* **8** (1968), 128–133), L. J. Lander–T. R. Parkin (*Math. Comp.* **21** (1967), 483–488), Stein–Stein [1].

Bateman and Horn [1] remark that H^* could just as well be formulated for integer-valued polynomials, with little modification; the same remark applies to all our polynomial results.

3. Other sieves: We might mention here also the "local" sieve of Selberg [3], Chang's sieve (see Chang [1] and Prachar [6], p. 157); the "trivial" sieve of Vinogradov–Linnik [1]; and the "larger" sieve of Gallagher [3].

Chapter 1

The Sieve of Eratosthenes: Formulation of the General Sieve Problem

1. Introductory Remarks

Let z be an integer $\geqslant 2$, and consider the integers n in the interval $I_z = [z, z^2)$. If we strike out from I_z first all the multiples of 2, then all the multiples of $3, 5, \ldots$ and so on for the multiples of all primes less than z, any integer remaining in I_z, or, we might say, any integer that has survived the "sifting" process, is itself a prime; for if n were a surviving (unsifted) integer having at least two prime factors, then n would be at least as large as z^2, and therefore n would lie outside I_z.

This sieve process (usually attributed to Eratosthenes, 3rd century B.C.) was devised in order to construct a table of prime numbers: given the first m ($\geqslant 2$) primes $p_1, \ldots, p_m$, the sieve of Eratosthenes enables us to extend the list of primes up to $(p_m + 2)^2$. The idea is quite simple, yet the process is, especially in conjunction with a computer, still a highly effective method for this purpose.

The process may be adapted in various ways to yield also other numerical information. Suppose for example, that we write down the numbers n of the interval $[z, z^2 - 2)$ in a column, and next to them, in a second column, the corresponding numbers $n + 2$. If the multiples of all the primes $<z$ are struck out from each column, we may find that some pairs $n, n + 2$ have survived the double sieve—that is, we may be left with some number pairs n, $n + 2$ in which each of $n, n + 2$ is a prime. Such pairs are called 'prime twins'; numerical evidence (or heuristic argument) suggests—but no one has proved yet—that there exist infinitely many prime twins. In the same way one may raise also, for example, the question of the existence of 'prime triplets', i.e. triads of primes of the form $n, n + 2, n + 6$; once again a sieve procedure— this time a triple sieve—affords a systematic means of constructing list of such triads. Etc.

We may describe the last two illustrations a little more simply, by avoiding talk of a double or triple sieve. Instead we form the sequence $\mathscr{A} = \{n(n + 2): z \leqslant n < z^2 - 2\}$ and sift $\mathscr{A}$ by the set $\mathfrak{P}$ of primes $< z$; or we form the sequence $\mathscr{A}' = \{n(n+2)(n+6): z \leqslant n < z^2 - 6\}$ and sift $\mathscr{A}'$ by the primes of $\mathfrak{P}$. In the case of $\mathscr{A}$, $n(n+2)$ survives the sifting process if, and only if, both n and $n + 2$ are primes; for if $n(n+2)$ has only prime factors $\geqslant z$, the same must be true of each of n and $n + 2$ and therefore, since $n + 2 < z^2$, neither n nor $n + 2$ can have more than one prime factor. In the same way, the only surviving members of $\mathscr{A}'$ yield prime triplets.

However extensive a table of primes, prime twins or prime triplets may be, we cannot infer from such a table that there exist infinitely many terms in any one of these three sequences, however plausible each proposition might seem from the frequent occurrence of primes, twins or triplets in the table itself. The foregoing remarks indicate that, while the sieve of Eratosthenes is a useful method from a numerical point of view, it does not automatically yield also theoretical information.

It is clear how we might hope to gain such theoretical information from the sieve: if we sift the integer sequence $\mathscr{A}$ (for example, think of $\mathscr{A}$ as one of the sequences mentioned above) by the set of primes, $\mathfrak{P}$, we might be able to devise a means of estimating the number of elements of $\mathscr{A}$ surviving the sifting procedure; thus an *upper* bound for this number would show that there cannot be too many primes (or prime twins, or prime triplets)—information that is often useful in arithmetical investigations—or, better still, a *positive lower bound* (as $z \to \infty$) would show that there exist infinitely many primes (a fact easily proved in other ways), or infinitely many prime twins (one of the notorious unsolved problems of prime number theory).

What we require, therefore, is a *theoretical* analogue of the Eratosthenes sieve. Evidently certain important questions in prime number theory can be formulated in terms of sifting a certain type of sequence $\mathscr{A}$ by a set $\mathfrak{P}$ of primes—the reader should have no difficulty now in expressing Goldbach's problem, or the problem of the infinitude of primes of the form $n^2 + 1$, as a sieve problem—and their answer depends on gaining significant information about the number of elements of $\mathscr{A}$ surviving the sifting procedure. Indeed, there are many such questions; moreover, for many of these a sieve approach appears to offer the most promising, and sometimes the only, mode of attack. The construction of a theoretical sieve is therefore of great interest, and such interest is heightened by the fact that the construction is, beyond a certain stage of effectiveness, exceedingly difficult. The modern sieve is, as yet, an imperfect method, and one may doubt whether it alone will ever succeed in settling difficult questions such as the Goldbach and prime twins conjectures. The reader will notice in later chapters that even now deep results from prime number theory can be harnessed effectively to existing sieve methods; on the other hand, the very generality of a sophisticated sieve theorem (cf. Chapters 8 and 9) tends to militate against its success in a specific problem.

Our aim in Chapter 1 is to formulate carefully the general sieve problem.

The wide diversity of arithmetical questions that can be interpreted in sieve language makes this a more difficult and delicate task than most writers on the sieve have acknowledged. Our method will be to build up the general picture by first examining in detail the particular kinds of sequences $\mathscr{A}$ that we shall want to sift (see the important examples in Section 3 to which we shall return again and again in later chapters when giving applications); and then (in Section 4) to introduce notation for the sifting sets $\mathfrak{P}$. The latter is necessary because even in respect of sifting sets we have to be prepared for considerable variety, as well as to exercise care in specifying the right $\mathfrak{P}$. Not only are there technical reasons why, sometimes, we have to exclude from $\mathfrak{P}$ the prime divisors of some number K characteristic of the problem; but the problem itself may require, if it is to be attacked by a sieve method, a rather different sifting set (different that is, compared with earlier examples), namely one consisting of a "thin" set of primes. For example, squarefree numbers expressible as sums of two integer squares are precisely those positive integers having all their odd prime factors $\equiv 1 \bmod 4$, that is, they would arise as a result of sifting the sequence of *all* positive squarefree integers† by the set $\mathfrak{P}$ of primes $\equiv 3 \bmod 4$. Actually, sieve problems involving thin sifting sets are beset by special difficulties.

In Section 5 we give Legendre's theoretical model of the Eratosthenes sieve (Theorem 1.1). Legendre used it in his famous heuristic investigation of the number $\pi(x)$ of primes not exceeding x, and by applying it to this very problem we shall see just how weak a theoretical tool the method of Eratosthenes–Legendre is. The first effective sieve method was devised by Viggo Brun only half a century ago, and a modern version of Brun's sieve is the subject of Chapter 2.

2. The Sequences $\mathscr{A}$

The sieve method deals with estimates for the number of elements in a (finite) sequence $\mathscr{A}$ that are not divisible by any prime number of a set $\mathfrak{P}$ of primes.

We consider a finite sequence

$$\mathscr{A} = \{a: \dots\}$$

of (not necessarily distinct and not necessarily positive) integers, where the defining properties of $\mathscr{A}$ will be entered in the space indicated by the dots. The problems we deal with require that $\mathscr{A}$ depends on certain parameters

† Indeed, the squarefree numbers themselves can be viewed as the result of sifting the sequence of all integers by the *squares* of primes! But we shall not consider such sifting sets in this account.

that may vary; it will always be the case that each admissible choice of the parameters is such as to keep $\mathscr{A}$ finite. Clearly it will be important to us, as far as possible, to make the parameters explicit in our calculations. For our purposes the most important are those that occur in the basic information we have to have about $\mathscr{A}$: thus the number of elements in

$$\mathscr{A}_d := \{a:\ a \in \mathscr{A}, a \equiv 0 \bmod d\},$$

i.e. the number of elements in $\mathscr{A}$ that are divisible by a squarefree integer d, we shall denote by $|\mathscr{A}_d|$. For $d = 1$ this is the number

$$|\mathscr{A}| = |\{a: \ldots\}|$$

of elements in $\mathscr{A}$.

It is a familiar experience in arithmetical investigations to use, instead of the exact number $|\mathscr{A}|$ of elements in $\mathscr{A}$, a convenient close approximation X to $|\mathscr{A}|$ (for example, $X = li\,x$ for $|\mathscr{A}| = \pi(x)$; see also Examples 1–6 below); we shall require

$$X > 1 \tag{2.1}$$

and we write

$$r_1 := |\mathscr{A}| - X \tag{2.2}$$

for the remainder. Next, for each prime p we choose $\omega_0(p)$ so that $(\omega_0(p)/p)X$, where X is the number introduced in (2.2), approximates to $|\mathscr{A}_p|$, and we write the remainder as

$$r_p := |\mathscr{A}_p| - \frac{\omega_0(p)}{p}X \qquad (\text{for all } p). \tag{2.3}$$

With these choices of X and $\omega_0(p)$ we now define, for each (squarefree) d,

$$\omega_0(1) := 1, \qquad \omega_0(d) := \prod_{p|d} \omega_0(p) \qquad (\mu(d) \neq 0), \tag{2.4}$$

so that $\omega_0(d)$ is a multiplicative function. Consistently with (2.2) and (2.3), we then introduce

$$r_d := |\mathscr{A}_d| - \frac{\omega_0(d)}{d}X \qquad (\mu(d) \neq 0). \tag{2.5}$$

Obviously these choices can be made in various ways, but, as one might expect, it will turn out that the smaller $|r_d|$ is (at any rate, on average), the better our results will be.

It will become clear from the examples below how these choices can be made in practice.

3. Basic Examples

Example 1. Let

$$\mathscr{A} = \{n: x - y < n \leqslant x\}$$

where x, y are real numbers satisfying

$$1 < y \leqslant x; \tag{3.1}$$

then

$$|\mathscr{A}_d| = \left|\left\{m: \frac{x-y}{d} < m \leqslant \frac{x}{d}\right\}\right| = \left[\frac{x}{d}\right] - \left[\frac{x-y}{d}\right] = \frac{y}{d} + \theta, \quad |\theta| \leqslant 1. \tag{3.2}$$

Therefore, in our notation, an approximation which is as good as it is convenient is provided by

$$X = y, \qquad \omega_0(p) = 1 \quad \text{for every } p, \tag{3.3}$$

and

$$\omega_0(d) = 1 \quad \text{for all (squarefree) } d;$$

and (3.2) then yields

$$|r_d| \leqslant 1, \qquad \mu(d) \neq 0. \tag{3.4}$$

Note the special case $y = x$, where the only change is that now $X = x$.

Example 2. We extend the preceding example to an arithmetic progression. We set

$$\mathscr{A} = \{n: x - y < n \leqslant x,\ n \equiv l \bmod k\},$$

where k and l are integers, and x and y are real numbers satisfying

$$1 \leqslant k < y \leqslant x. \tag{3.5}$$

Here

$$|\mathscr{A}_d| = \left|\left\{m: \frac{x-y}{d} < m \leqslant \frac{x}{d},\ dm \equiv l \bmod k\right\}\right|,$$

and we observe that the linear congruence

$$dm \equiv l \bmod k$$

is soluble in m if and only if $(d, k) \mid l$. Assuming that (d, k) is a factor of l, the congruence has exactly (d, k) solutions incongruent mod k. Hence

$$|\mathscr{A}_d| = \omega_0(d)\left(\frac{y}{dk} + \theta\right), \qquad |\theta| \leqslant 1, \tag{3.6}$$

where

$$\omega_0(d) = \begin{cases} 0 & \text{if } (d, k) \nmid l, \\ (d, k) & \text{if } (d, k) \mid l, \end{cases} \tag{3.7}$$

and so is a multiplicative function of d.

Further,

$$X = \frac{y}{k}, \qquad \omega_0(p) = 1 \quad \text{for} \quad p \nmid k, \tag{3.8}$$

and by (3.6) we always have

$$|r_d| \leqslant \omega_0(d). \tag{3.9}$$

In the special case $y = x$ the only change is, as before, $X = x/k$.

Note that in either case $X > 1$ and so satisfies (2.1).

Example 3. Let F be a polynomial of degree g with integer coefficients, and consider the sequence

$$\mathscr{A} = \{F(n) \colon x - y < n \leqslant x\},$$

where x and y are, as before, real numbers satisfying

$$1 < y \leqslant x. \tag{3.10}$$

Let us denote by $\rho(d) = \rho_F(d)$ the number of solutions of the congruence

$$F(n) \equiv 0 \bmod d \tag{3.11}$$

that are incongruent mod d. Here we find that

$$|\mathscr{A}_d| = |\{n \colon x - y < n \leqslant x, F(n) \equiv 0 \bmod d\}| = \rho(d)\left(\frac{y}{d} + \theta\right), |\theta| \leqslant 1; \tag{3.12}$$

therefore an appropriate choice is

$$X = y, \qquad \omega_0(d) = \rho(d), \tag{3.13}$$

and with this choice it follows from (3.12) that

$$|r_d| \leqslant \omega_0(d). \tag{3.14}$$

The arithmetic function ρ is, except in special circumstances (see Example 4 below), non-elementary in the sense that the individual numbers $\rho(d)$ are hard to determine.

However, it is well-known that ρ is a multiplicative function, and we have, in particular, that

$$\rho(d) = \prod_{p|d} \rho(p), \qquad \mu(d) \neq 0; \tag{3.15}$$

also, we know from Lagrange's theorem that $\rho(p) = p$ or $\rho(p) \leqslant g$, the degree of F. Hence if F has no fixed prime divisor—which is equivalent to the statement that $\rho(p) < p$ for all p—we have

$$\rho(p) \leqslant g \quad \text{if} \quad \rho(p) < p. \tag{3.16}$$

Apart from this we know that $\rho(p)$ is equal to k on average over the primes p, where k is the number of irreducible components of F. This can be expressed more precisely in various forms: for example, we know that as $x \to \infty$,

$$\sum_{p<x} \rho(p) \sim k \frac{x}{\log x}$$

and that

$$\sum_{p<x} \frac{\rho(p)}{p} \log p = k \log x + O_F(1), \tag{3.17}$$

where the O-constant may depend on the polynomial F, as is indicated by the notation O_F.

Example 4. Let us consider here a special case of the previous example, namely when F is the product of g linear factors (so that $k = g$). To be precise, let g be a natural number, and let

$$F_0(n) = \prod_{i=1}^{g} (a_i n + b_i) \tag{3.18}$$

where the coefficients a_i, b_i $(i = 1, \ldots, g)$ are integers satisfying

$$(a_i, b_i) = 1, \qquad i = 1, \ldots, g, \tag{3.19}$$

and let us assume that the discriminant of F_0 does not vanish, or that

$$E: = \prod_{i=1}^{g} a_i \prod_{1 \leqslant r < s \leqslant g} (a_r b_s - a_s b_r) \neq 0. \tag{3.20}$$

This condition ensures that the linear factors of F_0 are not constant and that they are distinct in the sense that none is a constant multiple of any other.

If

$$\mathscr{A} = \{F_0(n)\colon\ x - y < n \leqslant x\}, \tag{3.21}$$

where x and y are real and

$$1 < y \leqslant x, \tag{3.22}$$

then the number $|\mathscr{A}_d|$ is evaluated as in Example 3, so that here too we choose

$$X = y, \qquad \omega_0(d) = \rho(d), \tag{3.23}$$

and we have

$$|r_d| \leqslant \omega_0(d). \tag{3.24}$$

However, we are able here to make explicit the sense in which $\rho(p)$ is, on average, equal to g. First consider $\rho(p)$, i.e. consider the congruence

$$\prod_{i=1}^{g} (a_i n + b_i) \equiv 0 \bmod p. \tag{3.25}$$

The linear congruence

$$a_i n + b_i \equiv 0 \bmod p \tag{3.26}$$

has, in view of (3.19), no solution if $p \mid a_i$ and precisely one solution if $p \nmid a_i$. Hence our condition $(a_i, b_i) = 1$ for $i = 1, \ldots, g$ implies that each factor $a_i n + b_i$ contributes at most one solution to (3.25), so that

$$\rho(p) \leqslant g \qquad \text{for all } p. \tag{3.27}$$

Suppose that

$$p \nmid a_1 \ldots a_g,$$

so that each linear congruence (3.26) has exactly one solution. Two congruences may, of course, have the same solution, so that we cannot conclude at once that $\rho(p) = g$. However, suppose for example that

$$a_r n + b_r \equiv 0 \bmod p$$

and

$$a_s n + b_s \equiv 0 \bmod p$$

have the same solution (and that $r \neq s$). Multiplying the first congruence by a_s and the second by a_r, and subtracting the first from the second, it follows that

$$a_r b_s - a_s b_r \equiv 0 \bmod p,$$

i.e. that $p \mid E$; and putting all this information together we see that, at any rate,

$$\rho(p) = g \quad \text{if} \quad p \nmid E. \tag{3.28}$$

If $p \mid E$, $\rho(p)$ may assume all integer values intermediate between 0 (inclusive) and g; by Lagrange's theorem we know it cannot be larger, and it follows from above that

$$0 \leqslant \rho(p) < g \quad \text{if} \quad p \mid E.$$

The detailed enumeration of all possible cases would be of little value and excessively tedious.

Since ρ is multiplicative it follows from (3.28) that

$$\rho(d) = g^{\nu(d)} \quad \text{if} \quad \mu(d) \neq 0 \quad \text{and} \quad (d, E) = 1, \tag{3.29}$$

and from (3.27) that

$$\rho(d) \leqslant g^{\nu(d)} \quad \text{if} \quad \mu(d) \neq 0 \tag{3.30}$$

always; here and in the sequel, $\nu(d)$ denotes the numbers of prime factors of d.

Example 5. Let

$$\mathscr{A} = \{ap + b\colon\ p \leqslant x, p \equiv l \bmod k\} \tag{3.31}$$

where a, b, k, and l are integers, and x a real number, satisfying

$$ab \neq 0, \qquad (a, b) = 1, \qquad 2 \mid ab \tag{3.32}$$

(so that exactly one of the numbers a, b is even) and

$$(l, k) = 1, \qquad 1 \leqslant k < x. \tag{3.33}$$

Let d be a squarefree natural number and consider

$$|\mathscr{A}_d| = |\{p\colon\ p \leqslant x, p \equiv l \bmod k, ap + b \equiv 0 \bmod d\}|.$$

The linear congruence $an + b \equiv 0 \bmod d$ has no solution unless $(a, d) \mid b$; and if (a, d) is a factor of b, it has exactly (a, d) incongruent solutions. But since $(a, b) = 1$, $(a, d) \mid b$ if and only if $(a, d) = 1$, and the congruence has then the unique mod d solution.

$$n \equiv -ba' \bmod d \quad \text{where} \quad aa' \equiv 1 \bmod d.$$

If $(b, d) > 1$ and $n \equiv -ba' \bmod d$ then $(b, d) \mid n$. Hence there are at most $\nu(b)$ primes p such that $p \equiv -ba' \bmod d$. Summing up the discussion so far,

$$|\mathscr{A}_d| = \begin{cases} \vartheta\nu(b) & \text{if} \quad (a, d) = 1 \quad \text{and} \quad (b, d) > 1, \\ 0 & \text{if} \quad (a, d) > 1, \end{cases} \tag{3.34}$$

where $0 \leqslant \vartheta \leqslant 1$.

Let us now suppose that $(ab, d) = 1$. Then the arithemtic progressions $-ba' \bmod d$ and $l \bmod k$ intersect, by the Chinese Remainder Theorem, in some unique progression $l' \bmod [d, k]$, where $(l', [d, k]) = 1$, provided that $(d, k) \mid l + ba'$; and this progression contains infinitely many primes by Dirichlet's theorem. Thus

$$|\mathscr{A}_d| = |\{p\colon p \leqslant x, p \equiv l' \bmod [d, k]\}| = \pi(x, [d, k], l') \qquad \text{if} \quad (ab, d) = 1 \quad \text{and} \quad (d, k) \mid al + b. \tag{3.35}$$

There are many ways of approximating to $\pi(x; [d, k], l)$—for example,

$$\frac{1}{\phi([d, k])} \frac{x}{\log x}$$

would be a simple approximation—but we know from analytic number theory that

$$\frac{1}{\phi([d, k])} li\, x$$

is better. Now

$$\phi([d, k]) = \phi(d)\phi(k)/\phi((d, k)),$$

so that

$$\frac{1}{\phi([d, k])} li\, x = \frac{1}{d} \frac{d}{\phi(d)} \phi((d, k)) \frac{li\, x}{\phi(k)}.$$

It therefore seems reasonable to choose

$$X = \frac{li\, x}{\phi(k)}, \omega_0(d) = \begin{cases} 0, & (ab, d) > 1 \quad \text{or} \quad (d, k) \nmid al + b \\ \dfrac{d}{\phi(d)} \phi((d, k)), & (ab, d) = 1 \quad \text{and} \quad (d, k) \mid al + b; \end{cases} \tag{3.36}$$

with this choice of X and ω_0 (note that, in view of (2.1), the choice of X is admissible only provided that $\phi(k) < li\, x$) it follows from (3.34) and (3.35) that

$$r_d = |\mathscr{A}_d| - \frac{\omega_0(d)}{d} X \begin{cases} \leqslant \nu(b) \quad \text{if} \quad (ab, d) > 1 \quad \text{or} \quad (d, k) \nmid al + b, \\ = \pi(x; [d, k], l') - \dfrac{1}{\phi([d, k])} li\, x \\ \qquad \text{if} \quad (ab, d) = 1 \quad \text{and} \quad (d, k) \mid al + b. \end{cases} \tag{3.37}$$

If we restrict attention to those squarefree integers d for which

$$(d, kab) = 1 \tag{3.38}$$

we see that

$$\omega_0(d) = \frac{d}{\phi(d)} \tag{3.39}$$

and that

$$r_d = \pi(x; dk, l') - \frac{1}{\phi(dk)} li\, x. \tag{3.40}$$

Writing

$$E(x, q) = \max_{2 \leqslant y \leqslant x} \max_{\substack{1 \leqslant l' \leqslant q \\ (l', q) = 1}} \left| \pi(y; q, l') - \frac{1}{\phi(q)} li\, y \right|, \tag{3.41}$$

we see that

$$|r_d| \leqslant E(x, dk). \tag{3.42}$$

It is well-known (see Lemma 3.3) that $E(x, q)$ is, on average over q, (relatively) small.

Two specially interesting cases occur when

$$\text{(i)} \qquad a = 1,\ b = h\ \ (h \text{ even}) \tag{3.43}$$

$$\text{(ii)} \quad a = -1,\ b = N\ \ (N \text{ large and even}); \tag{3.44}$$

here (i) is connected with the prime twin conjecture, and (ii) with the Goldbach conjecture. In (i) we should like to know whether, as p runs through the odd primes, $p + h$ is infinitely often prime too, whilst in (ii) we should like to find at least one prime p, $2 < p < N$, such that $N - p$ is prime.

Example 6. Let F be a polynomial with integer coefficients, of degree g ($\geqslant 1$), and let ρ be defined as in Example 3, i.e. let $\rho(d)$ be the number of solutions of the congruence

$$F(n) \equiv 0 \bmod d. \tag{3.45}$$

We consider the sequence

$$\mathscr{A} = \{F(p){:}\ p \leqslant x, p \equiv l \bmod k\}$$

where l, k are integers and x is a real number satisfying

$$(l, k) = 1, \qquad 1 \leqslant k < x;$$

as in Example 5, we shall have to modify the second condition later to $\phi(k) < li\, x$.

Let d be a squarefree natural number. Then

$$|\mathscr{A}_d| = |\{p\colon\ p \leqslant x, p \equiv l \bmod k, F(p) \equiv 0 \bmod d\}|$$

$$= \sum_{\substack{m=1 \\ F(m) \equiv 0 \bmod d}}^{d} |\{p\colon\ p \leqslant x, p \equiv l \bmod k, p \equiv m \bmod d\}|,$$

and we observe that the sum over m has $\rho(d)$ terms. Of these, each term with $m \not\equiv l \bmod (d, k)$ is empty, and each term with $(m, d) > 1$ is at most 1 (since the progression $m \bmod d$ contains at most one prime if $(m, d) > 1$). Thus the sum on the right is equal to

$$\sum_{\substack{m=1 \\ (m,d)=1 \\ F(m) \equiv 0 \bmod d \\ m \equiv l \bmod (d,k)}}^{d} \pi(x;\ [k, d], h) + \vartheta\rho(d),$$

where $0 \leqslant \vartheta \leqslant 1$ and $h = h(l, m)$ is a unique (relatively prime) residue mod $[k, d]$. Hence

$$|\mathscr{A}_d| - \frac{\rho_1{}^*(d)}{\phi([k, d])}\, li\, x = \sum_{\substack{m=1 \\ (m,d)=1 \\ F(m) \equiv 0 \bmod d \\ m \equiv l \bmod (d,k)}}^{d} \{\pi(x;\ [k, d], h) = \frac{li\, x}{\phi([k, d])} + \vartheta\rho(d), \tag{3.46}$$

where $\rho_1{}^*(d)$ is the number of solutions of (3.45) which are $\equiv l \bmod (d, k)$ and coprime with d, i.e. of

$$F(m) \equiv 0 \bmod d, m \equiv l \bmod (d, k) \text{ for } (m, d) = 1;$$

and $\rho_1{}^*$ is equal to

$$\rho_1\left(\frac{d}{(d, k)}\right),$$

where $\rho_1(d)$ is the number of solutions of

$$F(m) \equiv 0 \bmod d \text{ for } (m, d) = 1. \tag{3.47}$$

Since $\rho_1(d)$ is a multiplicative function, so is $\rho_1{}^*(d)$. Proceeding as in Example 5, we may therefore choose

$$X = \frac{li\, x}{\phi(k)},\ \omega_0(d) = \rho_1{}^*(d)\phi((k, d))\frac{d}{\phi(d)}, \tag{3.48}$$

so that

$$X > 1 \quad \text{provided} \quad \phi(k) < li\, x, \tag{3.49}$$

and ω_0 is multiplicative; and we have, using the notation (3.41), that

$$|r_d| \leqslant \rho(d)\{E(x, [k, d]) + 1\}. \tag{3.50}$$

The situation is a little simpler whenever d is coprime with k. Then $(k, d) = 1$, $[k, d] = kd$ and therefore

$$\omega_0(p) = \frac{\rho_1(p)}{p-1}p \quad \text{if} \quad p \nmid k, \tag{3.51}$$

and

$$|r_d| \leqslant \rho(d)\{E(x, kd) + 1\} \quad \text{if} \quad \mu(d) \neq 0,\ (d, k) = 1. \tag{3.52}$$

Let us compare ρ_1 and ρ. We have

$$\rho_1(d) = \prod_{p|d} \rho_1(p), \quad \mu(d) \neq 0,$$

and $\rho_1(p)$ is the number of solutions of the congruence

$$F(m) \equiv 0 \bmod p, \quad p \nmid m.$$

Now $m = 0$ is a solution if and only if $p|F(0)$, so that

$$\rho_1(p) = \begin{cases} \rho(p) & \text{if} \quad p \nmid F(0), \\ \rho(p) - 1 & \text{if} \quad p \mid F(0). \end{cases} \tag{3.53}$$

We know about the function ρ, from (3.16), that

$$\rho(p) \leqslant g \quad \text{if } \rho(p) < p, \tag{3.54}$$

and hence

$$\rho_1(d) \leqslant \rho(d) \leqslant g^{\nu(d)} \quad \text{for} \quad \mu(d) \neq 0, \quad \text{if} \quad \rho(p) < p \quad \text{for all} \quad p \mid d. \tag{3.55}$$

4. The Sifting Set $\mathfrak{P}$ and the Sifting Function S

We now introduce a set

$$\mathfrak{P} = \{p: \ \ldots\}$$

of primes, and we denote its complement with respect to the set of all primes by

$$\overline{\mathfrak{P}}.$$

$\mathfrak{P}$ will usually be infinite, and often it simply consists of all primes not dividing some integer K; in this latter case we may write

$$\mathfrak{P} = \mathfrak{P}_K = \{p: p \nmid K\}$$

so that in this notation

$$\mathfrak{P}_1$$

is the set which consists of all primes.

In our theory a special rôle is played by *truncations* of $\mathfrak{P}$, i.e. sets of the type

$$\{p\colon\ p \in \mathfrak{P},\ p < z\}$$

where z is a real number satisfying

$$z \geqslant 2; \tag{4.1}$$

and in this connection we introduce

$$P(z)\colon = \prod_{\substack{p<z \\ p\in\mathfrak{P}}} p. \tag{4.2}$$

We are now in a position to explain what we mean by a *sieve problem*. The problem is to *sift* a certain sequence $\mathscr{A}$ (from the class of sequences described above) by a truncation (at z) of a certain set of primes, $\mathfrak{P}$; that is, to estimate the *sifting function*

$$S(\mathscr{A};\mathfrak{P},z)\colon\ = |\{a\colon\ a \in \mathscr{A}, (a, P(z)) = 1\}|. \tag{4.3}$$

In fact, we shall need to study also the more general sifting function

$$S(\mathscr{A}_q;\mathfrak{P},z)\colon = |\{a\colon\ a \in \mathscr{A}_q, (a, P(z)) = 1\}| \tag{4.4}$$

where q is a natural number satisfying

$$\mu(q) \neq 0,\ \ (q, P(z)) = 1,\ \ (q, \overline{\mathfrak{P}}) = 1; \tag{4.5}$$

here we mean by

$$(q, \overline{\mathfrak{P}}) = 1$$

that q is coprime with $\overline{\mathfrak{P}}$, that is, that none of the prime divisors of q is in $\overline{\mathfrak{P}}$. For example, if $\mathfrak{P} = \mathfrak{P}_K$ then $(q, \overline{\mathfrak{P}}) = 1$ means that $(q, K) = 1$.

Clearly

$$S(\mathscr{A};\mathfrak{P},z) = S(\mathscr{A}_1;\mathfrak{P},z).$$

It may seem surprising at first sight to regard $S(\mathscr{A}_q;\mathfrak{P},z)$ as being more general, or indeed as needing special consideration at all; one might reasonably argue that whatever results there are for a general sequence $\mathscr{A}$ may then be applied to any $\mathscr{A}_q$. The justification for singling out the sequences of type $\mathscr{A}_q$ will emerge as we develop the theory; but it may help if we offer some (necessarily vague) explanation now. In the course of finding estimates for $S(\mathscr{A};\mathfrak{P},z)$ we shall frequently make use of identities and iterative procedures (leading to successive improvements of results that are initially weak or very moderate in quality). When either procedure is applied to a particular sequence $\mathscr{A}$, many of the associated sequences $\mathscr{A}_q$ tend to appear quite naturally; one significant feature common to them all is that although they give rise to essentially the same X, for the allied calculations it will be important to choose the very *same* X (so that where $\mathscr{A}$ has X, $\mathscr{A}_q$ has $(\omega(q)/q)X$). The importance of the family of sequences $\mathscr{A}_q$ when studying $\mathscr{A}$ can be gauged by reference to Chapter 2 or, better still, to Chapters 8 and 9; and to single them out for special consideration is therefore quite natural. At the same time, their rôle is essentially a supporting one; whenever we come to state a result that is in some sense final, we shall do so with $q = 1$.

In order to see how a result for such a sifting function S can be related to a problem in prime number theory, we consider what is in a way the simplest possible example, i.e. we consider the sequence

$$\mathscr{A} = \{n\colon\ n \leqslant x\}$$

from Example 1, and we take for $\mathfrak{P}$ the set $\mathfrak{P}_1$ of all primes. Then, by (4.3) and (4.2), for any z ($\geqslant 2$)

$$S(\{n\colon\ n \leqslant x\};\ \mathfrak{P}_1, z) = \left|\left\{n\colon\ n \leqslant x, \left(n, \prod_{p<z} p\right) = 1\right\}\right|. \tag{4.6}$$

Let us assume that $\sqrt{x} < z(\leqslant x)$. Then, clearly, this number counts, apart from $n = 1$, no number less than z; however, in $[z, x]$ all primes are counted, but no composite number is, because it has a prime factor $\leqslant \sqrt{x} < z$. Hence

$$\begin{aligned} S(\{n\colon\ n \leqslant x\};\ \mathfrak{P}_1, z) &= 1 + |\{p\colon\ z \leqslant p \leqslant x\}| \\ &= \pi(x) + \theta z, \quad \text{if} \quad \sqrt{x} < z \leqslant x \end{aligned} \tag{4.7}$$

where $|\theta| \leqslant 1$.

Thus, to any result about this special function S there corresponds a result about $\pi(x)$.

In (4.6) the numbers $1 < n \leqslant \sqrt{x}$ were immediately discarded, so that we might have started with the sequence

$$\{n\colon\ \sqrt{x} < n \leqslant x\},$$

for which one finds in the same way as above that

$$S(\{n\colon\ \sqrt{x} < n \leqslant x\};\ \mathfrak{P}_1, z) = |\{p\colon\ z \leqslant p \leqslant x\}| \quad \text{if} \quad \sqrt{x}<z\leqslant x.$$

However, this is of no advantage, and in view of the knowledge we generally require about our sequences $\mathscr{A}$ it is easier to employ the simpler sequence from above.

Next let us consider what sifting functions can be related to the prime twin problem whose counting function is

$$|\{p\colon p \leqslant x,\ p+2 = p'\}|. \tag{4.8}$$

To this end we take the special case $g = 2$, $a_1 = 1$, $b_1 = 0$, $a_2 = 1$, $b_2 = 2$, $y = x$ of Example 4, i.e. we consider the sequence

$$\mathscr{A} = \{n(n+2)\colon n \leqslant x\}.$$

Here we see that

$$S(\{n(n+2)\colon n \leqslant x\};\ \mathfrak{P}_1, z) = \left|\left\{n(n+2)\colon n \leqslant x, \left(n(n+2), \prod_{p<z} p\right) = 1\right\}\right|$$

which can be written in various other ways, for example as

$$S(\{n(n+2)\colon n \leqslant x\};\ \mathfrak{P}_1, z) = \left|\left\{n\colon n \leqslant x, \left(n, \prod_{p<z} p\right) = 1, \left(n+2, \prod_{p<z} p\right) = 1\right\}\right|, \tag{4.9}$$

and from this, assuming that $x \geqslant 7$ and $\sqrt{(x+2)} < z \leqslant x$, we see immediately by the same arguments as above, that exactly those numbers n are counted which are primes in $[z, x]$ and for which simultaneously $n+2$ is also a prime number. Thus

$$S(\{n(n+2)\colon n \leqslant x\};\ \mathfrak{P}_1, z) = |\{p\colon z \leqslant p \leqslant x, p+2 = p'\}| \quad \text{if } \sqrt{(x+2)} < z \leqslant x,\ 7 \leqslant x, \tag{4.10}$$

and this deviates from the number in (4.8) by less than z.

In the case of $\pi(x)$ we gave a very simple example which showed that the way to link a problem from prime number theory with one of our sifting functions is by no means unique (and not necessarily obvious). We shall now give another, less trivial, example; this constitutes an important alternative approach to the prime twin problem which turns out to be here, as it will be in all the most delicate questions we shall raise later, even more advantageous than the previous one; a general discussion of this different type of approach is given in Section 5.5.

We consider

$$\mathscr{A} = \{p+2;\ p \leqslant x\}$$

which is the case $a = 1, b = 2, k = 1$ of the more general sequence we analysed in Example 5. Here we find that

$$S(\{p+2\colon p \leqslant x\};\ \mathfrak{P}_1, z) = |\{p\colon z-2 \leqslant p \leqslant x,\ p+2 = p'\}| \quad \text{if } \sqrt{(x+2)} < z \leqslant x. \tag{4.11}$$

As was pointed out above, this approximation to the prime twin problem is sometimes of advantage; however, the price we have to pay is having to work with a much less simple sequence, as can be seen by comparing in Examples 4 and 5 the data corresponding to the sequences we employed in (4.10) and (4.11).

There is still another point to be mentioned here. Satisfactory results for the sifting functions of our examples, at any rate for the range of z under consideration, are rather hard to achieve. Actually, for lower bounds the problems have even to be reformulated; but for upper bounds we can, and shall, use (4.7), (4.10) and (4.11) in conjunction with the trivial remark that $S(\mathscr{A};\ \mathfrak{P}, z)$ is decreasing when z increases (see e.g. (5.6) below).

The special importance of the primes of $\mathfrak{P}$ now carries with it the implication that, for a suitable description of the divisibility properties of $\mathscr{A}$ relative to $\mathfrak{P}$, it will be natural to switch from the function ω_0 to the arithmetic function ω defined by

$$\omega(p) = \begin{cases} \omega_0(p), & p \in \mathfrak{P}, \\ \\ 0, & p \in \overline{\mathfrak{P}}, \end{cases} \tag{4.12}$$

and extended to the set of all squarefree numbers by

$$\omega(1) := 1, \ \omega(d) := \prod_{p \mid d} \omega(p) \ \ (\mu(d) \neq 0); \tag{4.13}$$

so that ω is a multiplicative function, and ω depends on both $\mathscr{A}$ and $\mathfrak{P}$.
Correspondingly we introduce (c.f. (2.5))

$$R_d := |\mathscr{A}_d| - \frac{\omega(d)}{d} X \quad \text{if} \quad \mu(d) \neq 0, \tag{4.14}$$

so that, by virtue of (2.5) and (4.7),

$$\omega(d) = \omega_0(d) \quad \text{and} \quad R_d = r_d \quad \text{if} \quad \mu(d) \neq 0, \quad \text{and} \quad (d, \overline{\mathfrak{P}}) = 1. \tag{4.15}$$

At first sight definition (4.14) seems rather artificial, for R_d is plainly not a remainder term if $\omega(d) = 0$, which will be the case whenever $(d, \overline{\mathfrak{P}}) > 1$. However, it is actually more convenient not to introduce the condition $(d, \overline{\mathfrak{P}}) = 1$ into (4.14), first because this would give us a condition to verify before we could use (4.14), and then because, in fact, this condition is always satisfied when we want to use (4.14).

With the new function ω we form the product

$$W(z) := \prod_{p<z} \left(1 - \frac{\omega(p)}{p}\right). \tag{4.16}$$

We shall now require, as the first of a series of basic conditions we shall need to impose,

$$(\Omega_1) \qquad 0 \leqslant \frac{\omega(p)}{p} \leqslant 1 - \frac{1}{A_1}$$

for some suitable constant $A_1 \geqslant 1$. In most cases this will be used in the form

$$(\Omega_1) \qquad 1 \leqslant \frac{1}{1 - (\omega(p)/p)} \leqslant A_1.$$

Further, we introduce the multiplicative function

$$g(d) := \frac{\omega(d)}{d \prod_{p|d} (1 - (\omega(p)/p))}, \quad \mu(d) \neq 0; \tag{4.17}$$

note that, because of condition (Ω_1), g is well defined, moreover, the product in the denominator is uniformly bounded away from 0. By (4.12), we have that

$$g(d) = 0 \quad \text{if} \quad (d, \mathfrak{P}) > 1, \tag{4.18}$$

and also that

$$g(d) = 0 \quad \textit{if and only if} \quad \omega(d) = 0. \tag{4.19}$$

With g we form

$$G(z) := \sum_{d<z} \mu^2(d) g(d) \tag{4.20}$$

as well as

$$G(x, z) := \sum_{\substack{d<x \\ d|P(z)}} g(d) \tag{4.21}$$

where x is a real number satisfying

$$x > 0. \tag{4.22}$$

We note that, by (4.18)

$$G(x, z) = G(x) \quad \text{if} \quad z \geqslant x, \tag{4.23}$$

and in particular that

$$G(z, z) = G(z). \tag{4.24}$$

We complete the list of basic definitions be introducing the product

$$V(z) := \prod_{p<z} \left(1 - \frac{1}{p}\right). \tag{4.25}$$

The constants A_i, A etc. occurring in our conditions (c.f. (Ω_1)) are independent of X and z, and are assumed always to be $\geqslant 1$. The constants $B_i (\geqslant 1)$, and all the constants implied by the $\ll$- and O-symbols depend at most on the basic structure constants† A_i, κ, and α; where other parameters occur and dependence on them is to be kept explicit, we use e.g. $\ll_\varepsilon$ or O_ε to indicate that the constants implied here depend also on ε. The constants $C_i\,(\geqslant 1)$ may depend on the same parameters as the B_i's, as well as on various other parameters; in each case the nature of this additional dependence will be made explicit.

5. The Sieve of Eratosthenes-Legendre

One particularly simple, and indeed natural, approach to a sieve problem is through the use of the Möbius function, and this will lead us to the well-known Eratosthenes–Legendre formula (see (5.2) below). We shall derive from the formula a quantitative result (see (5.3) below) and to this end we shall now introduce two conditions, one on the remainder R_d and the other on the function ω, which are often satisfied in applications and are of a particularly simple as well as convenient form (for example, see the applications following Theorem 1.1 and some applications in Chapter 2): we suppose that

$$(R) \qquad |R_d| \leqslant \omega(d) \quad \text{if} \quad \mu(d) \neq 0,\ (d, \mathfrak{P}) = 1$$

and also that

$$(\Omega_0) \qquad \omega(p) \leqslant A_0.$$

If $(q, P(z)) = 1$, which implies that $(q, d) = 1$ in the calculation below, we have

$$S(\mathscr{A}_q;\ \mathfrak{P}, z) = \sum_{a \in \mathscr{A}_q} \sum_{\substack{d|a \\ d|P(z)}} \mu(d) = \sum_{d|P(z)} \mu(d) \sum_{\substack{a\in\mathscr{A} \\ a \equiv 0 \bmod qd}} 1 = \sum_{d|P(z)} \mu(d)\,|\mathscr{A}_{qd}|\,. \tag{5.1}$$

Hence, by (4.14), since q is squarefree,

$$S(\mathscr{A}_q;\ \mathfrak{P}, z) = \sum_{d|P(z)} \mu(d) \left\{\frac{\omega(qd)}{qd} X + R_{qd}\right\};$$

also, ω is a multiplicative function and therefore, by (4.12),

$$\sum_{d|P(z)} \mu(d)\frac{\omega(d)}{d} = \prod_{\substack{p<z \\ p\in\mathfrak{P}}} \left(1 - \frac{\omega(p)}{p}\right) = \prod_{p<z} \left(1 - \frac{\omega(p)}{p}\right).$$

† The important constants κ and α will make their appearance in later chapters.

These relations yield at once the formula (5.2) below. Furthermore, if we put $q = 1$ and impose the conditions (R) and (Ω_0), we have, for every R_d under consideration,

$$|R_d| \leqslant A_0^{\nu(d)},$$

and we infer that

$$\sum_{d|P(z)} |R_d| \leqslant \sum_{d|P(z)} A_0^{\nu(d)} = \prod_{p|P(z)} (1 + A_0) \leqslant (1 + A_0)^{\pi(z)} \leqslant (1 + A_0)^z.$$

This proves (5.3) below. We have now obtained

THEOREM 1.1. *We have*

$$S(\mathscr{A}_q;\ \mathfrak{P}, z) = \frac{\omega(q)}{q} XW(z) + \theta \sum_{d|P(z)} |R_{qd}|, \tag{5.2}$$

and, if conditions (R) *and* (Ω_0) *are satisfied, we have also that* †

$$(\Omega_0), (R)\colon\ S(\mathscr{A};\ \mathfrak{P}, z) = XW(z) + \theta(1 + A_0)^z, \tag{5.3}$$

where $|\theta| \leqslant 1$.

A noteworthy feature of (5.2) is that it holds without any conditions on $\mathscr{A}$ or $\mathfrak{P}$. We might note also that the condition $(q, \overline{\mathfrak{P}}) = 1$ (on q; see (4.5)) has not been used here.

Theorem 1.1 gives significant results only if z is *very* small compared with X, and so, though of historical interest, must be regarded as only a first, weak sieve result. It was the weakness of Theorem 1.1 which led in time to the powerful modern sieve methods which are the subject matter of this book.

We shall illustrate these remarks in a special case where we try to find via Theorem 1.1 an upper estimate for $\pi(x)$. To this end we consider, as before, the sequence $\mathscr{A} = \{n\colon n \leqslant x\}$ from Example 1, and with the set $\mathfrak{P}_1$ of all primes we have, by (4.6),

$$S(\{n\colon n \leqslant x\};\ \mathfrak{P}_1, z) = \left|\left\{n\colon n \leqslant x, \left(n, \prod_{p<z} p\right) = 1\right\}\right|. \tag{5.4}$$

By (3.3), (3.4) and (4.15) we may use

$$X = x,\ \omega(p) = 1 \quad \textit{for all } p, \quad |R_d| \leqslant 1 \quad \textit{if} \quad \mu(d) \neq 0.$$

Hence conditions (Ω_0) (with $A_0 = 1$) and (R) are satisfied, and (5.3) yields

$$S(\{n\colon n \leqslant x\};\ \mathfrak{P}_1, z) \leqslant x \prod_{p<z} \left(1 - \frac{1}{p}\right) + 2^z. \tag{5.5}$$

† "(Ω_0), (R):..." means that conditions (Ω_0) and (R) have been used to prove (5.3).

A direct use of (4.7) is out of the question since $z > \sqrt{x}$ would make the second term on the right far too big. However, for an upper estimate of $\pi(x)$ we may argue as follows: in (5.4) all primes of the interval $[z, x]$ are certainly counted so that

$$\pi(x) \leqslant S(\{n\colon\ n \leqslant x\};\ \mathfrak{P}_1, z) + z \tag{5.6}$$

for any z in $2 \leqslant z \leqslant x$.

Using the well-known estimate

$$\prod_{p<z}\left(1 - \frac{1}{p}\right) \ll \frac{1}{\log z} \tag{5.7}$$

we now obtain from (5.6) and (5.5) that

$$\pi(x) \ll \frac{x}{\log z} + 2^z,$$

and the choice of $z = \log x$ yields

$$\pi(x) \ll \frac{x}{\log\log x} \tag{5.8}$$

as $x \to \infty$.

We have seen that, in order to avoid a trivial deduction such as $\pi(x) \ll x$ or worse from Theorem 1.1, we had to choose z of order $\log x$. We could have estimated more carefully, namely replaced z by (essentially) $z/\log z$ in the second terms on the right of both (5.5) and (5.6), but this would have had very little effect.

When we have only $|R_d| \leqslant 1$ at our disposal—and it is a common experience that one cannot do essentially better—the error term in Theorem 1.1 cannot be better estimated than by $\ll X$ (as $X \to \infty$) already if

$$z > \frac{\log X}{\log 2}\log\log X.$$

On the other hand, by (2.2) and (4.8) we always have the trivial estimate

$$S(\mathscr{A};\ \mathfrak{P}, z) \leqslant |\mathscr{A}| \leqslant X + |R_1|. \tag{5.9}$$

This clarifies our earlier statement that in order to get a significant result from Theorem 1.1 z has to be very small, because usually we should like to have z a power of X (cf. (4.7), (4.10), (4.11)) rather than a power of $\log X$. However, because we had to choose z so small, our result (5.8) became rather weak; it is actually inferior to results which follow by other elementary methods in prime number theory. On the other hand—and this remark will be used several times later on—for *bounded* values of z Theorem 1.1 always gives a good result, and so it may serve as a starting point in further investigations of S in so far as it permits us then to assume immediately that

$$z \geqslant B_0,$$

where B_0 is sufficiently large.

NOTES

1.1. The sieve of Eratosthenes: see L. E. Dickson's "History of the theory of numbers" (Washington, 1919, Vol. I, Chapter XIII).

For an extensive table of prime numbers see D. N. Lehmer, "List of Prime Numbers from 1 to 10,006,721," Carnegie Institution of Washington No. 165 (1914).

Brun [9] attributes the term "prime twins" to Stäckel, who carried out some numerical calculations in connection with this and related problems. There are 1224 prime twins below 100,000 (Glaisher, *Mess. Math.* **8** (1878), 28–33) and 5328 below 600,000 (see Hardy–Littlewood [1]). Schinzel and Sierpinski [1] quote tables of prime "quadruplets" and "quintuplets" from V. A. Golubev, *Anzeigen Oesterr. Akad. Wiss.* (1956), 153–157; *ibid* (1957), 82–87 and Sexton, *ibid* (1955), 236–239. D. H. and E. Lehmer have carried out counts of various prime pairs, triplets and quadruplets up to 4×10^7, and their results are deposited in the Unpublished Math. Tables file of Math. tables and other aids to computation. Shen, *Nord. Tidskr. Inf. Beh.* **4** (1964), 243–245 reports that he has checked Goldbach's binary conjecture up to 33,000,000.

For earlier formulations of the general sieve problem see especially Ankeny and Onishi [1] and the important recent memoir of Selberg [5]. Of course, there are other accounts of sieve methods in which various degrees of generality and precision of formulation are achieved.

Erdös [11] uses a "squarefree" sieve, i.e. a sieve consisting of squares of primes, to show that the least square-free number in the arithmetic progression $l \bmod k$, $(l, k) = 1$, is $\ll k^{3/2}/\log k$. An "s-free" sieve occurs already in the works of Ricci [2, 4, 6, 7]; for some applications see also Cugiani [1, 2].

The earliest accounts of Brun's famous discovery are to be found in Brun [1, 2]. In point of time Merlin made the first serious attempt to go beyond Eratosthenes. Unfortunately he was killed in World War I; apart from an announcement in Comptes Rendus of 1911 (Merlin [1]) (communicated by Poincaré) there is only the one paper, Merlin [2], prepared by Hadamard for posthumous publication. It is rather obscure, but Brun obviously saw in it the basis for a viable approach and made handsome acknowledgement, in

Brun [5], of "Merlin's sieve". An analysis of Merlin's work can be found in R. Rotermund, Staatsexemensarbeit Marburg 1970.

1.2. The Möbius function μ is a multiplicative arithmetic function defined on the sequence of positive integers by

$$\mu(d) = \begin{cases} 1, \ d = 1 \\ 0 \ \text{if } d \text{ is not squarefree} \\ (-1)^r \text{ if } d \text{ is the product of } r \text{ distinct primes.} \end{cases}$$

It follows that if f is any multiplicative arithmetic function, then

$$\sum_{d|n} \mu(d)f(d) = \prod_{p|n} (1 - f(p));$$

in particular, taking $f \equiv 1$,

$$\sum_{d|n} \mu(d) = \begin{cases} 1, \ n = 1 \\ 0, \ n > 1. \end{cases}$$

If two arithmetic functions f and g (not necessarily multiplicative) are related by

$$f(n) = \sum_{d|n} g(d), \quad n = 1, 2, \ldots$$

then these relations may be inverted to give the Möbius inversion formula

$$g(n) = \sum_{d|n} \mu(d) f\left(\frac{n}{d}\right), \quad n = 1, 2, \ldots$$

(See H–W, Chapter XVI, especially Theorem 266.)

1.3. For a proof that ρ is multiplicative, see H–W, Theorem 122.
For a proof of Lagrange's theorem, see H–W, Theorem 107.
The asymptotic formula

$$\sum_{p<x} \rho(p) \sim \frac{x}{\log x}, \ x \to \infty,$$

derives from the Prime Ideal Theorem; for an account of this see Landau "Einführung in die elementare und analytische Theorie der algebraischen Zahlen und der Ideale," Chelsea, 1949. We shall have no occasion to use this deep result; all that we shall ever need is (1.3.17) which is due to Nagel [1].

For a statement and proof of the Chinese Remainder Theorem see H–W, Theorem 121.

Dirichlet's theorem for primes in arithmetic progression: a quantitative statement of this, derived from Dirichlet's argument by Mertens in 1874 (see *D*, Chapter 7), is that

$$\sum_{\substack{p<x \\ p\equiv l \bmod k}} \frac{1}{p} = \frac{1}{\phi(k)}\log\log x + O_k(1), \quad (l, k) = 1.$$

A much stronger statement of this is the Prime Number Theorem for primes in an arithmetic progression l mod k, $(l, k) = 1$, according to which, for fixed k,

$$\pi(x; k, l) \sim \frac{li\, x}{\phi(k)} + O(x \exp(-c\sqrt{\log x}))$$

(de la Vallée Poussin, *Mem. Acad. Belgique* **59** Nr. 1 (1899/1900)); the Prime Number Theorem itself is the special case $k = 1$. (See D, Chapter 20 or Prachar [6], Chapters III, IV; these books also describe how this result can be made uniform for $k \leqslant (\log x)^C$.)

The fact that ρ_1 is a multiplicative function on the sequence of squarefree numbers follows from the relation

$$\rho_1(d) = \sum_{\substack{t|d \\ t|F(0)}} \mu(t)\,\rho\,(d/t), \quad \mu(d) \neq 0,$$

since ρ also is multiplicative.

1.4. The functions ω_0 and ω: we could have introduced $\mathscr{A}$ and $\mathfrak{P}$ simultaneously at the outset and thus avoided the introduction of ω_0 by working with ω from the first. This might well be the appropriate procedure when planning a lecture course; but we have been at pains here to stress the initial independence of $\mathscr{A}$ and $\mathfrak{P}$. The same remark applies to the remainders r_d and R_d.

1.5. A.–M. Legendre, "Théorie des Nombres," 2 éd., Paris 1808, p.420.

Note that (5.1) (with $q = 1, \mathscr{A} = \{n: n \leqslant x\}, \mathfrak{P} = \mathfrak{P}_1, z = x^{\frac{1}{2}}$) states that

$$\pi(x) - \pi(x^{\frac{1}{2}}) + 1 = \sum_{d|P(x^{\frac{1}{2}})} \mu(d)\left[\frac{x}{d}\right],$$

a formula which Legendre used in his numerical studies of $\pi(x)$. For an account of various later improvements of Legendre's method (such as Meissel's) as a means of computing $\pi(x)$ exactly see D. H. Lehmer, *Illinois J. Math.* **3** (1959), 381–388.

(5.7): This estimate goes back to Euler's proof (in 1737) of the infinitude of prime numbers. It is easy to see that

$$1/V(z) = \prod_{p<z}\left(1 - \frac{1}{p}\right)^{-1} \geqslant \sum_{1 \leqslant n < z} \frac{1}{n} \geqslant \log z \qquad (z \geqslant 2),$$

so that $\ll$ in (5.7) can be replaced by $\leqslant$. Actually, Mertens proved (*J. für Math.* **78** (1874), 46–62) that

$$V(z) \sim \frac{e^{-\gamma}}{\log z}, \qquad z \to \infty.$$

(5.8): Čebyčev was the first to show that an elementary analysis of the prime decomposition of the binomial coefficient $\binom{2n}{n}$ leads to

$$\frac{x}{\log x} \ll \pi(x) \ll \frac{x}{\log x}$$

(H–W, Theorem 414 and Section 22.4). It is of interest that none of the standard sieve methods is good enough to prove the left-hand inequality.

One might say that the Eratosthenes–Legendre sieve rests on the use of the 'trivial' sieving factor $\sum_{d|n} \mu(d)$.

Nevertheless, there are numerous instances in the literature where use of this factor can be made to lead (when combined with other ideas) to highly non-trivial results. The most remarkable instance of this is I.M. Vinogradov's treatment of the exponential sum

$$\sum_{p \leqslant x} \exp(2\pi i \theta p)$$

in his celebrated proof of the three primes theorem (Goldbach's hypothesis for large enough odd integers)—see e.g. Prachar [6], Chapter VI. For other examples see e.g. Erdös [3, 14, 17], and Hooley [1]; for interesting applications see e.g. Vinogradov–Linnik [1] ,Prachar [9]; Ducci [1, 2]; Norton [2]. The corresponding sieving factor for s-free numbers is

$$\sum_{d^s|n} \mu(d) \qquad (s \geqslant 2).$$

Chapter 2

The Combinatorial Sieve

1. The General Method

The method of Eratosthenes–Legendre rests on the use of

$$\sum_{d \mid (a, P(z))} \mu(d)$$

as the characteristic function of the sequence of integers that are coprime with $P(z)$. It leads to a bad result because, unless z is very small, the "remainder" sum

$$\sum_{d \mid P(z)} |R_d|$$

has too many terms. (To keep the exposition simple, we shall now refer to Theorem 1.1 with $q = 1$.) The sieve methods considered by us set out to improve Theorem 1.1 by simulating

$$\sigma_0(n) := \sum_{d \mid n} \mu(d) \tag{1.1}$$

with an arithmetic function of type

$$\sigma(n) := \sum_{d \mid n} \mu(d)\,\chi(d), \qquad \sigma(1) = \chi(1) = 1, \tag{1.2}$$

where $\sigma(n)$ is sufficiently often near to 0 (if $n > 1$) for $\sigma\big((n, P(z))\big)$ to act rather like $\sigma_0\big((n, P(z))\big)$ in "picking out" integers n that are coprime with $P(z)$, while also being such as to keep the remainder in check.

When we invert (1.2) we obtain

$$\mu(d)\,\chi(d) = \sum_{\delta \mid d} \mu\left(\frac{d}{\delta}\right) \sigma(\delta); \tag{1.3}$$

and we use this relation to compare (see (1.5.1))

$$S(\mathscr{A}; \mathfrak{P}, z) = \sum_{d \mid P(z)} \mu(d)\,|\mathscr{A}_d| \tag{1.4}$$

with
$$\sum_{d|P(z)} \mu(d)\,\chi(d)\,|\mathscr{A}_d|.$$

To this end we use (1.3) in order to write

$$\sum_{d|P(z)} \mu(d)\,\chi(d)\,|\mathscr{A}_d| = \sum_{d|P(z)} |\mathscr{A}_d| \sum_{\delta|d} \mu\left(\frac{d}{\delta}\right) \sigma(\delta)$$
$$= \sum_{\delta|P(z)} \sigma(\delta) \sum_{t|P(z)/\delta} \mu(t)|\mathscr{A}_{\delta t}|$$

so that, by (1.4), on separating the term $\delta = 1$ from the rest, we obtain

$$S(\mathscr{A}; \mathfrak{P}, z) = \sum_{d|P(z)} \mu(d)\,\chi(d)\,|\mathscr{A}_d| - \sum_{1<d|P(z)} \sigma(d)\, S(\mathscr{A}_d; \mathfrak{P}^{(d)}, z), \tag{1.5}$$

where
$$\mathfrak{P}^{(d)} = \{p : p \in \mathfrak{P}, p \nmid d\}.$$

Although (1.5) gives us a basis for comparison, it suffers from the disadvantage that the second sum on the right hand side involves σ rather than χ. We shall therefore rewrite the second expression in a way (see identity (1.8) below) which, simultaneously, will provide us with a starting point for the combinatorial sieve. For this purpose we note the identity, valid for any squarefree $d > 1$ and any prime $p \mid d$,

$$\sigma(d) = \sum_{l|d/p} \mu(l)\,\chi(l) + \sum_{l|d/p} \mu(pl)\,\chi(pl) = \sum_{l|d/p} \mu(l)\,\{\chi(l) - \chi(pl)\}. \tag{1.6}$$

Let

$$P_{z_1,z} = \prod_{\substack{p\in\mathfrak{P}\\ z_1\leqslant p<z}} p = \frac{P(z)}{P(z_1)}, \qquad 2 \leqslant z_1 \leqslant z, \tag{1.7}$$

and introduce the notation $q(d)$ for the least prime factor of an integer $d > 1$, defining $q(1)$ to be ∞. Then we arrive easily at

$$S(\mathscr{A}; \mathfrak{P}, z_1) = \sum_{t|P_{z_1,z}} \sum_{\substack{a\in\mathscr{A}\\ (a,\,P(z_1))=1\\ (a,\,P_{z_1,z})=t}} 1 = \sum_{t|P_{z_1,z}} \sum_{\substack{a\in\mathscr{A}_t\\ (a,\,P(z)/t)=1}} 1$$
$$= \sum_{t|P_{z_1,z}} S(\mathscr{A}_t; \mathfrak{P}^{(t)}, z).$$

These remarks, using (1.6) with $p = q(d)$, enable us to write the second sum on the right of (1.5) in the form

$$\sum_{\delta|P(z)} \sum_{\substack{p|P(z)\\p<q(\delta)}} S(\mathscr{A}_{p\delta}; \mathfrak{P}^{(p\delta)}, z) \sum_{l|\delta} \mu(l)\{\chi(l) - \chi(pl)\}$$

$$= \sum_{l|P(z)} \sum_{\substack{p|P(z)\\p<q(l)}} \mu(l)\{\chi(l) - \chi(pl)\} \sum_{\substack{t|P(z)/l\\p<q(t)}} S(\mathscr{A}_{plt}; \mathfrak{P}^{(plt)}, z)$$

$$= \sum_{l|P(z)} \sum_{\substack{p|P(z)\\p<q(l)}} \mu(l)\{\chi(l) - \chi(pl)\} S(\mathscr{A}_{pl}; \mathfrak{P}^{(pl)}, p),$$

so that, on substituting in (1.5),

$$S(\mathscr{A}; \mathfrak{P}, z) = \sum_{d|P(z)} \mu(d)\chi(d)|\mathscr{A}_d| - \sum_{d|P(z)} \sum_{\substack{p|P(z)\\p<q(d)}} \mu(d)\{\chi(d) - \chi(pd)\} S(\mathscr{A}_{pd}; \mathfrak{P}, p). \tag{1.8}$$

It seems at first sight that there should be $\mathfrak{P}^{(pd)}$ in place of $\mathfrak{P}$ on the right of (1.8); however, what is implied there is the truncation of $\mathfrak{P}^{(pd)}$ at p, with $q(d) > p$, and this coincides with the truncation of $\mathfrak{P}$ at p. As has already been said, we shall need to return to (1.8) later; but we record at once one very special but very important case of (1.8), namely that corresponding to the choice $\chi(1) = 1, \chi(d) = 0$ for $d > 1$:

$$S(\mathscr{A}; \mathfrak{P}, z) = |\mathscr{A}| - \sum_{\substack{p<z\\p\in\mathfrak{P}}} S(\mathscr{A}_p; \mathfrak{P}, p). \tag{1.9}$$

Later on this identity will prove to be highly effective, especially when written in the "differenced" form

$$S(\mathscr{A}; \mathfrak{P}, z) = S(\mathscr{A}; \mathfrak{P}, z_1) - \sum_{\substack{z_1\leqslant p<z\\p\in\mathfrak{P}}} S(\mathscr{A}_p; \mathfrak{P}, p), \quad 2 \leqslant z_1 \leqslant z. \tag{1.10}$$

At first sight, the ideal choice of χ would be such as to make the second expression on the right of (1.5) (or (1.8)) of lower order of magnitude than the first, thus yielding an asymptotic formula for $S(\mathscr{A}; \mathfrak{P}, z)$. Unfortunately, it turns out that there is no method which will give such a formula for general $S(\mathscr{A}; \mathfrak{P}, z)$ unless z is small (see, for example, Theorem 1.1 or (2.16) below, or Theorem 2.5; there are, of course, special problems where it is possible to find asymptotic formulae), and so the utmost we can hope for is to find choices of χ which lead to good upper or lower estimates for $S(\mathscr{A}; \mathfrak{P}, z)$, i.e. bounds of the right, or nearly right, order of magnitude. It is at once clear from (1.5) that we obtain a bound of one kind or the other for $S(\mathscr{A}; \mathfrak{P}, z)$ if the second expression on the right is of constant sign.

We shall therefore shift our point of view and aim to obtain good upper as well as lower estimates of $S(\mathscr{A}; \mathfrak{P}, z)$ rather than an asymptotic formula (such as Theorem 1.1). (Nevertheless, combinatorial identities of the kind we have considered will continue to play an important part; for example, if we have an asymptotic formula or a lower (upper) estimate for $S(\mathscr{A}; \mathfrak{P}, z_1)$ and upper (lower) estimates for the $S(\mathscr{A}_p; \mathfrak{P}, p)$, then (1.10) yields a lower (upper) estimate of $S(\mathscr{A}; \mathfrak{P}, z)$.) Accordingly, we introduce functions σ_1 and σ_2 of type (1.2), that is,

$$\sigma_\nu(n) = \sum_{d|n} \mu(d)\,\chi_\nu(d), \qquad \sigma_\nu(1) = \chi_\nu(1) = 1 \qquad (\nu = 1, 2), \tag{1.11}$$

where σ_1, σ_2 satisfy also

$$\sigma_2(d) \leqslant \sigma_0(d) \leqslant \sigma_1(d) \quad \text{for all} \quad d \mid P(z); \tag{1.12}$$

then, as is equivalent to (1.12),

$$\sigma_2(d) \leqslant 0 \leqslant \sigma_1(d) \quad \text{for all} \quad d > 1, d \mid P(z), \tag{1.13}$$

and therefore, by (1.5),

$$\sum_{d|P(z)} \mu(d)\,\chi_2(d)\,|\mathscr{A}_d| \leqslant S(\mathscr{A}; \mathfrak{P}, z) \leqslant \sum_{d|P(z)} \mu(d)\,\chi_1(d)\,|\mathscr{A}_d|. \tag{1.14}$$

We remark that the inequalities (1.14) also follow from (1.8) with χ replaced by χ_1 and χ_2 in turn, provided that the functions χ_ν satisfy

$$(-1)^{\nu-1}\,\mu(d)\,\{\chi_\nu(d) - \chi_\nu(pd)\} \geqslant 0 \quad \text{whenever} \quad pd \mid P(z),$$
$$p < q(d) \qquad (\nu = 1, 2). \tag{1.15}$$

When we substitute (1.4.14) in (1.14) we obtain

$$X \sum_{d|P(z)} \mu(d)\,\chi_2(d)\frac{\omega(d)}{d} - \sum_{d|P(z)} |\chi_2(d)|\,|R_d| \leqslant S(\mathscr{A}; \mathfrak{P}, z)$$

$$\leqslant X \sum_{d|P(z)} \mu(d)\,\chi_1(d)\,\frac{\omega(d)}{d} + \sum_{d|P(z)} |\chi_1(d)|\,|R_d|. \tag{1.16}$$

The objective here then is to find efficient pairs of functions χ_ν (and hence σ_ν, $\nu = 1, 2$, satisfying (1.11) and (1.13)) for which the estimates in (1.16) yield a nontrivial *upper* bound when $\nu = 1$, and a non-trivial *lower* bound when $\nu = 2$; in the latter case it should be borne in mind that a *positive* lower bound may already represent a very substantial achievement.

We shall now make these remarks more explicit. We are faced with a pair of extremal problems: namely, to find functions χ_1 and χ_2 satisfying (1.11) and (1.12) such that

$$\text{(I)} \qquad \sum_{d|P(z)} \mu(d)\,\chi_1(d)\frac{\omega(d)}{d} \qquad \textit{is a minimum,}$$

and

$$\sum_{d|P(z)} \mu(d)\,\chi_2(d)\frac{\omega(d)}{d} \qquad \textit{is a maximum,}$$

subject to

$$\text{(II)} \qquad \sum_{d|P(z)} |\chi_\nu(d)|\,|R_d| \ (\nu = 1, 2) \ \textit{sufficiently small.}$$

We can, with advantage, clarify somewhat the requirements (I). If we apply (1.3) to χ_ν, write P for $P(z)$, and assume that (Ω_1) holds (so that $g(\delta)$ is well defined) we have

$$(\Omega_1)\colon \sum_{d|P} \mu(d)\,\chi_\nu(d)\frac{\omega(d)}{d}$$

$$= \sum_{d|P} \frac{\omega(d)}{d} \sum_{\delta|d} \mu\left(\frac{d}{\delta}\right)\sigma_\nu(\delta) = \sum_{\delta|P} \sigma_\nu(\delta)\frac{\omega(\delta)}{\delta} \sum_{t|(P/\delta)} \mu(t)\frac{\omega(t)}{t}$$

$$= \sum_{\delta|P} \sigma_\nu(\delta)\frac{\omega(\delta)}{\delta} \prod_{p|(P/\delta)} \left(1 - \frac{\omega(p)}{p}\right) = W(z) \sum_{\delta|P} \sigma_\nu(\delta) g(\delta)$$

$$= W(z)\,\{1 + \sum_{1<\delta|P(z)} \sigma_\nu(\delta)\,g(\delta)\}. \tag{1.17}$$

Accordingly, our problem is to minimize each of

$$\left|\sum_{1<\delta|P(z)} \sigma_\nu(\delta)\,g(\delta)\right| \qquad (\nu = 1, 2) \tag{1.18}$$

subject to (II); with the important additional requirement in the case of $\nu = 2$ that

$$1 + \sum_{1<\delta|P(z)} \sigma_2(\delta)\,g(\delta) > 0. \tag{1.19}$$

In this second formulation it becomes clear that the extremal problem for $\nu = 2$ (that is, the lower estimate problem) is more difficult, in confirmation of what one would expect on intuitive grounds.

In their full generality both problems appear to be very difficult; and, indeed, they remain still unsolved. In such circumstances it is clear what one must try: one has to impose further restrictions on χ_1 and χ_2 (or, what amounts to the same thing, on σ_1 and σ_2) which, while consistent with the basic conditions (1.11), (1.12), will by their nature make it possible to realize (I) subject to (II).

The experience of Theorem 1.1 suggests that the constraint (II) must be our first consideration, and here the indication is that we should require

$$\chi_\nu(d) = 0 \quad \text{for} \quad d \notin \mathscr{D}_\nu,$$

where $\mathscr{D}_1$ and $\mathscr{D}_2$ are certain sets of divisors of $P(z)$ which are *thin* enough to ensure both (II) and also that the expressions (1.18) are small (subject to the additional condition (1.19) in the case of $\nu = 2$). One obvious way of arranging for (II) to be satisfied is to take

$$\mathscr{D}_\nu = \{d : d \mid P(z), d < y_\nu\} \qquad (\nu = 1, 2)$$

where the parameters y_1, y_2 remain to be suitably chosen; although this step appears to be rather drastic, A. Selberg has shown that it can be used in the construction of an effective sieve method. The difficulty here is that the nature of $\mathscr{D}_\nu$ does not help in any obvious way with (I), and, in fact, this particular extremal problem too is still unsolved. What Selberg did was to introduce another restriction, on the nature of σ_1 rather than χ_1, which enabled him brilliantly both to satisfy the (right hand) inequality (1.12) trivially *and* to solve completely the (modified) minimum problem in (I). Selberg's approach to the case of $\nu = 2$ lacks the elegant simplicity of his upper bound method but this may be no more than a true reflection of the deep nature of the maximum problem in (I); his approach may be viewed as a compound of his upper bound ($\nu = 1$) method and the Buchstab identity (1.9). We shall begin to study Selberg's method in the next chapter.

A second way of constructing the sets $\mathscr{D}_1$ and $\mathscr{D}_2$ (which historically came first) depends on allowing the functions χ_ν to take the values 0 and 1 only, and choosing them to be respectively the characteristic functions of $\mathscr{D}_1$ and $\mathscr{D}_2$. We shall also impose on χ_1 and χ_2 the condition (1.15) which, as we have pointed out, ensures the truth of inequalities (1.14); and we shall refer to any method with these properties as a *combinatorial* sieve. We shall devote the remainder of Chapter 2 to such sieves. The structure of $\mathscr{D}_1$ and $\mathscr{D}_2$ in a combinatorial sieve is apt to be rather more complicated, but we are able to derive the basic features of a class of admissible structures from an analysis of the condition (1.15). Before we do so, we add one final remark about the general nature of a combinatorial sieve: it leads, more or less simultaneously, to effective upper *and* lower estimates of $S(\mathscr{A}; \mathfrak{P}, z)$. Attractive as this feature is, we may suspect for this very reason—such is the perversity of mathematics!—that the optimal sieve method will not be found in the class of combinatorial sieves. (But see Section 2.9).

Let us assume then from now on (to the end of this chapter) that

$$\chi_\nu(d) = 0 \quad \text{or} \quad \chi_\nu(d) = 1 \quad \text{if} \quad d \mid P(z) \qquad (\nu = 1, 2).$$

Suppose also that χ_ν is a function satisfying (1.15). If $pd \mid P(z)$ and $p < q(d)$, (1.15) can happen only in one of the following three ways:

(i) $\chi_\nu(d) = \chi_\nu(pd)$,

(ii) $\chi_\nu(d) = 1,\ \chi_\nu(pd) = 0,\ \mu(d) = (-1)^{\nu-1}$,

(iii) $\chi_\nu(d) = 0,\ \chi_\nu(pd) = 1,\ \mu(d) = (-1)^{\nu}$.

If (i) were *always* true, then χ_ν would be completely determined: we should have $\chi_\nu(d) = 1$ for *all* $d \mid P(z)$ and we should then be back to Theorem 1.1. But it is reasonable to try, in view of the difficulty of the basic problem, to ensure that (i) is true *often*. Now one way to bring this about—a way that is rather natural if we remember that χ_ν is the characteristic function of a set $\mathscr{D}_\nu$ of divisors of $P(z)$—is to require that $\mathscr{D}_\nu$ is *divisor-closed* in the sense of:

(iv) *if* $\chi_\nu(d) = 1$ *and* $t|d$, *then* $\chi_\nu(t) = 1$.

If χ_ν has property (iv), we see at once that (iii) can never arise; moreover, (i) fails to occur only when $\chi_\nu(pd) = 0$, and even then not unless $\chi_\nu(d) = 1$. In this last case we must, in view of (ii), guard against the possibility of having $\mu(d) = (-1)^\nu$; and we may do so by requiring $\chi_\nu(pd)$ to be 1 whenever $\chi_\nu(d) = 1$ and $\mu(d) = (-1)^\nu$.

To sum up: If $\chi_\nu(d)$ $(\nu = 1, 2)$ are defined for all positive integers d dividing $P(z)$, and if for either ν these functions have the properties†

$$\chi_\nu(d) = 0 \quad \textit{or} \quad \chi_\nu(d) = 1 \quad \textit{if} \quad d \mid P(z), \tag{1.20}$$

$$\chi_\nu(1) = 1, \tag{1.21}$$

$$\chi_\nu(d) = 1 \quad \textit{implies that} \quad \chi_\nu(t) = 1 \quad \textit{for all} \quad t \mid d,\ d \mid P(z), \tag{1.22}$$

$$\chi_\nu(t) = 1,\ \ \mu(t) = (-1)^\nu \quad \textit{imply that} \quad \chi_\nu(pt) = 1 \quad \textit{for all} \quad pt \mid P(z), p < q(t), \tag{1.23}$$

then (1.15) *and hence* (1.14) *are true*, and we may say that χ_1, χ_2 give rise to a combinatorial sieve.

† In terms of the sets $\mathscr{D}_\nu$ of which the χ_ν are characteristic functions, (1.22) implies that $\mathscr{D}_\nu$ is divisor-closed, and if (1.23) is satisfied then $\mathscr{D}_\nu$ is called *odd* when $\nu = 1$ and *even* when $\nu = 2$.

If we take $\nu = 2$ and $t = 1$ in (1.23), we see that (1.23) requires that

$$\chi_2(p) = 1 \quad \textit{for all} \quad p \mid P(z). \tag{1.24}$$

It is convenient to record here the identity

$$\chi_\nu(t) - \chi_\nu(pt) = (-1)^{\nu-1}\mu(t)\chi_\nu(t)\{1 - \chi_\nu(pt)\}$$
$$\textit{if} \quad pt \mid P(z),\ p < q(t) \qquad (\nu = 1, 2), \tag{1.25}$$

which will prove useful below. To check (1.25), we see that the relation is trivially true whenever $\chi_\nu(t) = \chi_\nu(pt)$, and in view of (1.22), this is always the case if $\chi_\nu(pt) = 1$. Hence the only non-trivial case is $\chi_\nu(pt) = 0$, $\chi_\nu(t) = 1$ (cf. (ii) above) and here we are able to infer from (1.23) that $\mu(t) = (-1)^{\nu-1}$, so that (1.25) is true in this case also.

We shall now bring the identity (1.17) to a form that is especially useful for combinatorial sieves. Denoting by p^+ the prime (of $\mathfrak{P}$) which succeeds p, and using the notation (1.7), we shall prove that

(Ω_1):

$$\sum_{d \mid P(z)} \mu(d)\chi_\nu(d)\frac{\omega(d)}{d}$$

$$= W(z)\left\{1 + (-1)^{\nu-1}\sum_{p<z}\frac{\omega(p)}{p}\frac{W(p)}{W(z)}\sum_{t \mid P_{p^+,z}}\frac{\chi_\nu(t)(1-\chi_\nu(pt))}{t}\omega(t)\right\}. \tag{1.26}$$

We could derive (1.26) more directly, but it is instructive to start from the identity†

$$\sum_{p \mid d}\{\chi_\nu((d, P_{p^+,z})) - \chi_\nu((d, P_{p,z}))\} = 1 - \chi_\nu(d), \qquad d \mid P(z).$$

If we substitute for $\chi_\nu(d)$, from this relation, on the left of (1.26), the sum becomes

$$W(z) + \sum_{d \mid P(z)}\sum_{p \mid d}\mu\left(\frac{d}{p}\right)\{\chi_\nu((d, P_{p^+,z})) - \chi_\nu((d, P_{p,z}))\}\frac{\omega(d)}{d};$$

† This identity is easily proved by noting that for $d = 1$ it is trivial, and for $d > 1$, $d = p_1 \ldots p_r$ with $p_1 < \ldots < p_r$, $p_i \in \mathfrak{P}$, the left hand side becomes

$$\sum_{i=1}^{r-1}\{\chi_\nu(p_{i+1}\ldots p_r) - \chi_\nu(p_i \ldots p_r)\} + 1 - \chi_\nu(p_r).$$

we now replace the variable of summation d by

$$d = \delta p t \quad \text{where} \quad \delta \mid P(p), t \mid P_{p^+, z},$$

and obtain, using (1.25) in the second step,

$$\sum_{d \mid P(z)} \mu(d)\chi_\nu(d)\frac{\omega(d)}{d}$$

$$= W(z) + \sum_{p<z} \frac{\omega(p)}{p} \sum_{\delta \mid P(p)} \mu(\delta)\frac{\omega(\delta)}{\delta} \sum_{t \mid P_{p^+, z}} \mu(t)\frac{\chi_\nu(t) - \chi_\nu(pt)}{t}\omega(t)$$

$$= W(z) + (-1)^{\nu-1} \sum_{p<z} \frac{\omega(p)}{p} W(p) \sum_{t \mid P_{p^+, z}} \frac{\chi_\nu(t)(1 - \chi_\nu(pt))}{t}\omega(t).$$

Identity (1.26) now follows at once if we assume that (Ω_1) is satisfied.

We end this section by exhibiting a particularly simple combinatorial sieve. Let r be an arbitrary positive integer, and define $\chi^{(r)}$ on the set of all divisors of $P(z)$ by

$$\chi^{(r)}(d) = \begin{cases} 1, & \nu(d) \leqslant r - 1, \\ 0 & \text{otherwise}, \end{cases} \tag{1.27}$$

so that $\chi^{(r)}$ is the characteristic function of the set

$$\mathscr{D}^{(r)} := \{d\colon\, d \mid P(z),\ \nu(d) \leqslant r - 1\}.$$

Then if $p < q(d)$,

$$\mu(d)\{\chi^{(r)}(d) - \chi^{(r)}(pd)\} = \begin{cases} 0, & \nu(d) \neq r - 1, \\ (-1)^{r-1}, & \nu(d) = r - 1; \end{cases} \tag{1.28}$$

and therefore if we write $\chi^{(r)}$ in place of χ in (1.8) we obtain†

$$S(\mathscr{A}; \mathfrak{P}, z) = \sum_{\substack{d \mid P(z) \\ \nu(d) \leqslant r-1}} \mu(d)|\mathscr{A}_d| - (-1)^{r-1} \sum_{\substack{\delta \mid P(z) \\ \nu(\delta) = r}} S(\mathscr{A}_\delta; \mathfrak{P}, q(d)). \tag{1.29}$$

If we now take r to be, in succession, $2s + 1$ and $2s + 2$, with s a non-negative integer, we can check easily that $\chi_1 = \chi^{(2s+1)}$ and $\chi_2 = \chi^{(2s+2)}$ generate a combinatorial sieve (i.e. satisfy (1.20) to (1.23)). In particular, they satisfy

† The case $r = 1$ of (1.29) yields (1.9).

(1.15) (cf. (1.28)); hence (1.14) is true and we obtain (as would follow immediately also from (1.29))

$$\sum_{\substack{d|P(z)\\ \nu(d)\leqslant 2s+1}} \mu(d)\,|\mathscr{A}_d| \leqslant S(\mathscr{A};\mathfrak{P},z) \leqslant \sum_{\substack{d|P(z)\\ \nu(d)\leqslant 2s}} \mu(d)\,|\mathscr{A}_d|; \tag{1.30}$$

or, if we do not wish to sacrifice equality, we have also from (1.29) and (1.5.9) that

$$S(\mathscr{A};\mathfrak{P},z) = \sum_{\substack{d|P(z)\\ \nu(d)\leqslant r-1}} \mu(d)\,|\mathscr{A}_d| + \theta \sum_{\substack{d|P(z)\\ \nu(d)=r}} |\mathscr{A}_d|,\ |\theta| \leqslant 1. \tag{1.31}$$

The class of functions $\chi^{(r)}$ generates what we call Brun's "pure" sieve, the subject matter of the next section. It yields, in (2.14) below, the first significant improvement of Theorem 1.1, (1.5.3). The pure sieve of Brun is the simplest (non-trivial) instance of a combinatorial sieve. Brun constructed more elaborate function pairs χ_ν along similar lines, and these will be described in Section 4, together with a simple account of the general Brun method. Sections 5–8 contain applications of Brun's sieve. Section 9 introduces Rosser's combinatorial sieve.

2. Brun's Pure Sieve

The subject of this section is Brun's pure sieve, which will lead us to the relation (2.14) below. We could set out from the inequalities (1.30), or from the relation (1.31), thus rendering Lemma 2.1 below unnecessary. However, we prefer to give an account which is independent of our discussion of combinatorial sieves and is based instead on the earlier relations (1.16) and (1.17).

What we have referred to as Brun's pure sieve can be made to depend on the following observation, which shows from first principles how one is lead to the functions $\chi^{(r)}$.

Lemma 2.1. *For any natural number n and for any non-negative integer s we have*

$$\sum_{\substack{d|n\\ \nu(d)\leqslant 2s+1}} \mu(d) \leqslant \sum_{d|n} \mu(d) \leqslant \sum_{\substack{d|n\\ \nu(d)\leqslant 2s}} \mu(d). \tag{2.1}$$

Proof. If $n = 1$ the three sums are obviously equal, so that we may suppose that $n > 1$. Writing $\nu(n) = v(\geqslant 1)$, we have

$$\sum_{\substack{d|n\\ \nu(d)=m}} \mu(d) = (-1)^m \binom{v}{m}$$

since $\binom{v}{m}$ is precisely the number of squarefree divisors of n having $v(d) = m$ (note that $\binom{v}{m} = 0$ for $v < m$). Hence, for any positive integer k,

$$\sigma^{(k)}(n) := \sum_{\substack{d|n \\ \nu(d) \leqslant k-1}} \mu(d) = \sum_{m=0}^{k-1} (-1)^m \binom{v}{m}. \tag{2.2}$$

We shall show that

$$\sigma^{(k)}(n) = (-1)^{k-1} \binom{v-1}{k-1}, \tag{2.3}$$

which, in view of (2.2) and the fact that $\sum_{d|n} \mu(d) = 0$ for $n > 1$, is more than is required for a proof of (2.1). We use induction with respect to k. The relation (2.3) is clearly true for $k = 1$; and assuming its truth for k it follows, because of $\binom{v}{k} - \binom{v-1}{k-1} = \binom{v-1}{k}$, that

$$\sigma^{(k+1)}(n) = (-1)^k \binom{v}{k} + \sigma^{(k)}(n) = (-1)^k \binom{v-1}{k}$$

i.e. (2.3) for $k + 1$.

Remark. It is worth noting that Lemma 2.1 follows at once, without any binomial calculations, from (1.30) on taking $P(z) = \prod_{p|n} p$ and $\mathscr{A} = \{n\}$.

Lemma 2.1 can also be put into a form where the parity of the bound for $\nu(d)$ does not matter, and which is sometimes more convenient.

COROLLARY. *If n and r are positive integers, then*

$$\sum_{d|n} \mu(d) = \sum_{\substack{d|n \\ \nu(d) \leqslant r-1}} \mu(d) + \vartheta \sum_{\substack{d|n \\ \nu(d) = r}} \mu(d), \tag{2.4}$$

where $0 \leqslant \vartheta \leqslant 1$.

Proof. For $n = 1$ (2.4) is clearly true; and for $n > 1$ (2.4) can be written as

$$0 = \sigma^{(r)}(n) + \vartheta(\sigma^{(r+1)}(n) - \sigma^{(r)}(n)),$$

which, by (2.3), is obviously true for a suitable ϑ in $0 \leqslant \vartheta \leqslant 1$.

Lemma 2.1 enables us now to derive an improved sieve result from (1.16) and (1.17). Let us suppose, for the sake of argument, that ω is subject to (Ω_0) and (Ω_1), and that the remainders R_d satisfy (R).

Let r be an arbitrary positive integer, and let $\chi^{(r)}$ be as defined in (1.27). Then

$$\sigma^{(r)}(n) = \sum_{d|n} \mu(d)\chi^{(r)}(d) = \sum_{\substack{d|n \\ \nu(d) \leqslant r-1}} \mu(d)$$

is, by Lemma 2.1, a function of type σ_1 if r is odd and of type σ_2 if r is even. Moreover, by (2.3)

$$|\sigma^{(r)}(n)| = \binom{\nu(n)-1}{r-1} \leqslant \binom{\nu(n)}{r} \quad \text{if} \quad n > 1.$$

Making use of the hypotheses, it follows that the expression corresponding to (1.18) is

(Ω_1):

$$\left|\sum_{1<d|P(z)} \sigma^{(r)}(d)g(d)\right| \leqslant \sum_{1<d|P(z)} \binom{\nu(d)}{r} g(d)$$

$$= \sum_{m=r}^{\nu(P(z))} \binom{m}{r} \sum_{\substack{1<d|P(z) \\ \nu(d)=m}} g(d) \leqslant \sum_{m=r}^{\infty} \binom{m}{r} \frac{1}{m!}\left(\sum_{p<z} g(p)\right)^m$$

$$= \frac{1}{r!}\left(\sum_{p<z} g(p)\right)^r \exp\left\{\sum_{p<z} g(p)\right\}, \tag{2.5}$$

and also that

$$(R): \qquad \sum_{d|P(z)} \chi^{(r)}(d)\,|R_d| \leqslant \sum_{\substack{d|P(z) \\ \nu(d)\leqslant r-1}} \omega(d) \leqslant \left(1 + \sum_{p<z} \omega(p)\right)^{r-1}. \tag{2.6}$$

Thus, by (1.16) and (1.17), we obtain

(Ω_1), (R):

$$S(\mathscr{A}; \mathfrak{P}, z) = XW(z)\left(1 + \theta\frac{1}{r!}\left(\sum_{p<z} g(p)\right)^r \exp\left\{\sum_{p<z} g(p)\right\}\right)$$

$$+ \theta'\left(1 + \sum_{p<z} \omega(p)\right)^{r-1}, \tag{2.7}$$

for any positive integer r, where $|\theta| \leqslant 1$, $|\theta'| \leqslant 1$.

To take the estimate in (2.7) further we could use (Ω_0) in

$$(\Omega_1): \qquad g(p) \leqslant A_1\frac{\omega(p)}{p}, \tag{2.8}$$

so that

$$(\Omega_0), (\Omega_1): \qquad g(p) \leqslant \frac{A_0 A_1}{p}; \tag{2.9}$$

or we could require of ω simply that

$$\frac{\omega(p)}{p} \leqslant \frac{A}{p+A}, \tag{2.10}$$

which is also very convenient, because it implies both (Ω_0) and (Ω_1) as well as

$$g(p) \leqslant \frac{A}{p}.$$

Let us now impose condition (Ω_0) and suppose that $z \geqslant B_1 = B_1(A_0)$. Then

$$\sum_{p<z} \frac{1}{p} \leqslant \log\log z + 1,$$

and hence, by (2.9),

$$\sum_{p<z} g(p) \leqslant A_0 A_1 (\log\log z + 1).$$

We put

$$r = \left[\frac{A_0 A_1}{\lambda}(\log\log z + 1)\right] + 1,$$

where λ is a parameter satisfying

$$0 < \lambda e^{1+\lambda} \leqslant 1,$$

so that

$$\sum_{p<z} g(p) \leqslant \lambda r.$$

Using the well-known estimate

$$\frac{1}{r!} \leqslant \left(\frac{e}{r}\right)^r, \quad r = 1, 2, \ldots, \tag{2.11}$$

we obtain

$$\frac{1}{r!}\left(\sum_{p<z} g(p)\right)^r \exp\left\{\sum_{p<z} g(p)\right\} \leqslant \left(\frac{e}{r}\right)^r (\lambda r)^r e^{\lambda r} = (\lambda e^{1+\lambda})^r. \tag{2.12}$$

Also

$$\left(1 + \sum_{p<z} \omega(p)\right)^{r-1} \leqslant z^{r-1} \leqslant z^{(A_0 A_1/\lambda)(\log\log z + 1)}. \tag{2.13}$$

The estimates (2.12) and (2.13) now yield, when combined with (2.7),

$(\Omega_0), (\Omega_1), (R)$:

$$S(\mathscr{A};\mathfrak{P},z) = XW(z)\{1 + \theta\,(\lambda e^{1+\lambda})^{(A_0A_1/\lambda)\,(\log\log z+1)}\} + \theta' z^{(A_0A_1/\lambda)\,(\log\log z+1)}, \quad (\text{as } x \to \infty) \tag{2.14}$$

where $|\theta| \leqslant 1$, $|\theta'| \leqslant 1$ and

$$0 < \lambda e^{1+\lambda} \leqslant 1. \tag{2.15}$$

We have omitted the condition $z \geqslant B_1$ from (2.14), because otherwise (2.14) follows from (1.5.3).

Let us compare this consequence of the simplest case of Brun's sieve with (1.5.3). In the example following Theorem 1.1 (cf. (1.5.8)) we saw that there we could not choose z to be a larger power of $\log X$ than the first in order to obtain a significant result. Now, however, we are able to formulate a substantial improvement of (1.5.3). If we impose the much weaker constraint $\log z \leqslant \sqrt{\log X}$, and choose λ by

$$\frac{1}{\lambda} = \frac{1}{2A_0A_1}\,\frac{\log X}{\log z(\log\log z + 1)},$$

then λ clearly satisfies (2.15) if X is large, and (2.14) leads readily to the special relation

$(\Omega_0), (\Omega_1), (R)$:

$$S(\mathscr{A};\mathfrak{P},z) = XW(z)\{1 + \theta e^{-\sqrt{\log X}}\} + \theta' X^{\frac{1}{2}},$$
$$\log z \leqslant \sqrt{\log X},\ |\theta| \leqslant 1,\ |\theta'| \leqslant 1. \tag{2.16}$$

The simplest Brun sieve ((2.14) or (2.16)) thus gives us a considerable improvement of (1.5.3)—for example, we could allow z to be any power of $\log X$; however, it still has the defect that no power of X, however small, for z would give a satisfactory result, and, as we have pointed out before, this would be necessary for the most interesting cases.

Next, as another illustration of (2.14), consider the number of prime twins given in (1.4.8). A direct use of (1.4.10) is impossible, but for an upper estimate it suffices to remark that by the same argument which lead to (1.4.10) we see that the sifting function S there counts at least all twin-pairs $\geqslant z$, so that

$$|\{p\colon p \leqslant x,\ p+2 = p'\}| \leqslant S(\{n(n+2) : n \leqslant x\};\mathfrak{P}_1, z) + z. \tag{2.17}$$

By (1.3.23), (1.3.24) and (1.4.15) we may use

$$X = x,\ \omega(2) = 1,\ \omega(p) = 2 \quad \text{for} \quad p > 2,\ |R_d| \leqslant \omega(d) \quad \text{for} \quad \mu(d) \neq 0,$$

so that (Ω_0) with $A_0 = 2$, (Ω_1) with $A_1 = 3$, and (R) are satisfied; incidentally, we remark that it would be slightly more convenient to work with $\mathfrak{P}_2$ instead of $\mathfrak{P}_1$. Therefore, by (2.14), the right-hand side of (2.17) is, choosing $\lambda = \frac{1}{4}$,

$$\ll x \prod_{2<p<z} \left(1 - \frac{2}{p}\right) + z^{24(\log\log z + 1)}.$$

By (1.5.7) we find that

$$\prod_{2<p<z} \left(1 - \frac{2}{p}\right) \leqslant \prod_{2<p<z} \left(1 - \frac{1}{p}\right)^2 \ll \frac{1}{\log^2 z},$$

and with the choice of

$$\log z = \frac{1}{25} \frac{\log x}{\log\log x} \tag{2.18}$$

we now obtain

$$|\{p\colon p \leqslant x,\ p + 2 = p'\}| \ll \frac{x}{\log^2 x} (\log\log x)^2, \tag{2.19}$$

as $x \to \infty$.

Actually, this estimate was the first result Brun obtained for the prime twin problem. He put it into the striking form, which is an easy consequence of (2.19), that the sum

$$\sum_{\substack{p \\ p+2=p'}} \frac{1}{p}$$

is, at any rate, convergent. Later, by more refined methods, he was able to remove the factor $(\log\log x)^2$. The best result which is known so far in this direction is contained in Theorem 3.11 (see Chapter 3).

We illustrated Theorem 1.1 by deriving from it an upper estimate (cf. (1.5.8)) for $\pi(x)$: when (2.14) is applied to this problem, one obtains

$$\pi(x) \ll \frac{x}{\log x} \log\log x.$$

Although still inferior to what results from other elementary methods, this is a considerable improvement of (1.5.8). If we had applied Theorem 1.1 to the counting number on the left of (2.19) we should have obtained only a trivial estimate.

3. Technical Preparation

We shall devote this section to some technical preparation, with the main object of deriving several basic properties of the function $W(z)$, which are needed for Brun's sieve and are also of importance in some later chapters.

We start with a few remarks about our ω-conditions.

First we recall from (2.10) that it is sometimes convenient to require of ω simply that

$$\frac{\omega(p)}{p} \leqslant \frac{A}{p+A} \tag{3.1}$$

because this inequality implies both (Ω_0) and (Ω_1).

Actually, we shall prove Lemma 2.3 below under yet another condition on ω, of an "average" kind. We shall suppose that, for suitable constants $\kappa(>0)$ and $A_2(\geqslant 1)$,

$$(\Omega_2(\kappa)) \qquad \sum_{w \leqslant p < z} \frac{\omega(p)\log p}{p} \leqslant \kappa \log\frac{z}{w} + A_2 \quad \text{if} \quad 2 \leqslant w \leqslant z.$$

Let us clarify the connection between $(\Omega_2(k))$ and (Ω_0). It is known that

$$\sum_{w \leqslant p \leqslant z} \frac{\log p}{p} \leqslant \log\frac{z}{w} + 1 \quad \text{if} \quad 2 \leqslant w \leqslant z; \tag{3.2}$$

and this implies at once that

Lemma 2.2. (Ω_0) :

$$\sum_{w \leqslant p < z} \frac{\omega(p)\log p}{p} \leqslant A_0 \log\frac{z}{w} + A_0 \quad \textit{if} \quad 2 \leqslant w \leqslant z;$$

in other words, (Ω_0) *implies* $(\Omega_2(\kappa))$ *with* $\kappa = A_2 = A_0$.

Thus (Ω_0) is the stronger of the two conditions; however it is much simpler and much more convenient to check.

Lemma 2.3. $(\Omega_2(\kappa))$:

$$\sum_{w \leqslant p < z} \frac{\omega(p)}{p} \leqslant \kappa \log\frac{\log z}{\log w} + \frac{A_2}{\log w} \quad \textit{if} \quad 2 \leqslant w \leqslant z. \tag{3.3}$$

If $\omega(p)$ satisfies both (Ω_1) and $(\Omega_2(\kappa))$, then if

$$2 \leqslant w \leqslant z,$$

we have

$$\sum_{w \leqslant p < z} g(p) \leqslant \kappa \log \frac{\log z}{\log w} + O\left(\frac{1}{\log w}\right), \tag{3.4}$$

$$\frac{W(w)}{W(z)} \leqslant \left(\frac{\log z}{\log w}\right)^{\kappa} \left\{1 + O\left(\frac{1}{\log w}\right)\right\} = O\left(\frac{\log^{\kappa} z}{\log^{\kappa} w}\right), \tag{3.5}$$

and

$$\frac{1}{W(z)} = O(\log^{\kappa} z). \tag{3.6}$$

Proof. By partial summation $(\Omega_2(\kappa))$ implies that

$$\sum_{w \leqslant p < z} \frac{\omega(p)}{p} \leqslant \frac{\kappa \log (z/w) + A_2}{\log z} + \int_w^z \frac{\kappa \log (t/w) + A_2}{t \log^2 t}\, dt.$$

A simple calculation shows that the right-hand side equals

$$\kappa \log \frac{\log z}{\log w} + \frac{A_2}{\log w},$$

and this proves (3.3).

By another application of partial summation we obtain

$$\sum_{w \leqslant p < z} \frac{\omega(p)}{p \log p} \leqslant \frac{1}{\log w} \left\{\kappa + \frac{A_2}{\log w}\right\}. \tag{3.7}$$

If, in particular, we take $w = p$ and $z = p + \varepsilon$ in $(\Omega_2(\kappa))$ we see that $(\Omega_2(\kappa))$ also implies that†

$$\frac{\omega(p)}{p} \log p \leqslant A_2. \tag{3.8}$$

Hence, using (2.8) and (3.7),

$$\sum_{w \leqslant p < z} \frac{\omega(p)}{p} g(p) \leqslant A_1 A_2 \sum_{w \leqslant p < z} \frac{\omega(p)}{p \log p} \leqslant \frac{A_1 A_2}{\log w} \left\{\kappa + \frac{A_2}{\log w}\right\}. \tag{3.9}$$

† This appears to be a very weak result, but we cannot derive anything better from $(\Omega_2(\kappa))$ for a single term. If (Ω_0) were available, we could obviously do much better.

By definition

$$1 + g(p) = \frac{1}{1 - \dfrac{\omega(p)}{p}} \tag{3.10}$$

and

$$g(p) = \frac{\omega(p)}{p} + \frac{\omega(p)}{p} g(p).$$

Hence it follows from (3.3) and (3.9) that

$$\sum_{w \leqslant p < z} g(p) \leqslant \kappa \log \frac{\log z}{\log w} + \frac{A_2}{\log w} + \frac{A_1 A_2}{\log w} \left\{ \kappa + \frac{A_2}{\log w} \right\}, \tag{3.11}$$

which proves (3.4), and it also yields

$$\frac{W(w)}{W(z)} = \frac{1}{\prod\limits_{w \leqslant p < z} \left(1 - \dfrac{\omega(p)}{p}\right)}$$

$$= \prod_{w \leqslant p < z} (1 + g(p)) \leqslant \exp \left\{ \sum_{w \leqslant p < z} g(p) \right\}$$

$$\leqslant \exp \left\{ \kappa \log \frac{\log z}{\log w} + \frac{A_2}{\log w} \left(1 + A_1 \kappa + \frac{A_1 A_2}{\log 2}\right) \right\}. \tag{3.12}$$

This proves (3.5), and (3.6) follows from (3.5) on putting $w = 2$.

LEMMA 2.4. $(\Omega_2(\kappa))$: *We have*

$$\sum_{p < z} \omega(p) \leqslant (\kappa + A_2)\, li\, z + \frac{2A_2}{\log 2} \leqslant A(2\, li\, z + 3), \qquad A := \max(\kappa, A_2). \tag{3.13}$$

Proof. By (3.3) we have

$$\sum_{p<z} \omega(p) = 2 \sum_{p<z} \frac{\omega(p)}{p} + \int_2^z \sum_{w \leqslant p < z} \frac{\omega(p)}{p} \, dw$$

$$\leqslant 2\kappa \log \left(\frac{\log z}{\log 2}\right) + \frac{2A_2}{\log 2} + \int_2^z \left(\kappa \log \frac{\log z}{\log w} + \frac{A_2}{\log w}\right) dw$$

$$= \frac{2A_2}{\log 2} + (\kappa + A_2)\, li\, z,$$

and this proves (3.13).

We end this section by pointing out that use of the weaker condition $(\Omega_2(\kappa))$ in place of (Ω_0) leads to more general analogues of (1.5.3) and (2.16).

Taking (1.5.3) first, we deduce at once from (1.5.2) (with $q = 1$) that if (R) is imposed then

$$S(\mathscr{A}; \mathfrak{P}, z) = XW(z) + \theta \exp\left(\sum_{p<z} \omega(p)\right), \qquad |\theta| \leqslant 1;$$

and it follows from Lemma 2.4 that

$(\Omega(\kappa)), (R)$:

$$S(\mathscr{A}; \mathfrak{P}, z) = XW(z) + \theta e^{A(2 \operatorname{li} z + 3)}, \qquad A = \max(\kappa, A_2), \qquad |\theta| \leqslant 1. \tag{3.14}$$

This is a more general version of Eratosthenes–Legendre than (1.5.3).

We turn next to (2.16). In view of (3.14), we may assume, as we did at the corresponding stage in the previous section, that

$$z \geqslant B_2$$

where $B_2 = B_2(\kappa, A_2)$. We set out from (2.7) but now make use of (3.13) and

$$\sum_{p<z} g(p) \leqslant \kappa \log\log z + c_0, \tag{3.15}$$

$$c_0 = \kappa \log\left(\frac{1}{\log 2}\right) + \frac{A_2}{\log 2}\left\{1 + A_1\left(\kappa + \frac{A_2}{\log 2}\right)\right\},$$

the latter being a direct consequence of (3.11) (with $w = 2$). With λ satisfying (2.15) we choose

$$r = \left[\frac{1}{\lambda}(\kappa \log\log z + c_0)\right] + 1$$

so that

$$\sum_{p<z} g(p) \leqslant \lambda r$$

as before, and, arguing now in a closely similar manner we arrive at (cf. (2.14))

$(\Omega_1), (\Omega_2(\kappa)), (R)$:

$$\begin{aligned} S(\mathscr{A}; \mathfrak{P}, z) = XW(z)\{1 + \theta(\lambda e^{1+\lambda})^{(\kappa \log\log z + c_0)/\lambda}\} \\ + \theta' z^{(\kappa \log\log z + c_0)/\lambda}, \quad |\theta| \leqslant 1, |\theta'| \leqslant 1. \end{aligned} \tag{3.16}$$

Finally, if we choose

$$\frac{1}{\lambda} = \frac{1}{2}\frac{\log X}{\log z(\kappa \log\log z + c_0)},$$

so that (2.15) is satisfied provided that $\log z \leqslant \sqrt{(\log X)}$ and X is large enough, we arrive easily at (cf. (2.16))

$(\Omega_1), (\Omega_2(\kappa)), (R)$:

$$S(\mathscr{A}; \mathfrak{P}, z) = XW(z)\{1 + \theta e^{-\sqrt{\log X}}\} + \theta' X^{\frac{1}{2}}, \qquad (3.17)$$
$$\log z \leqslant \sqrt{(\log X)}, |\theta| \leqslant 1, |\theta'| \leqslant 1,$$

as $X \to \infty$.

4. Brun's Sieve

In Sections 2 and 3 we made it clear that if we take the sets $\mathscr{D}_1$ and $\mathscr{D}_2$ of Section 1 to be of the type $\mathscr{D}^{(r)}$ we already obtain a result vastly superior to Theorem 1.1. But there are other encouraging signs to be found there. Let us consider, for example, the calculation in (2.6)—the estimation of the remainder term. The point here is that r has to be taken rather large to obtain an effective result, but this appears to compel us to choose z smaller than we might wish. However, we can easily see how to overcome this difficulty. Let us decide that the prime divisors of d from $\mathscr{D}^{(r)}$ should, apart from being at most $r - 1$ in number, also be restricted in size in the sense that at most D_1, say, of them can come from an interval $z_1 \leqslant p < z$. In this event we should have at once the superior estimate

$$\sum_{d|P(z)} \chi^{(r)}(d)|R_d| \leqslant \left(1 + \sum_{p<z} \omega(p)\right)^{D_1} \left(1 + \sum_{p<z_1} \omega(p)\right)^{r-1-D_1};$$

and if that did not suffice we could introduce a further parameter, $z_2 < z_1$, and require not more than D_2 prime divisors of d in $\mathscr{D}^{(r)}$ to come from the interval $z_2 \leqslant p < z_1$; and so forth.

Of course, the introduction of such additional restrictions on the members of $\mathscr{D}^{(r)}$ complicates the structure of $\mathscr{D}^{(r)}$, and hence the estimation of the associated expression in (1.14); and from the expositions available in the literature Brun's method does, indeed, appear to be exceedingly involved. However, a recent suggestion by Levin has enabled us to formalise Brun's technique in a very simple way so that his method can be seen, almost for the first time, to be both exceptionally elegant and unexpectedly powerful.

Our object in this section then is to describe Brun's general method and, in particular, to prove Theorem 2.1 below, our main Brun result. Although the proof is relatively simple, and even though no effort is made to obtain the best possible result, Theorem 2.1 is already rather powerful, as will be evident from our applications.

THEOREM 2.1.† (Ω_1), $(\Omega_2(\kappa))$, (R): *Let b be a positive integer, and let λ be a real number satisfying*

$$0 < \lambda e^{1+\lambda} < 1. \tag{4.1}$$

Then

$$S(\mathscr{A};\mathfrak{P},z) \leqslant XW(z)\left\{1 + 2\frac{\lambda^{2b+1}e^{2\lambda}}{1-\lambda^2 e^{2+2\lambda}}\exp\left((2b+3)\frac{c_1}{\lambda\log z}\right)\right\} + O(z^{2b+(2\cdot 01/(e^{2\lambda/\kappa}-1))}) \tag{4.2}$$

and

$$S(\mathscr{A};\mathfrak{P},z) \geqslant XW(z)\left\{1 - 2\frac{\lambda^{2b}e^{2\lambda}}{1-\lambda^2 e^{2+2\lambda}}\exp((2b+2)\frac{c_1}{\lambda\log z}\right\} + O(z^{2b-1+(2\cdot 01/(e^{2\lambda/\kappa}-1))}) \tag{4.3}$$

where

$$c_1 = \frac{A_2}{2}\left\{1 + A_1\left(\kappa + \frac{A_2}{\log 2}\right)\right\}.$$

Remark 1. Note that the constants implied by the use of the O-notation do not depend on b and λ.

Remark 2. It is easy to check that replacement of condition (R) by the more general

$$|R_d| \leqslant L\omega(d)$$

changes Theorem 2.1 only to the extent of introducing a factor L into the last error term in each of (4.2) and (4.3).

Proof. We may assume that

$$z \geqslant B_3, \tag{4.4}$$

where B_3 is sufficiently large, since otherwise our theorem follows from (3.14).

Following our discussion in Section 1 of the combinatorial sieve we recall that if $\chi_1(d)$, $\chi_2(d)$ are functions defined on the set of divisors of $P(z)$ and possess properties (1.20) to (1.23), then, by (1.16),

$$(-1)^\nu S(\mathscr{A};\mathfrak{P},z) \geqslant (-1)^\nu X\sum_{d|P(z)}\mu(d)\chi_\nu(d)\frac{\omega(d)}{d} - \sum_{d|P(z)}\chi_\nu(d)\,|R_d| \qquad (\nu = 1,2) \tag{4.5}$$

† We recall that by Theorem 2.1. (Ω_1), $(\Omega_2(\kappa))$, (R):. . .' we mean that the stated conclusions of Theorem 2.1 hold under the conditions (Ω_1), $(\Omega_2(\kappa))$ and (R).

where, by (1.26),

$$\frac{1}{W(z)}\sum_{d|P(z)}\mu(d)\chi_\nu(d)\frac{\omega(d)}{d}$$

$$= 1 + (-1)^{\nu-1}\sum_{p<z}\frac{\omega(p)}{p}\frac{W(p)}{W(z)}\sum_{t|P_{p^+,z}}\frac{\chi_\nu(t)(1-\chi_\nu(pt))}{t}\omega(t) \qquad (\nu = 1, 2). \tag{4.6}$$

We shall now construct a class of functions $\chi_\nu (\nu = 1, 2)$ conforming to the above requirements. To this end we introduce a positive integer r and real numbers z_n such that

$$2 = z_r < z_{r-1} < \ldots < z_1 < z_0 = z, \tag{4.7}$$

and let b be a positive integer. Then, for either ν and for each n, $1 \leqslant n \leqslant r$, we put

$$\chi_\nu(d) = 1 \quad \text{if} \quad \nu((d, P_{z_n,z})) \leqslant 2b - \nu + 2n - 1 \quad \text{for} \quad n = 1, \ldots, r \tag{4.8}$$

and for the remaining divisors d of $P(z)$ we put $\chi_\nu(d) = 0$. With this definition, (1.20), (1.21) and (1.22) are clearly satisfied. As for (1.23), consider a t satisfying $\chi_\nu(t) = 1$, $p < q(t)$ and $z_m \leqslant p < z_{m-1}$, say; clearly we have to check (4.8) (with $d = pt$) only for $n = m$, i.e. we have to confirm that $\nu(pt) \leqslant 2b - \nu + 2m - 1$. But $\nu(t) \leqslant 2b - \nu + 2m - 1$; if also $\mu(t) = (-1)^{\nu(t)} = (-1)^\nu$, then $\nu(t) = 2b - \nu + 2m - 1$ is impossible. Hence $\nu(t) < 2b - \nu + 2m - 1$, and therefore $\nu(pt) \leqslant 2b - \nu + 2m - 1$. Hence $\chi_\nu(pt) = 1$ and so (1.23) too is satisfied.

Now the sum on the right of (4.6) is

$$\sum_{n=1}^{r}\sum_{z_n \leqslant p < z_{n-1}}\frac{\omega(p)}{p}\frac{W(p)}{W(z)}\sum_{t|P_{p^+,z}}\frac{\chi_\nu(t)(1-\chi_\nu(pt))}{t}\omega(t)$$

$$\leqslant \sum_{n=1}^{r}\frac{W(z_n)}{W(z)}\sum_{z_n \leqslant p < z_{n-1}}\frac{\omega(p)}{p}\sum_{t|P_{p^+,z}}\frac{\chi_\nu(t)(1-\chi_\nu(pt))}{t}\omega(t)$$

since $W(p) \leqslant W(z_n)$ if $z_n \leqslant p < z_{n-1}$; moreover, for each t making a contribution to the innermost sum on the right we have, necessarily, that $\chi_\nu(t) = 1$, $\chi_\nu(pt) = 0$, which, in the light of (4.8), implies that $\nu(t) = 2b - \nu + 2n - 1$. It follows that the sum is at most†

$$\sum_{n=1}^{r}\frac{W(z_n)}{W(z)}\sum_{\substack{d|P_{z_n,z}\\ \nu(d)=2b-\nu+2n}}\frac{\omega(d)}{d},$$

† In taking this step we have disregarded some conditions on the prime decomposition of t arising from (4.8). For certain kinds of applications, these conditions may be retained, with a gain in precision.

so that we arrive at the estimate

$$\sum_{p<z} \frac{\omega(p)}{p} \frac{W(p)}{W(z)} \sum_{t|P_{p+,z}} \frac{\chi_\nu(t)(1-\chi_\nu(pt))}{t}\omega(t)$$

$$\leqslant \sum_{n=1}^{r} \frac{W(z_n)}{W(z)} \frac{1}{(2b-\nu+2n)!} \left(\sum_{z_n \leqslant p<z} \frac{\omega(p)}{p}\right)^{2b-\nu+2n}. \tag{4.9}$$

Now let λ be a real number satisfying (4.1), and let us suppose that

$$\frac{W(z_n)}{W(z)} \leqslant e^{2(n\lambda+c)} \quad \text{for} \quad n = 1, \ldots, r, \tag{4.10}$$

where

$$c = \frac{c_1}{\log z}; \tag{4.11}$$

the inequalities (4.10) have yet to be verified for a suitable choice of the numbers z_n from (4.7). Then

$$\sum_{z_n \leqslant p<z} \frac{\omega(p)}{p} \leqslant \sum_{z_n \leqslant p<z} \log \frac{1}{1-\omega(p)/p} = \log \frac{W(z_n)}{W(z)} < 2(n\lambda + c)$$
$$\text{for} \quad n = 1, \ldots, r,$$

and it follows that the sum on the right of (4.9) is at most

$$\sum_{n=1}^{r} e^{2n\lambda+2c} \frac{(2n\lambda+2c)^{2b-\nu+2n}}{(2b-\nu+2n)!}$$

$$\leqslant e^{2c}(\lambda+c)^{2b-\nu} \sum_{n=1}^{r} \frac{(2ne^{-1})^{2n}}{(2n)!} \left(1+\frac{c}{n\lambda}\right)^{2n} (\lambda e^{1+\lambda})^{2n}$$

since $(2b-\nu+2n)! \geqslant (2n)!\,(2n)^{2b-\nu}$. We observe that $(ne^{-1})^n/n!$ is decreasing and that $(1+c/n\lambda)^{2n} \leqslant e^{2c/\lambda}$. Hence the sum under consideration is at most

$$e^{2c}(\lambda+c)^{2b-\nu} 2\, e^{-2}\, e^{2c/\lambda} \sum_{n=1}^{\infty} (\lambda\, e^{1+\lambda})^{2n}$$

$$= \frac{2\lambda^{2b-\nu+2}\, e^{2\lambda}}{1-\lambda^2 e^{2+2\lambda}} \left(1+\frac{c}{\lambda}\right)^{2b-\nu} e^{2c(1+1/\lambda)} \leqslant 2\, \frac{\lambda^{2b-\nu+2}\, e^{2\lambda}}{1-\lambda^2 e^{2+2\lambda}}\, e^{(2b-\nu+4)c/\lambda}.$$

Thus we obtain from (4.6) and (4.9) that

$$\sum_{d|P(z)} \mu(d)\,\chi_\nu(d)\,\frac{\omega(d)}{d}$$

$$= W(z)\left(1 + \theta 2\,\frac{\lambda^{2b-\nu+2}\,e^{2\lambda}}{1-\lambda^2\,e^{2+2\lambda}}\,e^{(2b-\nu+4)c/\lambda}\right), \qquad \nu = 1, 2, \qquad (4.12)$$

where $|\theta| \leqslant 1$.

We turn to the second sum on the right of (4.5). Using (*R*), (4.8), $\Omega_2(\kappa)$ and (3.13) we find that

$$\sum_{d|P(z)} \chi_\nu(d)\,|R_d| \leqslant \sum_{d|P(z)} \chi_\nu(d)\,\omega(d)$$

$$\leqslant \left(1 + \sum_{p<z}\omega(p)\right)^{2b-\nu+1}\prod_{n=1}^{r-1}\left(1 + \sum_{p<z_n}\omega(p)\right)^2$$

$$\leqslant (1 + A(2liz + 3))^{2b-\nu+1}\prod_{n=1}^{r-1}(1 + A(2liz_n + 3))^2,$$

$$\nu = 1, 2. \qquad (4.13)$$

We shall now choose our numbers z_n. To this end let Λ be a real number satisfying

$$\Lambda > 0, \qquad (4.14)$$

and then define z_n by

$$\log z_n = e^{-n\Lambda}\log z \quad \text{for} \quad n = 1, \ldots, r-1, \quad z_r = 2; \qquad (4.15)$$

here, in order to satisfy (4.7), r is chosen such that

$$\log z_{r-1} = e^{-(r-1)\Lambda}\log z > \log 2,$$

but

$$e^{-r\Lambda}\log z \leqslant \log 2,$$

so that

$$e^{(r-1)\Lambda} < \frac{\log z}{\log 2} \leqslant e^{r\Lambda}. \qquad (4.16)$$

Then, with a suitable constant B_4, it follows from (4.13) that

$$\sum_{d|P(z)} \chi_\nu(d)\,|R_d| \leqslant \left(\frac{B_4 z}{\log z}\right)^{2b-\nu+1}\prod_{n=1}^{r-1}\left(\frac{B_4 z_n e^{n\Lambda}}{\log z}\right)^2, \qquad \nu = 1, 2. \qquad (4.17)$$

We note that

$$\prod_{n=1}^{r-1} \left(\frac{B_4 e^{n\Lambda}}{\log z}\right) = \left(\frac{B_4 e^{\frac{1}{2}r\Lambda}}{\log z}\right)^{r-1},$$

and by (4.16) and (4.4) we have

$$\frac{B_4 e^{\frac{1}{2}r\Lambda}}{\log z} \leqslant \frac{B_4 e^{\Lambda/2}}{\log z} \sqrt{\frac{\log z}{\log 2}} < 1.$$

Furthermore, by (4.15),

$$\prod_{n=1}^{r-1} z_n{}^2 = \exp\left\{2 \log z \sum_{n=1}^{r-1} e^{-n\Lambda}\right\} \leqslant z^{2/(e\Lambda - 1)},$$

and using these estimates in (4.17), it follows that

$$\sum_{d \mid P(z)} \chi_\nu(d)|R_d| = O(z^{2b-\nu+1+2/(e\Lambda-1)}), \qquad \nu = 1, 2. \tag{4.18}$$

We still have to verify (4.10). By (3.12) and (4.11) it follows that

$$\frac{W(z_n)}{W(z)} \leqslant \exp\left\{n \Lambda \kappa + \frac{2c_1 e^{n\Lambda}}{\log z}\right\}$$

$$= e^{2c} \exp\left\{n\left(\Lambda \kappa + \frac{2c_1}{\log z} \cdot \frac{e^{n\Lambda} - 1}{n}\right)\right\}, \qquad n = 1, \ldots, r, \tag{4.19}$$

where

$$c_1 = \frac{A_2}{2}\left(1 + A_1 \kappa + \frac{A_1 A_2}{\log 2}\right) \tag{4.20}$$

(originally for $n = 1, \ldots, r - 1$ only but, in view of (4.16), also for $n = r$). Since $\Lambda > 0$, we have

$$\frac{e^{n\Lambda} - 1}{n} \leqslant \frac{e^{r\Lambda} - 1}{r},$$

and the latter expression is, by (4.16), at most

$$\Lambda \frac{e^{r\Lambda}}{r\Lambda} \leqslant \Lambda \frac{e^{\Lambda}}{\log 2} \frac{\log z}{\log(\log z/\log 2)}.$$

Hence we obtain from (4.19)

$$\frac{W(z_n)}{W(z)} \leqslant e^{2c} \exp\left\{n \Lambda \kappa \left(1 + \frac{2c_1 e^{\Lambda}}{\kappa \log 2} \frac{1}{\log(\log z/\log 2)}\right)\right\}, \qquad n = 1, \ldots, r.$$

Therefore, in view of (4.4), we meet the requirement (4.10) (and (4.14)) if we simply put

$$\Lambda = \frac{2\lambda}{\kappa}\frac{1}{1+\varepsilon}, \qquad \varepsilon = \frac{1}{200\, e^{1/\kappa}}.$$

Finally, we have

$$e^{2\lambda/\kappa} - e^{\Lambda} \leqslant \left(\frac{2\lambda}{\kappa} - \Lambda\right)e^{2\lambda/\kappa} \leqslant \varepsilon^{\Lambda} e^{1/\kappa},$$

so that, using that $e^{\Lambda} - 1 \geqslant \Lambda$,

$$\frac{e^{2\lambda/\kappa} - 1}{e^{\Lambda} - 1} \leqslant 1 + \frac{\varepsilon\Lambda\, e^{1/\kappa}}{e^{\Lambda} - 1} \leqslant 1 + \varepsilon\, e^{1/\kappa} = \frac{2{\cdot}01}{2}.$$

If we now insert this in (4.18) we obtain

$$\sum_{d|P(z)} \chi_\nu(d)\,|R_d| = O(z^{2b-\nu+1+(2\cdot01/(e^{2\lambda/\kappa}-1))}), \qquad \nu = 1, 2. \tag{4.21}$$

Our theorem now follows from (4.5), (4.12), (4.11) and (4.21).

As a first application we return to the prime twin problem. Although we shall obtain in Chapter 9 a much stronger result, it already shows the power of Theorem 2.1. We have (see (1.4.9))

$$\begin{aligned} &S(\{n(n+2): n \leqslant x; \mathfrak{P}_1, z) \\ &= \left|\left\{n: n \leqslant x, \left(n, \prod_{p<z} p\right) = 1, \left(n+2, \prod_{p<z} p\right) = 1\right\}\right|, \end{aligned} \tag{4.22}$$

and in the proof of (2.19) we saw that we may use here

$$X = x, \qquad \omega(2) = 1, \qquad \omega(p) = 2 \quad \text{for} \quad p > 2,$$

and that then (Ω_0) with $A_0 = 2$, (Ω_1) with $A_1 = 3$, and (R) are satisfied. Therefore (since, by Lemma 2.2, (Ω_0) implies $(\Omega_2(\kappa))$ with $\kappa = A_2 = A_0$) Theorem 2.1 can be applied, and we obtain by Lemma 2.2 and (3.6) from (4.3), choosing $b = 1$,

$$\begin{aligned} &S(\{n(n+2): n \leqslant x\}; \mathfrak{P}_1, z) \\ &\geqslant \tfrac{1}{2}x \prod_{2<p<z}\left(1 - \frac{2}{p}\right)\left\{1 - \frac{2\lambda^2 e^{2\lambda}}{1 - \lambda^2 e^{2+2\lambda}} \exp\left(\frac{4c_1}{\lambda \log z}\right)\right. \\ &\qquad \left. + O\left(\frac{\log^2 z \,.\, z^{1+2\cdot01/(e^{\lambda}-1)}}{x}\right)\right\}. \end{aligned} \tag{4.23}$$

We can find constants λ (satisfying (4.1)) and u such that

$$1 - \frac{2\lambda^2 e^{2\lambda}}{1-\lambda^2 e^{2+2\lambda}} > 0 \qquad \text{and} \qquad 1 + \frac{2{\cdot}01}{e^\lambda - 1} < u < 8. \tag{4.24}$$

If we then put $z = x^{1/u}$ the right hand side of (4.23) tends to infinity as $x \to \infty$ since

$$\exp\left(\frac{4c_1}{\lambda \log z}\right) = 1 + O\left(\frac{1}{\log z}\right),$$

which shows that then the number of n's in (4.22) tends to infinity as $x \to \infty$.

Each prime divisor of each n counted in (4.22) is $\geqslant z$. Hence for the total number Ω of prime divisors we find

$$x \geqslant z^{\Omega(n)} = x^{\Omega(n)/u} \quad \text{and} \quad x \geqslant x^{\Omega(n+2)/u},$$

so that by (4.24) both $\Omega(n)$ and $\Omega(n+2)$ are $\leqslant u < 8$, i.e. $\leqslant 7$. Thus we have proved that *there are infinitely many numbers n such that both n and $n + 2$ have at most 7 prime factors.* Similarly it follows from Theorem 2.1 that *every sufficiently large even integer can be written as a sum of two numbers which both have at most 7 prime factors.*

In order to show how (4.24) can be realised we note that the first condition there is equivalent to $1 > \lambda^2 e^{2\lambda}(2 + e^2)$ or

$$\lambda e^\lambda < \frac{1}{\sqrt{(2 + e^2)}},$$

and for the second it suffices to find a constant λ such that $e^\lambda - 1 > (2{\cdot}01)/7$, or

$$\lambda e^\lambda > \frac{9{\cdot}01}{7} \log \frac{9{\cdot}01}{7};$$

and now we see that it suffices to check (using accurate log tables!) that

$$\frac{9{\cdot}01}{7} \log \frac{9{\cdot}01}{7} < 1{\cdot}288 \log(1{\cdot}288) < 0{\cdot}326 < \frac{1}{\sqrt{(2 + e^2)}}.$$

Before proceeding to other applications of Theorem 2.1 we shall show that there is one significant way in which this result can be extended: while the conditions (Ω_1) and $(\Omega_2(\kappa))$ cannot be weakened in any sensible way, we shall now find that the rather over-simple condition (R) can be relaxed appreciably at little cost, and that this enhances the power of the Brun theory to a remarkable extent.

If we omit condition (R) from the statement of Theorem 2.1, but then follow precisely all the steps of the proof of Theorem 2.1 so far as the dominant terms are concerned (cf. (4.6)), choosing all the various parameters in precisely the same way, we arrive again at the results (4.2) and (4.3) except that the two error terms have now to be left in the form

$$\sum_{d|P(z)} \chi_\nu(d)|R_d|$$

with $\nu = 1$ and 2 respectively. With (R) no longer available, we cannot then argue as we did in (4.13); but if we review our earlier treatment of the sums

$$\sum_{d|P(z)} \chi_\nu(d)\omega(d)$$

and inspect (4.13) and the argument leading up to (4.18) more closely, we see at once that our choice of χ_1, χ_2 does at any rate allow us to draw from these same arguments the conclusion that, for any constant $A \geqslant 1$, we have, for all sufficiently large z, that

$$\sum_{d|P(z)} \chi_\nu(d)A^{\nu(d)} = O(z^{2b+1-\nu+(2\cdot 01/(e^{2\lambda/\kappa}-1))}), \quad \nu = 1, 2, \tag{4.25}$$

where the implied O-constant, while it may depend on A_1, A_2, κ as well as A does not depend on b and λ.

With these estimates at our disposal we now introduce two new conditions† on the remainder terms R_d to replace the single condition (R): We assume that there exists a real number $L \geqslant 1$ and a constant $A_0{}' \geqslant 1$ such that

$$(R_0) \qquad |R_d| \leqslant L\left(\frac{X \log X}{d} + 1\right)A_0'^{\nu(d)} \quad \text{for} \quad \mu(d) \neq 0, \qquad (d, \overline{\mathfrak{P}}) = 1;$$

and we assume also that for some constant $\alpha\,(0 < \alpha \leqslant 1)$ there exists corresponding to any given constant $U \geqslant 1$ a positive constant C_0 such that

$$(R_1\,(\kappa, \alpha)) \qquad \sum_{\substack{d < X^\alpha \log^{-C_0} X \\ (d, \overline{\mathfrak{P}})=1}} \mu^2(d)\,|R_d| = O_U\left(\frac{X}{\log^{\kappa+U} X}\right).$$

Clearly (R_0) is, in general, much weaker than (R) (take for example the common case when (Ω_0) is valid), while (R) implies a condition of type $(R_1\,(\kappa, 1))$. Moreover, $(R_1\,(\kappa, \alpha))$ is an *average* condition of the sort indicated

† We shall not use these conditions again outside Chapter 2, although we shall have to introduce something similar in the Selberg Theory later on (cf. $(R(\kappa, \alpha))$ in Section 7.6).

in Chapter 1 (see Example 5). With these conditions and (4.25) we have for each of $\nu = 1, 2$, that

$$\sum_{d|P(z)} \chi_\nu(d)\,|R_d|$$

$$\leqslant \sum_{\substack{d < X^\alpha \log^{-C_0} X \\ (d,\mathfrak{P})=1}} |R_d| + L \sum_{\substack{d|P(z) \\ d \geqslant X^\alpha \log^{-C_0} X}} \left(\frac{X \log X}{d} + 1\right) A_0'^{\nu(d)} \chi_\nu(d)$$

$$\leqslant O_U\left(\frac{X}{\log^{\kappa+U} X}\right) + 2LX^{1-\alpha} \log^{C_0+1} X \sum_{d|P(z)} A_0'^{\nu(d)} \chi_\nu(d)$$

$$= O_U\left(\frac{X}{\log^{\kappa+U} X} + LX^{1-\alpha} z^{2b+1-\nu+(2\cdot 01/(e^{2\lambda/\kappa}-1))} \log^{C_0+1} X\right).$$

If we adopt the convenient notation $u = \log X/\log z$ and apply (3.6) we derive

$$\sum_{d|P(z)} \chi_\nu(d)\,|R_d| \ll_U XW(z)$$

$$\times \left\{\frac{u^{-\kappa}}{\log^U X} + Lz^{-\alpha u+2b+1-\nu+(2\cdot 01/(e^{2\lambda/\kappa}-1))} u^{C_0+1} \log^{C_0+\kappa+1} z\right\}, \quad \nu = 1, 2,$$

and hence the more general†

THEOREM 2.1′. (Ω_1), $(\Omega_2(\kappa))$, (R_0), $(R_1(\kappa, \alpha))$: *Let b be a positive integer, let λ be a real number satisfying* (4.1) *and c_1 the constant defined in* (4.20), *and write*

$$u = \frac{\log X}{\log z}. \tag{4.26}$$

Then

$$S(\mathscr{A}; \mathfrak{P}, z) \leqslant XW(z)\left\{1 + 2\frac{\lambda^{2b+1} e^{2\lambda}}{1-\lambda^2 e^{2+2\lambda}} \exp\left((2b+3)\frac{c_1}{\lambda \log z}\right)\right.$$

$$\left. + O(Lz^{-\alpha u+2b+(2\cdot 01/(e^{2\lambda/\kappa}-1))} u^{C_0+1} \log^{C_0+\kappa+1} z) + O_U(u^{-\kappa} \log^{-U} X)\right\} \tag{4.27}$$

† In proving this theorem we have assumed tacitly that z is large enough. As was the case in the proof of Theorem 2.1 (cf. (4.4)) we may make this assumption since it is an easy exercise to deduce from Theorem 1.1 with (R_0) and $(R_1(\kappa, \alpha))$ a result that is at least as good as Theorem 2.1′ for bounded z.

and

$$S(\mathscr{A};\mathfrak{P},z) \geqslant XW(z)\left\{1 - 2\frac{\lambda^{2b}e^{2\lambda}}{1-\lambda^2 e^{2+2\lambda}}\exp\left((2b+2)\frac{c_1}{\lambda\log z}\right)\right.$$

$$\left. + O(Lz^{-\alpha u+2b-1+(2\cdot 01)/(e^{2\lambda/\kappa}-1)}u^{C_0+1}\log^{C_0+\kappa+1}z) + O_U(u^{-\kappa}\log^{-U}X)\right\}, \tag{4.28}$$

where the O-constants, while they may depend on $A_0', A_1, A_2, \kappa, \alpha$ and U, do not depend on λ or b.

We shall illustrate the strength of Theorem 2.1′, as well as its status relative to Theorem 2.1, in terms of the prime twin problem. This time we take as starting point the formulation (1.4.11), and accordingly we choose

$$\mathscr{A} = \{p+2:\ p \leqslant x\}, \qquad \mathfrak{P} = \{p:\ p > 2\} = \mathfrak{P}_2,$$

so that

$$\overline{\mathfrak{P}} = \{2\}.$$

Here Example 5 (with $a = 1, b = 2, k = 1$) applies, and so we choose

$$X = lix, \quad \omega(2) = 0, \quad \omega(p) = \frac{p}{p-1} \quad \text{for} \quad p > 2,$$

whence (cf. (1.3.42))

$$|R_d| \leqslant E(x,d) \quad \text{if} \quad \mu(d) \neq 0.$$

It is easy to confirm that then (Ω_1) holds with $A_1 = 2$, $(\Omega_2(\kappa))$ with $\kappa = 1 = A_2$, (R_0) with $L = 2$ and $A_0' = 1$; and that $(R_1(1,\alpha))$ holds with $\alpha = 1/2$ by virtue of Bombieri's theorem (see Lemma 3.3—sharp versions of this result allow us even to take $C_0 = U + 13$). With this choice of parameters we take $b = 1$ in (4.28) and obtain (since now $1 \leqslant c_1 < 2$ by (4.20))

$$S(\{p+2;\ p \leqslant x\};\mathfrak{P}_2,z)$$

$$\geqslant (li\,x)\prod_{2<p<z}\left(1-\frac{1}{p-1}\right)\left\{1 - 2\frac{\lambda^2 e^{2\lambda}}{1-\lambda^2 e^{2+2\lambda}}\exp\left(\frac{8}{\lambda\log z}\right)\right.$$

$$\left. + O(z^{-\frac{1}{2}u+1+(2\cdot 01/(e^{2\lambda}-1))}u^{U+14}\log^{U+15}z) + O(u^{-1}\log^{-U}x)\right\}. \tag{4.29}$$

We can find numerical constants λ (satisfying (4.1)) and u such that (cf. (4.24))

$$1-\frac{2\lambda^2 e^{2\lambda}}{1-\lambda^2 e^{2+2\lambda}}>0 \quad\text{and}\quad 2+\frac{4{\cdot}02}{e^{2\lambda}-1}<u<9; \tag{4.30}$$

actually $e^{\lambda}=1{\cdot}288$ is an admissible choice. Arguing as before, we may now conclude that *there exist infinitely many primes p such that $p+2$ is a number having at most* 8 *prime factors.*

This is a striking improvement on what we were able to deduce in this connection from Theorem 2.1; but we can do even a little better. The following technique for sharpening our result applies quite generally to all uses of Theorems 2.1 and 2.1′ and, indeed, has been standard procedure in the Brun method from the earliest times.

Let us turn back to the sum (4.9) and to the arguments we used there (for the most part these were simple inequalities involving binomial and exponential terms) to arrive at a reasonably accurate upper estimate in a reasonably simple form: now leave the term corresponding to $n=1$ (or even the first two or three or four terms) alone and apply the various simplifying devices to the later terms only. We shall illustrate this procedure in the prime twins problem: we are interested therefore only in the dominant term in (4.28) (or (4.3) for that matter) and so take $v=2$, $b=1$ and treat a as zero (cf. (4.11)). The sum under consideration now reads, with the first term separated off,

$$\tfrac{1}{2}e^{2\lambda}(2\lambda)^2+\sum_{n=2}^{r}e^{2n\lambda}\frac{(2n\lambda)^{2n}}{(2n)!}=2\lambda^2e^{2\lambda}+\sum_{n=2}^{r}\frac{(2ne^{-1})^{2n}}{(2n)!}(\lambda e^{1+\lambda})^{2n}$$

$$\leqslant 2\lambda^2e^{2\lambda}+\frac{(4e^{-1})^4}{4!}\sum_{n=2}^{\infty}(\lambda e^{1+\lambda})^{2n}=2\lambda^2e^{2\lambda}+\frac{32}{3}\frac{\lambda^4e^{4\lambda}}{1-\lambda^2e^{2+2\lambda}},$$

so that the dominant term in curly brackets on the right of (4.28) is (cf. (4.30))

$$1-2\lambda^2e^{2\lambda}\left(1+\frac{16}{3}\frac{\lambda^2e^{2\lambda}}{1-\lambda^2e^{2+2\lambda}}\right).$$

This expression is positive (actually it exceeds 0·04) when $e^{\lambda}=1{\cdot}293$; and this choice of λ allows us to find a u satisfying (cf. (4.30) again)

$$2+\frac{4{\cdot}02}{e^{2\lambda}-1}<u<8.$$

Hence we may conclude that $p+2$ (*p prime*) *is infinitely often a number having at most* 7 *prime factors* (*counted according to multiplicity*).

A closely similar argument would show that *every sufficiently large even integer is the sum of a prime and a number having at most 7 prime factors.*

We could improve our method further still in the context of this and many other related questions in prime number theory (for a systematic presentation of these see Chapters 9 and 10) by taking advantage of the combinatorial improvement mentioned in the footnote on p. 58, and by combining Theorems 2.1, 2.1′ with iterative procedures (such as are enshrined in the Buchstab formulae (1.10)) or weighting devices (see Chapters 9 and 10 again). We shall not do so because, starting in Chapter 3, we shall develop a theory which, for these proposes, is appreciably more powerful.

Nevertheless, enough has been said to indicate the power of the Brun method in this context; and in Section 8 we shall give some applications for which Theorems 2.1 and 2.1′ are singularly well suited.

5. A General Upper Bound O-Result

The following application of the upper estimate (4.2) in Theorem 2.1, although very simple, has, because of its weak general conditions, a wide range of applications. Whenever a O-estimate suffices (or is the only thing we are able to obtain) it is the most appropriate result. It is usually this theorem which is implied in the literature when one states (or reads) "Brun's sieve method gives...".

THEOREM 2.2. (Ω_1), $(\Omega_2(\kappa))$, (R): *For any*† $A(>0)$,

$$S(\mathscr{A};\mathfrak{P},z) \leqslant B_5\, X \prod_{p<z}\left(1-\frac{\omega(p)}{p}\right) \quad \textit{if} \quad z \leqslant X^A, \tag{5.1}$$

and

$$S(\mathscr{A};\mathfrak{P},z) \leqslant B_5\, X \prod_{p<X}\left(-\frac{\omega(p)}{p}\right) \quad \textit{if} \quad z \geqslant X^{1/A}. \tag{5.2}$$

Remark. By virtue of Lemma 2.2, condition $(\Omega_2(\kappa))$ may be replaced by (Ω_0). This facilitates the use of Theorem 2.2 for many applications.

Proof. We begin with the trivial estimate, provided by (1.4.14), namely

$$S(\mathscr{A};\mathfrak{P},z) \leqslant |\mathscr{A}| \leqslant X + |R_1| \leqslant X + 1;$$

here we have used condition (R) and the fact that $\omega(1) = 1$. If $X < 2^A$ this estimate implies both (5.1) and (5.2). We may suppose therefore that $X \geqslant 2^A$.

† Note that $B_5 = B_5\,(A, A_1, A_2, \kappa)$.

Next we take $b = 1$ and $\lambda = (2{\cdot}01)/8$ in Theorem 2.1.
Then

$$\frac{2{\cdot}01}{e^{2\lambda/\kappa} - 1} \leqslant \frac{2{\cdot}01}{2\lambda}\kappa = 4\kappa$$

and we obtain from (4.2)

$$S(\mathscr{A}; \mathfrak{P}, z) \leqslant XW(z)\left\{1 + O(1) + O\left(\frac{z^{2+4\kappa}}{XW(z)}\right)\right\}.$$

Now suppose that $z \leqslant X^{1/A}$; then by (3.6) and using for example (and without loss of generality), $A \geqslant 3 + 4\kappa$,

$$\frac{z^{2+4\kappa}}{XW(z)} \ll \frac{z^{2+4\kappa}\log^{\kappa} z}{X} \ll X^{(2+4\kappa)/A-1}\log^{\kappa} X \ll 1,$$

and it follows that

$$S(\mathscr{A}; \mathfrak{P}, z) \ll XW(z) \quad \text{for} \quad z \leqslant X^{1/A}. \tag{5.3}$$

The rest of the proof of Theorem 2.2 is an easy consequence of (5.3). We may now assume in both (5.1) and (5.2) that $z \geqslant X^{1/A}$.
Then trivially

$$S(\mathscr{A}; \mathfrak{P}, z) \leqslant S(\mathscr{A}; \mathfrak{P}, X^{1/A}),$$

and so, by (5.3),

$$S(\mathscr{A}; \mathfrak{P}, z) = O\left(XW(X^A)\frac{W(X^{1/A})}{W(X^A)}\right) = O_A(XW(X^A)) \tag{5.4}$$

since, by (3.5),

$$\frac{W(X^{1/A})}{W(X^A)} = O_A(1).$$

But since $\omega(p) \geqslant 0$ for all primes p, we have

$$W(X^A) \leqslant W(z) \quad \text{if} \quad z \leqslant X^A,$$

and this proves (5.1). Finally, since

$$W(X^A) \leqslant W(X) \quad \text{for} \quad A \geqslant 1,$$

(5.2) follows at once, and completes the proof of Theorem 2.2.

6. Sifting by a Thin set of Primes

The next two sections are devoted to applications of Theorem 2.2.

THEOREM 2.3. *Let g be a natural number, let a_i, b_i $(i = 1, \dots, g)$ be pairs of integers satisfying*

$$(a_i, b_i) = 1, \qquad i = 1, \dots, g,$$

and define

$$E := \prod_{i=1}^{g} a_i \prod_{1 \leqslant r < s \leqslant g} (a_r b_s - a_s b_r) \neq 0.$$

Let y and x be real numbers satisfying

$$1 < y \leqslant x.$$

Further, let $\mathfrak{P}$ be a set of primes for which there exist constants δ and A such that

$$\sum_{\substack{p<y \\ p \in \mathfrak{P}}} \frac{1}{p} \geqslant \delta \log\log y - A. \tag{6.1}$$

Then

$$|\{n:\ x - y < n \leqslant x,\ ((a_i n + b_i), \mathfrak{P}) = 1 \quad \text{for} \quad i = 1, \dots, g\}|$$
$$\ll \prod_{\substack{p \mid E \\ p \in \mathfrak{P}}} \left(1 - \frac{1}{p}\right)^{\rho(p) - g} \frac{y}{\log^{\delta g} y}, \tag{6.2}$$

where $\rho(p)$ denotes the number of solutions of

$$\prod_{i=1}^{g} (a_i n + b_i) \equiv 0 \bmod p, \tag{6.3}$$

and where the constant implied by the $\ll$-notation depends on g and A only.

We offer a number of comments in order to elucidate the rather long statement of this theorem.

Remark 1. The expression on the left of (6.2) counts all integers in $x - y < n \leqslant x$ for which $\prod_{i=1}^{g} (a_i n + b_i)$ is not divisible by any prime of $\mathfrak{P}$.

Remark 2. The product on the right of (6.2) is always finite since $E \neq 0$.

Remark 3. We have stressed the nature of the $\ll$-constant (its dependence on g and A only), because for some applications it is important that this constant should be independent of other parameters; in particular, that it should not depend on the coefficients a_i and b_i.

Remark 4. It will be clear from the proof (c.f. (6.7)) that the product on the right of (6.2) may have attached to it the further condition $p < y$.

Proof. We write

$$F_0(n) = \prod_{i=1}^{g} (a_i n + b_i),$$

and consider (in relation to $\mathfrak{P}$) the sequence

$$\mathscr{A} = \{F_0(n):\ x - y < n \leqslant x\},$$

which was analysed in Example 4. By (1.4.12) and (1.3.23)

$$\omega(p) = \begin{cases} \rho(p) & \text{if} \quad p \in \mathfrak{P}, \\ 0 & \text{if} \quad p \in \overline{\mathfrak{P}}, \end{cases} \tag{6.4}$$

and by (1.4.15) and (1.3.24)

$$|R_d| \leqslant \omega(d) \quad \text{if} \quad \mu(d) \neq 0 \quad \text{and} \quad (d, \overline{\mathfrak{P}}) = 1;$$

thus condition (R) is satisfied.

We may assume that

$$\rho(p) < p \quad \text{for all} \quad p \in \mathfrak{P}, \tag{6.5}$$

since otherwise the left side of (6.2) is zero and our theorem is trivially true.

By (1.3.27) we see that (Ω_0) is satisfied with $A_0 = g$. Moreover,

$$\frac{\omega(p)}{p} \leqslant \frac{g}{p} \leqslant 1 - \frac{1}{g+1} \quad \text{if} \quad p \geqslant g + 1$$

and, by (6.5),

$$\frac{\omega(p)}{p} \leqslant \frac{p-1}{p} \leqslant 1 - \frac{1}{g+1} \quad \text{if} \quad p \leqslant g + 1;$$

thus (Ω_1) holds with $A_1 = g + 1$. We have now checked that the conditions of Theorem 2.2 are satisfied, and that the constants arising from these conditions depend on g only. If we take

$$A = 1 \quad \text{and} \quad z = X = y,$$

we obtain by Theorem 2.2 and (6.4)

$$\begin{aligned}|\{n\colon\ x-y<n\leqslant x,\ (F_0(n),\mathfrak{P})=1\}|&\\ \leqslant|\{n\colon\ x-y<n\leqslant x,\ (F_0(n),P(y))=1\}|&\\ =S(\mathscr{A};\mathfrak{P},y)\ll y\prod_{\substack{p<y\\ p\in\mathfrak{P}}}\left(1-\frac{\rho(p)}{p}\right),&\qquad(6.6)\end{aligned}$$

where the $\ll$-constant depends on g only.

It remains to discuss the product on the right. We have

$$\prod_{\substack{p<y\\ p\in\mathfrak{P}}}\left(1-\frac{\rho(p)}{p}\right)\leqslant\prod_{\substack{p<y\\ p\in\mathfrak{P}}}\left(1-\frac{1}{p}\right)^{\rho(p)}=\prod_{\substack{p<y\\ p\in\mathfrak{P}}}\left(1-\frac{1}{p}\right)^{g}\prod_{\substack{p<y\\ p\in\mathfrak{P}}}\left(1-\frac{1}{p}\right)^{\rho(p)-g}.$$

The first product on the right is, by (6.1), at most

$$\exp\left\{-g\sum_{\substack{p<y\\ p\in\mathfrak{P}}}\frac{1}{p}\right\}\leqslant\frac{e^{Ag}}{\log^{\delta g}y};$$

whereas the second product is, by (1.3.28) and (1.3.27), equal to

$$\prod_{\substack{p<y\\ p|E\\ p\in\mathfrak{P}}}\left(1-\frac{1}{p}\right)^{\rho(p)-g}\leqslant\prod_{\substack{p|E\\ p\in\mathfrak{P}}}\left(1-\frac{1}{p}\right)^{\rho(p)-g}.\qquad(6.7)$$

Hence Theorem 2.3 now follows from (6.6).

Next we shall give some of the more important special cases of Theorem 2.3 as corollaries. We start with the case $g = 1$ and take

$$a_1 = 1,\qquad b_1 = 0.$$

Then $E = 1$, and Theorem 2.3 yields at once

COROLLARY 2.3.1. *If* $1 < y \leqslant x$ *and* $\mathfrak{P}$ *is a set of primes such that, with some constants* δ *and* A,

$$\sum_{\substack{p<y\\ p\in\mathfrak{P}}}\frac{1}{p}\geqslant\delta\log\log y-A.\qquad(6.8)$$

Then

$$|\{n\colon\ x-y<n\leqslant x,\ (n,\mathfrak{P})=1\}|\ll\frac{y}{\log^{\delta}y},$$

where the constant implied by the $\ll$*-symbol depends on* A *only.*

In most cases where Theorem 2.3 is applied, $\mathfrak{P}$ consists of all primes belonging to one or more residue classes to some modulus k; or $\mathfrak{P}$ is in some other sense a "thin" set of primes. If $\mathfrak{P}$ were not of this kind, other more precise theorems would be available. So let us take as our $\mathfrak{P}$ the set

$$\mathfrak{P}_{l,k} := \{p : p \equiv l \bmod k\}, \qquad (l, k) = 1, \tag{6.9}$$

or, more generally, if we wish to exclude the primes of several residue classes, the set

$$\bigcup_{i=1}^{r} \mathfrak{P}_{l_i,k}, \qquad (l_i, k) = 1 \quad \text{for} \quad i = 1, \ldots, r \tag{6.10}$$

where the l_i's belong to r $(\leqslant \phi(k))$ distinct residue classes mod k. (It is clear from the Chinese Remainder Theorem that it suffices to consider one modulus k only.) The density of such a set $\mathfrak{P}$ (in the sense of (6.1)) derives from Mertens' result

$$\sum_{\substack{p<x \\ p \equiv l \bmod k}} \frac{1}{p} = \frac{1}{\phi(k)} \log\log x + O_k(1), \qquad (l, k) = 1, \tag{6.11}$$

where, in the general case (6.10), there would be a factor $r/(\phi(k))$ on the right-hand side. In particular, our condition (6.1) is satisfied in the case (6.10) with

$$\delta = \frac{r}{\phi(k)} \quad \text{and} \quad A = O_k(1).$$

We now derive from Corollary 2.3.1. the following result.

COROLLARY 2.3.2. *Suppose that* $1 < y \leqslant x$, *and let* $l_1, \ldots, l_r$ $(1 \leqslant r \leqslant \phi(k))$ *represent* r *distinct residue classes modulo* k *with* $(l_i, k) = 1$ *for* $i = 1, \ldots, r$. *Then*

$$|\{n : x - y < n \leqslant x, (n, \mathfrak{P}_{l_i,k}) = 1 \quad \text{for} \quad i = 1, \ldots, r\}| \ll \frac{y}{\log^{r/\phi(k)} y};$$

in particular, for any pair of integers l, k *with* $(l, k) = 1$, *we have*

$$|\{n :\ x - y < n \leqslant x,\ (n, \mathfrak{P}_{l,k}) = 1\}| \ll \frac{y}{\log^{1/\phi(k)} y}.$$

In both cases the constants implied by the $\ll$*-symbol depend on* k *only.*

Because of its connection with the number of representations by sums of two squares, the case $k = 4$ is of particular interest. Here $\phi(k) = 2$ and Corollary 2.3.2 yields

COROLLARY 2.3.3. *Suppose that $1 < y \leqslant x$. Then for each of $l = 1$ and $l = 3$ we have*

$$|\{n: \ x - y < n \leqslant x, \ (n, \mathfrak{P}_{l,4}) = 1\}| \ll \frac{y}{\sqrt{(\log y)}},$$

where the constant implied by the $\ll$- symbol is absolute.

There is a useful comment to be made in connection with these special results. Sometimes a problem requires $\mathfrak{P}$ to be not exactly *a* $\mathfrak{P}_{l,k}$, but a slight modification of $\mathfrak{P}_{l,k}$; for example, the appropriate $\mathfrak{P}$ might be

$$\{2\} \cup \mathfrak{P}_{3,4} \quad \text{or} \quad \{2, 3\} \cup \mathfrak{P}_{1,6}.$$

In such cases we should note that a change of $\mathfrak{P}$ brought about by inserting or omitting a fixed number of primes does not alter δ (in (6.8)) and therefore does not affect our results. Alternatively, we may observe that if we replace $\mathfrak{P}_{l,k}$ by some $\mathfrak{P}$ such that

$$\mathfrak{P} \supset \mathfrak{P}_{l,k},$$

the number of n's that survive the sifting process is not increased, so that we may still use the estimate obtained with $\mathfrak{P}_{l,k}$.

Finally, let us go back to Theorem 2.3 and take

$$g = 2, \qquad y = x = N, \qquad \mathfrak{P} = \mathfrak{P}_{l,k},$$

where $N(\geqslant 2)$ is an integer, $(l, k) = 1$ and

$$a_1 = 1, \qquad b_1 = 0, \qquad a_2 = 1, \qquad b_2 = h, \tag{6.12}$$

where h is some non-zero integer. Then, making use of (6.11),

$$E = h, \qquad \delta = \frac{1}{\phi(k)}, \qquad \rho(p) = 1 \quad \text{if} \quad p \mid E.$$

We obtain in this case the following result.

COROLLARY 2.3.4. *Let $N(\geqslant 2)$ and $h(\neq 0)$ be integers and let l and k be coprime. Then*

$$|\{n:\ n \leqslant N,\ (n(n+h), \mathfrak{P}_{l,k}) = 1\}| \ll \prod_{\substack{p|h \\ p \equiv l \bmod k}} \left(1 - \frac{1}{p}\right)^{-1} \frac{N}{\log^{2/\phi(k)} N},$$

where the constant implied by the $\ll$-symbol depends on k only.

If instead of (6.12) we choose

$$a_1 = 1, \quad b_1 = 0, \qquad a_2 = -1, \qquad b_2 = N,$$

it follows that

$$E = -N, \qquad \delta = \frac{1}{\phi(k)}, \qquad \rho(p) = 1 \quad \text{if} \quad p \mid E,$$

and we obtain

COROLLARY 2.3.5. *Let $N(\geqslant 2)$ be an integer, and let l and k be coprime. Then*

$$|\{n:\ n \leqslant N,\ (n(N-n), \mathfrak{P}_{l,k}) = 1\}| \ll \prod_{\substack{p|N \\ p \equiv l \bmod k}} \left(1 - \frac{1}{p}\right)^{-1} \frac{N}{\log^{2/\phi(k)} N},$$

where the constant implied by the $\ll$-symbol depends on k only.

In these last results the left hand side is the number of integers $n \leqslant N$ such that both n and $n + h$ in one case, and both n and $N - n$ in the other, do not contain any prime factor $p \equiv l \bmod k$. The extension to r progressions is obvious.

7. FURTHER APPLICATIONS

Without encountering new difficulties, we could use Theorem 2.3 to estimate the number of n's (in some interval) such that each $a_i n + b_i$ is a prime number; and we could even require n to lie in an arithmetic progression. We leave this application as an exercise for the reader, and turn instead to the problem where n is restricted to prime numbers. The great advantage of the result we obtain (Theorem 2.4 below) is its uniformity: the $\ll$-constant will depend only on g. Its disadvantage is that the numerical factor (i.e. the $\ll$-constant) is not explicit. Actually, Theorem 2.4 can also be derived from Theorem 2.3; it may even be regarded as a special case. However, a direct application of Theorem 2.2 introduces an illuminating procedure.

THEOREM 2.4 *Let g and k be natural numbers, and let l, a_i, b_i $(i = 1, \ldots, g)$ be integers satisfying*

$$E := \prod_{i=1}^{g} a_i \prod_{1 \leqslant r < s \leqslant g} (a_r b_s - a_s b_r) \neq 0, \qquad b_1 \ldots b_g \neq 0. \tag{7.1}$$

Let y and x be real numbers satisfying

$$k < y \leqslant x. \tag{7.2}$$

Then

$$|\{p\colon x - y < p \leqslant x,\ p \equiv l \bmod k,\ a_i p + b_i \text{ prime for } i = 1, \ldots, g\}|$$

$$\ll \prod_{\substack{p \mid E \\ p \nmid k}} \left(1 - \frac{1}{p}\right)^{\rho(p) - g} \prod_{p \mid kb_1 \ldots b_g} \left(1 - \frac{1}{p}\right)^{-1} \left(\frac{k}{\phi(k)}\right)^{g} \frac{y/k}{\log^{g+1}(y/k)}, \tag{7.3}$$

where $\rho(p)$ denotes the number of solutions of

$$\prod_{i=1}^{g} (a_i n + b_i) \equiv \bmod p, \tag{7.4}$$

and where the constant implied by the $\ll$-symbol depends on g only.

Proof. Theorem 2.4 contains some trivial cases. We could exclude these by introducing into the theorem some further conditions. However, since we are interested in only those non-trivial cases where these conditions are satisfied, it is more convenient to discuss them (the conditions) here, before embarking on the main proof, rather than to incorporate them in the statement of the theorem.

First of all, we may clearly assume that the left side of (7.3) is not zero. Then if $(a_j, b_j) > 1$ for some j, and $a_j p + b_j = p_j$, a prime, it follows that $(a_j, b_j) = p_j$ and the congruence $a_j n + b_j \equiv 0 \bmod p$ has at most one solution for each prime $p \neq p_j$. Hence $\rho(p) \leqslant g$ for all but at most g primes p, which shows that the right hand side of (7.3) is not less than some positive constant (depending on g only) whereas $a_j p + b_j = p_j$ implies that the left hand side is at most 1. Therefore we may assume that

$$(a_i, b_i) = 1 \quad \text{for} \quad i = 1, \ldots, g.$$

Then each factor $a_i n + b_i$ contributes at most one solution to (7.4), so that

$$\rho(p) \leqslant g \quad \text{for all } p. \tag{7.5}$$

Now the first three expressions on the right of (7.3) are not less than 1, and this shows that all cases where the left side (of (7.3)) is $O_g(1)$ may be excluded. Thus we may assume that

$$\rho(p) < p \quad \text{for all} \quad p, \tag{7.6}$$

and also that

$$\rho(p) < p - 1 \quad \text{for} \quad p \nmid b_1 \ldots b_g. \tag{7.7}$$

To see that (7.7) is justified, let us suppose that there is a prime, $\bar{p}$, with $\rho(\bar{p}) = \bar{p} - 1$ and $\bar{p} \nmid b_1 \ldots b_g$. The latter expresses the fact that $n \equiv 0 \bmod \bar{p}$ is not a solution of (7.4) (with $p = \bar{p}$). Putting

$$F_0(n) = \prod_{i=1}^{g} (a_i n + b_i),$$

it follows that $\bar{p} \mid F_0(n)$ for all $n \not\equiv 0 \bmod \bar{p}$, and in particular that $\bar{p} \mid F_0(p)$ for all $p \neq \bar{p}$.

Finally, when we come to choose

$$z = \sqrt{(y/k)},$$

we may assume that our standard condition $z \geqslant 2$ is satisfied.

We now take

$$\mathscr{A} = \{nF_0(n): \ x - y < n \leqslant x, \ n \equiv l \bmod k\}$$

and

$$\mathfrak{P} = \mathfrak{P}_k = \{p: \ p \nmid k\}.$$

Then whenever $(d, k) = 1$, which is here equivalent to $(d, \mathfrak{P}) = 1$, we have

$$|\mathscr{A}_d| = |\{n: \ x - y < n \leqslant x, \ n \equiv l \bmod k, \ nF_0(n) \equiv 0 \bmod d\}|$$

$$= \rho'(d)\left(\frac{y}{dk} + \theta\right), \quad |\theta| \leqslant 1,$$

where $\rho'(d)$ is the number of solutions of

$$nF_0(n) \equiv 0 \bmod d, \tag{7.8}$$

and ρ' is a multiplicative function. Hence, choosing

$$X = \frac{y}{k}, \quad \omega(p) = \begin{cases} \rho'(p) & \text{if} \quad p \nmid k \\ 0 & \text{if} \quad p \mid k, \end{cases} \tag{7.9}$$

we have

$$|R_d| \leqslant \omega(d) \quad \text{if} \quad (d, k) = 1,$$

so that condition (R) of Theorem 2.2 is satisfied.

If now we compare (7.8) with (7.4) we see that $\rho'(p) = \rho(p) + 1$ if $n \equiv 0 \bmod p$ is not a solution of (7.4), i.e. if $p \nmid b_1 \dots b_g$; otherwise $\rho'(p) = \rho(p)$. Hence

$$\rho'(p) = \begin{cases} \rho(p) + 1 & \text{if } p \nmid b_1 \dots b_g \\ \rho(p) & \text{if } p \mid b_1 \dots b_g, \end{cases} \quad (p \nmid k). \tag{7.10}$$

By (7.5) condition (Ω_0) holds with $A_0 = g + 1$. Thus

$$\frac{\omega(p)}{p} \leqslant \frac{g+1}{p} \leqslant 1 - \frac{1}{g+2} \quad \text{if} \quad p \geqslant g + 2;$$

on the other hand, by (7.7) and (7.6),

$$\frac{\omega(p)}{p} \leqslant \frac{p-1}{p} \leqslant 1 - \frac{1}{g+1} \quad \text{if} \quad p \leqslant g + 1,$$

so that (Ω_1) holds with $A_1 = g + 2$. Hence the conditions of Theorem 2.2 are satisfied, and the constants are seen to depend on g only.

We now take

$$z = \sqrt{(y/k)} \quad \text{and} \quad A = 2,$$

and obtain by (5.2) and (7.9) that

$$S(\mathscr{A}; \mathfrak{P}, \sqrt{(y/k)}) \ll \frac{y}{k} \prod_{\substack{p < y/k \\ p \nmid k}} \left(1 - \frac{\rho'(p)}{p}\right), \tag{7.11}$$

where the $\ll$-constant depends on g only. Recall that, by definition,

$$S(\mathscr{A}; \mathfrak{P}, \sqrt{(y/k)})$$
$$= \left|\left\{n\colon x - y < n \leqslant x,\ n \equiv l \bmod k,\ \left(nF_0(n), \prod_{\substack{p < \sqrt{(y/k)} \\ p \nmid k}} p\right) = 1\right\}\right|.$$

Let us consider the numbers n in $x - y < n \leqslant x$, $n \equiv l \bmod k$ such that n and each $a_i n + b_i$ $(i = 1, \dots, g)$ are prime numbers. If these primes are at least $\sqrt{(y/k)}$, then such an n is counted in $S(\mathscr{A}; \mathfrak{P}, \sqrt{(y/k)})$. Otherwise $n < \sqrt{(y/k)}$ or

$$(0 <)\, a_i n + b_i < \sqrt{(y/k)} \tag{7.12}$$

for at least one i. Since each a_i is non-zero by (7.1), (7.12) is satisfied by less than $\sqrt{(y/k)}$ numbers, and it follows from (7.11) that

$$|\{p\colon x - y < p \leqslant x,\ p \equiv l \bmod k,\ a_i p + b_i \text{ prime for } i = 1, \ldots, g\}| \ll \frac{y}{k} \prod_{\substack{p < y/k \\ p \nmid k}} \left(1 - \frac{\rho'(p)}{p}\right) + \sqrt{\frac{y}{k}}, \tag{7.13}$$

where the $\ll$-constant depends only on g.

It remains to deal with the product on the right of (7.13), and this can be estimated in the following way. We have

$$\prod_{\substack{p < y/k \\ p \nmid k}} \left(1 - \frac{\rho'(p)}{p}\right) \leqslant \prod_{\substack{p < y/k \\ p \nmid k}} \left(1 - \frac{1}{p}\right)^{\rho'(p)}$$

$$= \prod_{\substack{p < y/k \\ p \nmid k}} \left(1 - \frac{1}{p}\right)^{\rho'(p) - g - 1} \prod_{\substack{p < y/k \\ p \mid k}} \left(1 - \frac{1}{p}\right)^{-g-1} \prod_{p < y/k} \left(1 - \frac{1}{p}\right)^{g+1}, \tag{7.14}$$

and we deal separately with the first of the three products on the right. We proved in (1.3.28) that

$$\rho(p) = g \quad \text{if} \quad p \nmid E,$$

whence

$$\prod_{\substack{p < y/k \\ p \nmid k}} \left(1 - \frac{1}{p}\right)^{\rho'(p) - g - 1}$$

$$\leqslant \prod_{\substack{p \mid E \\ p \nmid b_1 \ldots b_g \\ p \nmid k}} \left(1 - \frac{1}{p}\right)^{\rho(p) - g} \prod_{\substack{p \mid E \\ p \mid b_1 \ldots b_g \\ p \nmid k}} \left(1 - \frac{1}{p}\right)^{\rho(p) - g - 1} \prod_{\substack{p \nmid E \\ p \mid b_1 \ldots b_g \\ p \nmid k}} \left(1 - \frac{1}{p}\right)^{-1}$$

$$= \prod_{\substack{p \mid E \\ p \nmid k}} \left(1 - \frac{1}{p}\right)^{\rho(p) - g} \prod_{\substack{p \mid b_1 \ldots b_g \\ p \nmid k}} \left(1 - \frac{1}{p}\right)^{-1}; \tag{7.15}$$

if we incorporate this in (7.14) we find that

$$\prod_{\substack{p < y/k \\ p \nmid k}} \left(1 - \frac{\rho'(p)}{p}\right) \leqslant \prod_{\substack{p \mid E \\ p \nmid k}} \left(1 - \frac{1}{p}\right)^{\rho(p) - g}$$

$$\times \prod_{p \mid k b_1 \ldots b_g} \left(1 - \frac{1}{p}\right)^{-1} \prod_{p \mid k} \left(1 - \frac{1}{p}\right)^{-g} \prod_{p < y/k} \left(1 - \frac{1}{p}\right)^{g+1}.$$

Since, by (1.5.7),

$$\prod_{p < y/k} \left(1 - \frac{1}{p}\right)^{g+1} = O_g\left(\frac{1}{\log^{g+1}(y/k)}\right),$$

we arrive finally, in view of (7.13), at (7.3).

We can see from (7.14) and (7.15) that the estimate in (7.3) can be somewhat sharpened by retaining the condition $p < y/k$ everywhere. If we used (5.1) instead of (5.2) in our proof we should find, as can easily be checked, that we could add even the condition $p < z\ (\leqslant y/k)$ and then replace the last factor on the right of (7.3) by

$$\frac{y/k}{\log^{g+1} z}.$$

We shall now give as corollaries some of the most important special cases of Theorem 2.4. First of all consider the case $g = 1$, so that

$$E = a.$$

We eliminate $\rho(p)$ from (7.3) simply by using the fact that $\rho(p) \geqslant 0$, and observe that if $(a, b) = 1$ (if $(a, b) > 1$, Corollary 2.4.1 below is trivial),

$$\prod_{\substack{p \mid a \\ p \nmid k}} \left(1 - \frac{1}{p}\right)^{-1} \prod_{p \mid kb} \left(1 - \frac{1}{p}\right)^{-1} = \prod_{p \mid kab} \left(1 - \frac{1}{p}\right)^{-1}.$$

COROLLARY 2.4.1. *Let k be a natural number and let l, a and b be integers with $ab \neq 0$. Let y and x be real numbers satisfying*

$$k < y \leqslant x.$$

Then

$$|\{p\colon\ x - y < p \leqslant x,\ p \equiv l \bmod k,\ ap + b = p'\}|$$
$$\ll \prod_{p \mid kab} \left(1 - \frac{1}{p}\right)^{-1} \frac{y}{\phi(k) \log^2(y/k)},$$

and the constant implied by the $\ll$-symbol is absolute.

Apart from the non-explicitness of the constant, this result is a considerable extension of Theorem 3.12 below (and so of Theorem 3.11 which corresponds to the special cases $a = 1, b = h$ and $a = -1, b = N$ of Theorem 3.12); first because the present result is valid for an arbitrary interval, but also because the range of k is now practically unrestricted.

Next we take

$$k = 1 \quad \text{and} \quad a_i = 1 \qquad (i = 1, \ldots, g)$$

in Theorem 2.4, and write h_i in place of $b_i\ (i = 1, \ldots, g)$. Then condition (7.1) is satisfied if

$$h_1 < \ldots < h_g \quad \text{and} \quad h_i \neq 0 \qquad (i = 1, \ldots, g).$$

Also, since $a_i = 1$, each factor in (7.4) contributes a solution

$$n \equiv -h_i \bmod p,$$

so that $\rho(p)$ is the number of numbers among $h_1, \ldots, h_g$ that are distinct mod p.

COROLLARY 2.4.2. *Let g be a natural number and let $h_1, \ldots, h_g$ be integers satisfying*

$$h_1 < \ldots < h_g \quad \text{and} \quad h_i \neq 0 \qquad (i = 1, \ldots, g).$$

Let y and x be real numbers satisfying

$$1 < y \leqslant x.$$

Then

$$|\{p\colon\ x - y < p \leqslant x,\ p + h_i \text{ prime for } i = 1, \ldots, g\}|$$
$$\ll \prod_{\substack{p \mid \Pi(h_s - h_r) \\ 1 \leqslant r < s \leqslant g}} \left(1 - \frac{1}{p}\right)^{\rho(p) - g} \prod_{p \mid h_1 \ldots h_g} \left(1 - \frac{1}{p}\right)^{-1} \frac{y}{\log^{g+1} y},$$

where $\rho(p)$ denotes the number of modulo p distinct numbers among the h_i's, and where the constant implied by the $\ll$-notation depends only on g.

Note that the integers h_i are not necessarily all of the same sign.

Finally we take

$$g = 2, \qquad k = 1, \qquad a_1 = a_2 = -1, \qquad b_1 = M, \qquad b_2 = N,$$

and deduce from Theorem 2.4

COROLLARY 2.4.3. *Let M and N be natural numbers satisfying*

$$M > N \quad \text{and} \quad 1 < y \leqslant x.$$

Then

$$|\{p\colon\ x - y < p \leqslant x,\ M - p = p',\ N - p = p''\}|$$
$$\ll \prod_{p \mid M - N} \left(1 - \frac{1}{p}\right)^{-1} \prod_{p \mid MN} \left(1 - \frac{1}{p}\right)^{-1} \frac{y}{\log^3 y},$$

and the constant implied by the $\ll$-symbol is absolute.

What we are doing here is to count the number of solutions of

$$p_1 + p_2 = M, \qquad p_2 + p_3 = N;$$

the reader should have no difficulty in deriving from Theorem 2.4 the generalization of this problem to the case with $g > 2$.

Since $M > N$, the most important special case of Corollary 2.4.3 corresponds to the choice $y = x = N$.

8. Fundamental Lemma

We turn now to a very effective application of the simple version of Brun's sieve given in Theorems 2.1 and 2.1′. In the more recent literature each of Theorems 2.5, 2.5′ below is often referred to as a "fundamental lemma". Such results are of interest only for z small in comparison with X; however, in this respect they constitute remarkable extensions of (3.17). At the same time Theorem 2.5 is also a considerable refinement of Theorem 2.2.

Theorem 2.5. (Ω_1), $(\Omega_2(\kappa))$, (R): *Let $X \geqslant z$ and set*

$$u = \frac{\log X}{\log z}.$$

Then

$$S(\mathscr{A}; \mathfrak{P}, z) = XW(z)\{1 + O(\exp(-u(\log u - \log\log 3u - \log\kappa - 2))) + O(\exp - (\sqrt{\log X}))\}.$$

Remark. We remarked after the statement of Theorem 2.1 that condition (R) could be replaced there by

$$|R_d| \leqslant L\omega(d)$$

with only a slight consequent modification of the result of Theorem 2.1. Similarly, (3.17) and also Theorem 2.5 remain valid under this more general condition if a factor L is inserted into the second error term in each result.

Proof. Our condition $X \geqslant z$ is equivalent to $u \geqslant 1$; it may be weakened, or again it could be replaced by a stronger condition. In fact, for small values of u Theorem 2.5 is merely a restatement of (5.1) in Theorem 2.2, and we may therefore, in the light of the latter result, assume throughout the rest of the proof that

$$u \geqslant B_6 \tag{8.1}$$

where B_6 is sufficiently large. Moreover, in view of (3.17) and (3.6), we may assume that

$$\log z > u.$$

We choose

$$b = \left[\frac{u}{2} - \frac{u}{2\log u}\right] \quad \text{and} \quad \lambda = \frac{e\kappa \log u}{u}.$$

Then Theorem 2.1 yields, in view of (3.6),

$$S(\mathscr{A}; \mathfrak{P}, z) = XW(z)\left\{1 + O\left(\exp\left\{-2b\log\frac{1}{\lambda} + 2b\frac{c_1}{\lambda\log z}\right\}\right)\right.$$
$$\left. + O\left(\exp\left\{-\log z\left(u - 2b - \frac{e\kappa}{2}\right) + \kappa\log\log z\right\}\right)\right\}, \quad (8.2)$$

and the error terms are

$$O\left(\exp\left\{-\left(u - \frac{u}{\log u}\right)\log\left(\frac{u}{e\,\kappa\log u}\right) + O\left(\frac{u}{\log u}\right)\right\}\right)$$
$$+ O\left(\exp\left\{-\log z\frac{u}{2\log u} + \kappa\log\log z\right\}\right)$$
$$= O(\exp\{-u(\log u - \log\log u - \log\kappa - 2)\}) + O(\exp(-\sqrt{\log X}))$$

because $\log z > u$. Therefore our Theorem follows from (8.2).

In the same way we deduce from Theorem 2.1′

THEOREM 2.5′. (Ω_1), $(\Omega_2(\kappa))$, (R_0), $(R_1(\kappa, \alpha))$: *Let* $X \geqslant z$ *and write*

$$u = \frac{\log X}{\log z}.$$

Then

$$S(\mathscr{A}; \mathfrak{P}, z) = XW(z)\{1 + O(\exp\{-\alpha u(\log u - \log\log 3u - \log(x/\alpha) - 2)\})$$
$$+ O_U(L\log^{-U} X)\},$$

where the O-constants may depend on U as well as on the usual constants A_0', A_1, A_2, κ *and* α.

Proof. We proceed as in the proof of Theorem 2.5. The above result is of interest only as $u \to \infty$ and we shall concentrate therefore on the case of u large. (For bounded u there would be no difficulty in proving an analogue of Theorem 2.2 valid under the weaker R-conditions.)

Furthermore, for $u \geqslant \log z$, that is to say, for $\log z \leqslant \sqrt{(\log X)}$, the reader would find it a simple exercise to check the truth of the following analogues of (3.14) and (3.17):

Subject to (Ω_1), $(\Omega_2(\kappa))$, (R_0) and $(R_1(\kappa, \alpha))$,

$$S(\mathscr{A}; \mathfrak{P}, z) = XW(z)\{1 + O_U(\log^{-U} X) + O(LX^{-\alpha} \log^{C_0+\kappa} X(1 + A_0')^{\pi(z)})\}$$

and

$$S(\mathscr{A}; \mathfrak{P}, z) = XW(z)\{1 + O_U(\log^{-U} X)\} + O(LX^{1-\frac{1}{2}\alpha} \log^{C_0} X);$$

and to confirm that both these are better in their limited ranges of effectiveness than the general result being proved.

These last remarks allow us to suppose that

$$\log z > u,$$

and here an application of Theorem 2.1′ with

$$b = \left[\frac{\alpha}{2} u - \frac{\alpha}{2} \frac{u}{\log u}\right], \qquad \lambda = \frac{e\kappa}{\alpha} \frac{\log u}{u}$$

leads readily (via an analogue of (8.2)) to the result.

Theorems 2.5 and 2.5′ show how remarkably successful Brun's method is when the size of z is governed by the condition

$$\frac{\log z}{\log X} \to 0 \quad \text{as} \quad X \to \infty.$$

Indeed, for this range of values of z, Selberg's method cannot quite match the precision of Brun's (cf. Theorem 7.2).

Fundamental lemmas have important applications in those problems where one needs precise information about the distribution of numbers (belonging to some integer sequence) which have no "very" small prime factors. For example, such problems arise in the study of additive arithmetic functions. We end this chapter with two general results giving very sharp asymptotic formulae for the distribution of such numbers in polynomial sequences.

Let u and x be real numbers such that $u \geqslant 1$ and $x^{1/u} \geqslant 2$; and let $q = q(x, u)$ (with or without suffices) denote for the remainder of this chapter a number having no prime divisors less than $x^{1/u}$ and satisfying

$$\frac{\log q}{\log x} \ll 1$$

(the implied constant being independent of x and u). Such numbers q may clearly be regarded as free of very small prime factors and are sometimes referred to as *quasi-primes* (relative to x and u). If we let $x \to \infty$ and take $u = u(x)$ to be a positive function tending arbitrarily slowly to infinity with x, then, by Theorem 2.5, for any number X such that

$$1 \ll \frac{\log X}{\log x} \ll 1,$$

we have

$$|\{q\colon q \leqslant X\}| \sim \left(ue^{-\gamma}\frac{\log X}{\log x}\right)\pi(X),$$

so that quasi-primes are, in a "local" sense, hardly more dense than the primes and are actually less dense than numbers having at most two prime factors (provided that $u(x)$ tends to infinity more slowly than log log x). Indeed, Theorem 2.5 yields a high quality "quasi-prime number theorem". The following two theorems provide, in very precise form, far-reaching generalizations of this last remark.

THEOREM 2.6 *Let $F_1(n), \ldots, F_k(n)$ be distinct irreducible polynomials with integer coefficients and write*

$$F(n) = F_1(n) \ldots F_k(n).$$

Let $\rho(p)$ denote the number of solutions of the congruence

$$F(n) \equiv 0 \bmod p,$$

and assume that

$$\rho(p) < p \text{ for all primes } p. \tag{8.3}$$

Let u and x be real numbers such that

$$u \geqslant 1 \text{ and } x^{1/u} \geqslant 2.$$

Then

$$|\{n\colon 1 \leqslant n \leqslant x,\ F_i(n) = q_i \quad for \quad i = 1, \ldots, k\}|$$

$$= x \prod_{p < x^{1/u}} \left(1 - \frac{\rho(p)}{p}\right) \{1 + O_F(\exp(-u(\log u - \log\log 3u - \log k - 2))$$

$$+ O_F(\exp(-\sqrt{\log x}))\} \tag{8.4}$$

Moreover the expression on the right is

$$(ue^{-\gamma})^k \prod_p \left(1 - \frac{\rho(p) - 1}{p - 1}\right)\left(1 - \frac{1}{p}\right)^{-k+1} \frac{x}{\log^k x}$$

$$\times \left\{1 + O_F(\exp\{-u(\log u - \log\log 3u - \log k - 2)\}) + O_F\left(\frac{u}{\log x}\right)\right\}, \tag{8.5}$$

where all the O_F-constants depend at most on the coefficients and degrees of $F_1, \ldots, F_k$.

Remark. Although we have not required the polynomials F_i to have positive degrees, the theorem is, in fact, of interest only when this is the case; for if one of the F_i's had zero degree it would, by (8.3), have to be equal to ± 1.

Proof. We take $\mathscr{A} = \{F(n): 1 \leqslant n \leqslant x\}$ and $\mathfrak{P} = \mathfrak{P}_1$ (so that $\bar{\mathfrak{P}}$ is empty), and we now refer back to the analysis in Example 3 (with $y = x$) of Chapter 1. It will then be evident that we should choose

$$X = x, \qquad \omega(d) = \rho(d)$$

and that then (R) is satisfied, as are (Ω_1) with $A_1 = g + 1$ (g being the degree of F; cf. (1.3.16)) and $(\Omega_2(\kappa))$ with $\kappa = k$ and $A_2 = O_F(1)$ (cf.(1.3.17)). We may therefore apply Theorem 2.5 with $z = x^{1/u}$, and (8.4) follows at once. Next, we have

$$\prod_{p<x^{1/u}} \left(1 - \frac{\rho(p)}{p}\right) = \prod_p \left(1 - \frac{\rho(p)}{p}\right)\left(1 - \frac{1}{p}\right)^{-k} \frac{e^{-\gamma k} u^k}{\log^k x} \left\{1 + O_F\left(\frac{u}{\log x}\right)\right\}$$

(as can be seen from (5.6.5), and (5.2.5) of Lemma 5.3—see the proof of Theorem 5.3) where the infinite product is convergent. The identity

$$\left(1 - \frac{\rho(p)}{p}\right)\left(1 - \frac{1}{p}\right)^{-1} = 1 - \frac{\rho(p) - 1}{p - 1} \tag{8.6}$$

then leads at once to (8.5).

There are situations in which the dependence of the O-constants on the coefficients of the F_i leads to difficulties. It is therefore worth remarking that, by (1.3.16), (Ω_0) holds (with $A_0 = g$) in the situation of Theorem 2.6 so that, by Lemma 2.2, $(\Omega_2(\kappa))$ is satisfied with $\kappa = A_2 = g$. Hence (8.4) holds with both O-constants depending at most on g, provided we replace the k occurring in the first error term (as $\log k$, in the exponent) by g.

THEOREM 2.6′. *With the notation of Theorem 2.6, suppose in addition that*

$$F_i(n) \neq \pm n \qquad (i = 1, \ldots, k), \tag{8.7}$$

and that $\rho(p)$ satisfies (as well as (8.3)) the condition

$$\rho(p) < p - 1 \quad \text{if} \quad p \nmid F(0). \tag{8.8}$$

Let g be the degree of F, let u and x be real numbers such that

$$u \geqslant 3 \quad \text{and} \quad x^{1/u} \geqslant 2;$$

and define

$$\rho_1(p) = \begin{cases} \rho(p) - 1, & p \mid F(0), \\ \rho(p) & , \; p \nmid F(0). \end{cases} \tag{8.9}$$

Then

$$|\{p\colon p \leqslant x,\ F_i(p) = q_i \text{ for } i = 1, \ldots, k\}| = (li\,x) \prod_{p < x^{1/u}} \left(1 - \frac{\rho_1(p)}{p-1}\right)$$

$$\times \left\{1 + O_F(\exp\{-\tfrac{1}{3}u(\log u - \log\log 2u - \log 3k - 2)\}) + O_F\left(\frac{1}{\log x}\right)\right\}. \tag{8.10}$$

Moreover, the expression on the right is equal to

$$(ue^{-\gamma})^k \prod_p \left(1 - \frac{\rho_1(p)+1}{p}\right)\left(1 - \frac{1}{p}\right)^{-k-1} \frac{x}{\log^{k+1} x}$$

$$\times \left\{1 + O_F(\exp\{-\tfrac{1}{3}u(\log u - \log\log 2u - \log 3k - 2)\}) + O_F\left(\frac{u}{\log x}\right)\right\}. \tag{8.11}$$

Proof. We take $\mathscr{A} = \{F(p)\colon p \leqslant x\}$, $\mathfrak{P} = \mathfrak{P}_1$ (so that again $\overline{\mathfrak{P}}$ is empty) and this time hark back to Example 6 (with $k = 1$) of Chapter 1. In line with the analysis carried out there of $|\mathscr{A}_d|$, we choose

$$X = li\,x \quad \text{and} \quad \omega(d) = \frac{d\rho_1(d)}{\phi(d)}$$

and we then find that

$$|R_d| \leqslant \{E(x,d) + 1\}\,\rho(d) \quad \text{if} \quad \mu(d) \neq 0, \tag{8.12}$$

where

$$E(x,d) = \max_{\substack{1 \leqslant l \leqslant d \\ (l,d)=1}} \left|\pi(x,d,l) - \frac{li\,x}{\phi(d)}\right|.$$

We find also that (Ω_1) holds with $A_1 = g + 1$ ($g = \deg F$) using (1.3.51), (1.3.53) (identical with (8.9)) and (1.3.54); that $(\Omega_2(\kappa))$ holds with $\kappa = k$ and $A_2 = O_F(1)$ for the same reasons and because of (1.3.17); and so far as (R_0) and $(R_1\,(\kappa, \alpha))$ are concerned, we have by (8.12) and (1.3.55) that

$$|R_d| \leqslant \{E(x, d) + 1\}g^{\nu(d)} \quad \text{if} \quad \mu(d) \neq 0.$$

This inequality implies on the one hand that, trivially,

$$\begin{aligned} |R_d| &\leqslant \left\{\frac{x}{d} + 2\right\}g^{\nu(d)} \\ &\leqslant 2\left\{\frac{X \log X}{d} + 1\right\}g^{\nu(d)} \quad \text{if} \quad \mu(d) \neq 0, \end{aligned}$$

so that (R_0) holds with $L = 2$ and $A_0' = g$. On the other hand, this same inequality leads via Bombieri's Theorem (see Lemma 3.5 (with $k = 1, h = g$) and Lemma 3.4 below) to $(R_1\,(\kappa, \alpha))$ with $\kappa = k, \alpha = \frac{1}{2}$ and taking (as we may as well do) $U = 1$.

We may therefore apply Theorem 2.5′, and we take $z = x^{1/u}$. Now $x \geqslant 2^u \geqslant 8$ by hypothesis, and therefore the "u" in the statement of Theorem 2.5′ (which is not out u here but $\log X/\log z$) satisfies

$$\frac{\log X}{\log z} = u\frac{\log (li\, x)}{\log x} \geqslant u\frac{\log (x/\log x)}{\log x} = u\left(1 - \frac{\log\log x}{\log x}\right) \geqslant \frac{2}{3}u.$$

Hence (8.10) follows after a little simple arithmetic from Theorem 2.5′.

As for the last statement of the theorem, we have by (8.6),

$$1 - \frac{\rho_1(p)}{p - 1} = \left(1 - \frac{\rho_1(p) + 1}{p}\right)\left(1 - \frac{1}{p}\right)^{-1}$$

and therefore the product on the right of (8.10) is equal to

$$\prod_p \left(1 - \frac{\rho_1(p) + 1}{p}\right)\left(1 - \frac{1}{p}\right)^{-k-1} \frac{e^{-\gamma k}\, u^k}{\log^k x}\left\{1 + O_F\left(\frac{u}{\log x}\right)\right\}$$

as we may readily check by referring ahead to Chapter 5 (see the references given at the corresponding stage in the proof of Theorem 2.6).

This completes the proof of Theorem 2.6′.

As an illustration of Theorem 2.6′ let us take the special case of $k = 1$, $F(n) = F_1(n) = n + 2$, so that $\rho_1(2) = 0$ and $\rho_1(p) = 1$ if $p > 2$. We obtain from (8.11) that, uniformly in u,

$$|\{p\colon p \leqslant x,\ p + 2 = q\}| = 2(ue^{-\gamma}) \prod_{p>2} \left(1 - \frac{1}{(p-1)^2}\right) \frac{x}{\log^2 x}$$

$$\times \left\{1 + O(\exp\{-\tfrac{1}{3}u(\log u - \log\log 2u - \log 3 - 2)\}) + O\left(\frac{u}{\log x}\right)\right\},$$

$$(u \geqslant 3,\ x^{1/u} \geqslant 2),$$

which confirms the analogue for quasi-primes of the prime twin conjecture; apart from the factor $ue^{-\gamma}$, the dominant term is precisely the conjectured dominant term in the prime twin case!

9. Rosser's Sieve

In this last section we shall mention briefly another realization of a combinatorial sieve, due to Rosser. Specifically, we exhibit his choice of the functions $\chi_1(d)$, $\chi_2(d)$ which are defined on the set of divisors d of $P(z)$ and possess properties (1.20) to (1.23).

Let r be a natural number (in practice r will be large), and require, as a first step, that

$$\chi_\nu(1) = 1, \quad \chi_\nu(d) = 0 \quad \text{if} \quad \nu(d) > 2r + 1 - \nu \qquad (d \mid P(z);\ \nu = 1, 2). \tag{9.1}$$

Next, let $\beta = \beta_\kappa$ be a fixed positive number > 1. We write a typical divisor $d > 1$ of $P(z)$ in the form

$$d = p_i p_{i-1} \cdots p_1 \qquad (p_i < \cdots < p_1 < z) \tag{9.2}$$

and we shall now concern ourselves only with those d's for which

$$i = \nu(d) \leqslant 2r + 1 - \nu. \tag{9.3}$$

Let

$$\chi_\nu(d) = 1 \quad \text{whenever} \quad \nu(d) \leqslant 1 + \frac{1 - (-1)^\nu}{2} \qquad (\nu = 1, 2); \tag{9.4}$$

so that, if $\mathscr{D}_1$ and $\mathscr{D}_2$ are the subsets of the set of divisors of $P(z)$ of which χ_1 and χ_2 are, respectively, the characteristic functions, then $\mathscr{D}_1$ contains 1, all primes p and all products pp' of primes that divide $P(z)$, and $\mathscr{D}_2$ contains 1 and all primes p dividing $P(z)$.

Now suppose that d is as in (9.2), with

$$1 + \frac{1 - (-1)^{\nu}}{2} < i \leqslant 2r + 1 - \nu.$$

Then let $\chi_{\nu}(d) = 1$ if, and only if, the prime factors of d satisfy the inequalities

$$p_{2j+\nu}^{\beta} p_{2j+\nu-1} \cdots p_1 < X, \qquad j = 0, \ldots, \left[\frac{i - \nu}{2}\right]. \tag{9.5}$$

Let us first make the case $\nu = 1$ explicit. Here $\chi_1(d) = 1$ if $d = 1$ or d is a prime or d is the product of two primes (all dividing $P(z)$), and, if $i \geqslant 3$, $\chi_1(d) = 1$ if, and only if,

$$\begin{aligned} &p_1^{\beta} < X \\ &p_3^{\beta} p_2 p_1 < X \\ &\cdots \\ &p_{2j_0+1}^{\beta} p_{2j_0} \cdots p_1 < X, \qquad j_1 = \left[\frac{i-1}{2}\right], \end{aligned} \tag{9.6}$$

for a d given by (9.2).

Next we clarify the case $\nu = 2$. Here we have $\chi_2(d) = 1$ if $d = 1$ or if d is a prime (dividing $P(z)$); and, if d is again as in (9.2) with $i \geqslant 2$, $\chi_2(d) = 1$ if, and only if,

$$\begin{aligned} &p_2^{\beta} p_1 < X \\ &p_4^{\beta} p_3 p_2 p_1 < X \\ &\cdots \\ &p_{2j_1+2}^{\beta} p_{2j_1+1} \cdots p_1 < X, \qquad j_1 = \left[\frac{i-2}{2}\right]. \end{aligned} \tag{9.7}$$

We now confirm that χ_1 and χ_2 so defined give rise to a combinatorial sieve: of the four basic requirements to be checked, (1.20) and (1.21) hold trivially; a simple verification shows that (1.22) is satisfied, i.e. that $\mathscr{D}_1$ and $\mathscr{D}_2$ are divisor-closed. (For example, take χ_2: if d is given by (9.2) and $i_{2l}, i_{2l-1}, \ldots, i_1$ is a subset of $i, i-1, \ldots, 1$ with $i_{2l} < \ldots < i_1$, then

$$p_{i_{2l}}^{\beta} p_{i_{2l-1}} \cdots p_{i_1} < p_{2l}^{\beta} p_{2l-1} \cdots p_1 < X$$

by construction; etc.) It remains to check (1.23). Suppose $\nu = 1$, and $\chi_1(d) = 1$ with $\mu(d) = -1$. Then i is odd and therefore $i \leqslant 2r - 1$. Take a prime p of $\mathfrak{P}$ so that $p < p_i$ and consider $\chi_1(pd)$. Clearly $\nu(pd) = i + 1$ is even and $\leqslant 2r$; writing $i = 2l - 1$ and $p = p_{2l}$ it is clear that the conditions for χ_1 (pd) to be 1 are precisely the conditions (9.6) with $j_0 = l - 1$, and these do indeed hold since $\chi_1(d) = 1$. A similar argument confirms (1.23) for $\nu = 2$.

The Rosser functions χ_ν may now be substituted into the fundamental combinatorial identity (1.26) and they lead, in principle at least, to upper and lower sieve estimates in the manner of Brun. In practice the accurate estimation of the sums involved presents technical difficulties; it turns out that there is interplay between the sums corresponding to upper and lower estimates of a kind we shall encounter in Chapter 8, and that the correct choice of β_κ should emerge from this process. Only in the cases $\kappa = 1$ and $\kappa = \frac{1}{2}$ have these difficulties been successfully resolved; but for these values of κ Rosser's sieve is, in a certain sense, optimal.

NOTES

2.1. Formulae (1.9), (1.10) are known in the literature as Buchstab's identities (see Buchstab [1]). In fact, earlier workers had used such identities, for example to refine Legendre's numerical study of $\pi(x)$ (for references, e.g. to the papers of Meissel, see D. H. Lehmer, *Illinois J. Math.* **3** (1959), 381–388, and they are to be found already in the works of Brun himself (Brun [5] p. 5, [7] pp. 32–33) and Rademacher [1] p. 24. However, to Buchstab must go the credit for exploiting these identities in various novel and highly effective ways in the context of sieve theory.

For the earliest account of Selberg's sieve see Selberg [1].

The nomenclature associated with the divisor sets $\mathscr{D}_\nu$ $(\nu = 1, 2)$ appears to derive from Vinogradov [2] and Levin [9].

Identity (1.26) is taken from Levin [9].

2.2. For Brun's pure sieve see Brun [2], also Landau [3], Rademacher [3].

Lemma 2.1, Corollary: see Hooley [1].

Condition (2.10) occurs e.g. in Lavrik [3].

According to a classical result of Mertens (see Rosser–Schoenfeld [1])

$$\sum_{p<z} \frac{1}{p} = \log\log z + B + O\left(\frac{1}{\log z}\right)$$

where B is an absolute constant,

$$B = \gamma + \sum_p \left\{\log\left(1 - \frac{1}{p}\right) + \frac{1}{p}\right\} = 0{\cdot}26149\ldots.$$

(2.16): not required for later application, but is an intermediate result of the fundamental lemma type (see Section 2.8).

(2.19) and the convergence of $\sum_{p+2=p'} 1/p$ were established in Brun [6]. These results represent the earliest theoretical information bearing on the prime twins problem. Note that here (2.14) must be used and not (2.16); the latter would lead to a much worse result $-$ $x/(\log x)$ in (2.19) in place of $x(\log\log x/\log x)^2$. For the record, the convergence of $\sum_{p+h=p'} (1/p)$ $(2 \mid h)$ was first probed by Segal [1] (see also Bays [1]).

2.3. (3.2) follows from Theorems 6 and 21 if the immensely useful Rosser–Schoenfeld [1]. From elementary prime number theory (see H–W, Theorem 425) we know that

$$\sum_{p<z} \frac{\log p}{p} = \log z + O(1),$$

but this would not suffice to prove (3.2).

(3.3), (3.4): as we saw in Section 2, estimates of this kind are much simpler to derive on the basis of (Ω_0).

(3.17) is of exactly the same form as (2.16) but holds under a weaker condition $((\Omega_2\ (\kappa))$ in place of $(\Omega_0))$ (cf. Section 2.8).

2.4. Brun carried out his researches on the sieve between the years 1915 and 1924 (Brun [1, 2, 3, 4, 5, 6, 7, 8, 10]; Brun [9] is expository). Rademacher [1] gave an important improved version of Brun's sieve and Estermann [1] sharpened it further still. Rényi used Brun's sieve in the Estermann form in his classic contribution to Goldbach's conjecture. Our account resembles Estermann's, but derives directly from the formulation of Levin [9].

According to Erdös [16] "Brun's method is perhaps our most powerful elementary tool in number theory"; we shall not attempt, in these Notes or in the Bibliography, to list all the instances where Brun's sieve has been used in an auxiliary rôle as an arithmetical aid (Theorem 2.2 of Section 5 covers most of these applications). As an approach towards the famous problems described in the Introduction, Linnik [3] remarks that "the path to the solution of [these] problems was opened up by Viggo Brun". All the problems in question have an elementary formulation, Gelfond and Linnik [1] make the point that, for this very reason, it is natural to search for an arithmetical (as opposed to a transcendental) method of solution. Such a method, it if exists, will frequently give (they say) a simple and natural view of the problem and the reason underlying its existence. Frequently elementary methods succeed where analytic methods fail. This is the case with binary additive problems of the Goldbach type. . . .

Landau [2] describes Brun's upper bound sieve; Trost [1] and Gelfond–Linnik [1] describe Brun's sieve. A curious feature of sieve literature is that while there is frequent use of Brun's *method* there are only a few attempts to formulate a general Brun *theorem* (such as Theorem 2.1); as a result there are surprisingly many papers which repeat in considerable detail the steps of Brun's argument. We refer the reader f.e. to the papers of Ricci, James, Wang and Pan, all of whom have given striking applications of Brun's method.

Theorem 2.1: See Halberstam–Richert [4]. The introduction of the parameter b derives from yet a further refinement of Brun's method by Tartakovskiĭ [1, 2]. The constant 2·01 can be replaced by any constant >2, as is evident from the proof.

Footnote on p. 58: the argument leading to (4.9) may, in certain circumstances, be refined. Such a refinement has been carried out by Hagedorn (Diplomarbeit, Ulm 1972) on the basis of the following combinatorial identity

$$\sum_{\substack{j_1+\ldots+j_n=2n-1\\ j_1+\ldots+j_k\leqslant 2k-1(k=1,\ldots,n-1)}} \frac{1}{j_1!\ldots j_n!} = \frac{1}{n}\sum_{j_1+\ldots+j_n=2n-1}\frac{1}{j_1!\ldots j_n!}$$

where each sum extends over all n-tuples $(j_1,\ldots,j_n)$ of non-negative integers satisfying the stated conditions; and the expression on the right is equal to $n^{2n-2}/(2n-1)!$. From this identity one may derive also that (for $n>1$)

$$\sum_{\substack{j_1+\ldots+j_n=2n\\ j_1+\ldots+j_k\leqslant 2k-1(k=1,\ldots,n-1)}} \frac{1}{j_1!\ldots j_n!} \leqslant \frac{1}{3}\frac{n^{2n-2}}{(2n-1)!}$$

By P_r we denote an integer having at most r prime factors, counted according to multiplicity. In later chapters we refer to such a number as an almost-prime of order r (a term apparently introduced by Levin [10]). Then we derive here, from (4.23) and (4.24), that there exist infinitely many almost-primes P_7 such that $P_7+2=P_7'$; and in the same way we can get

$$N = P_7 + P_7' \qquad (2\mid N,\ N\geqslant N_0)$$

as an approximation to Goldbach's conjecture; let us describe this pair of results in "short-hand" as (7, 7): in this terminology, Brun [7] was the first to derive a result of this kind, namely (9, 9), to be followed by (7, 7) of Rademacher [1]; (6, 6) of Estermann [1]; (5, 7), (4, 9), (3, 15), (2, 366) of Ricci [6, 7]; (5, 5) of Buchstab [2]; (4, 4) of Buchstab [3], Tartakovskiĭ [1, 2]; and (a, b) with $a+b\leqslant 6$ of Kuhn [1, 2, 3]. All these are based on Brun's sieve; but the better results require in addition use of other devices such as Buchstab identities and/or Kuhn's weights. Further improvements will be cited later.

Bombieri's theorem: For a version permitting the choice $U+13$ for C_0 see Montgomery [2]. According to the latest version, $C_0=U+\frac{7}{2}$ is admissible (see Vaughan, *J. London Math. Soc.*, to appear). Taking advantage of this deep result, which renders the use of *GRH* in earlier papers unnecessary (see references below) we derive here (in the notation from above) the superior results (1, 8) and even (1, 7); moreover, Hagedorn's combinatorial refinement (*loc. cit.*) leads to (1, 6)—this without use of Buchstab iteration or Kuhn weights. Rényi [1, 2] was the first to obtain a result of type $(1, b)$; in addition to Brun's sieve (in Estermann form) he used the large sieve in combination with analytical ideas in an early—in fact, the earliest—version of the Bombieri result.

Previously, Estermann [1] had proved (1, 6) on the basis of a weak form of *GRH*. Subsequently Wang [2], [11] attained (1, 4), (1, 3) respectively, both on the basis of *GRH*.

No further comment is necessary to underline the power of Brun's method when harnessed with other combinatorial devices and with additional information about the distribution of primes in arithmetical progressions. (James [3] gives a clear account of the way in which Brun's method can be improved systematically by successive iterations of Buchstab identities.)

(5.3): In the argument leading up to this inequality we assume $A \geqslant 3 + 4\kappa$ and claim that this involves no loss of generality. Indeed, there is none: for if (5.1) and (5.3) hold for $A = 3 + 4\kappa$ they are certainly true for any positive $A < 3 + 4\kappa$, since then $X^A < X^{3+4\kappa}$ and $X^{1/A} > X^{1/(3+4\kappa)}$.

2.6. Theorem 2.3: The "thin set" condition (6.1) occurs in S. Selberg [3], where the special case $g = 2, a_1 = a_2 = 1, b_1 = 0$ is discussed. For a statement in case $g = 3$ see Rieger [10]. For $\delta = 1$, i.e. when $\mathfrak{P}$ is not thin, cf. Prachar [6], Sätze 4.2, 4.7.

Corollary 2.3.1: see Hooley [2] for an application with $\mathfrak{P} = \{p\colon (D/p) = -1\}$ (so that $\delta = \frac{1}{2}$) and $y = x$.

(6.11): see appropriate Note attached to Section 1.3.

Corollary 2.3.2: Stated in Hooley [4], with $r = 1$, $l = 5$, $k = 6$, $y = x$. For $y = x$ the result is connected with the so-called Lehmer problem which asks, in this context, for an asymptotic formula (see Landau "Handbuch der Lehre von der Verteilung der Primzahlen", I, Berlin, 1909, pp. 643 *et seq.*, and the reference on p. 147 of Prachar [6]; the conjectured formula as stated in the latter lacks a constant factor on the right).

Corollary 2.3.3: cf. Rieger [12], who has $l = 3$.

Corollary 2.3.4: cf. Rieger [13], who has $l = 3$, $k = 4$.

Corollary 2.3.5: cf. Rieger [10], who has $l = 3$, $k = 4$ and hints at other results of this kind (see also Rieger [11, 12]).

Note that Corollaries 2.3.4, 2.3.5 yield bad upper estimates for

$$|\{p\colon p \leqslant N,\ (p + h, \mathfrak{P}_{l,k}) = 1\}| \quad \text{and} \quad |\{p\colon p \leqslant N,\ (N - p, \mathfrak{P}_{l,k}) = 1\}|$$

unless $\phi(k) = 1$.

For Brun lower sieve applications with thin $\mathfrak{P}$ see James [1, 3]. As a specimen result from the first of these, his Theorem 4 states: For all sufficiently large $n \equiv 2 \bmod 4$, n is representable in the form $n = a + b$, where a, b have all but at most two of their prime factors congruent 1 mod 4. See also James [2] for an application to representation by ternary quadratic forms. Here he obtains a good lower bound for

$$|\{n\colon n \leqslant N^{\frac{1}{2}}, n = \bmod Dk, (N - n^2, P(N^{\frac{1}{2}})) = 1\}|,$$

where

$$\mathfrak{P} = \{p\colon p > c, \left(\frac{D}{p}\right) = -1\}.$$

Schinzel [3] (see also Schinzel–Wang [1], Lemma 7) gives another remarkable "thin $\mathfrak{P}$" lower sieve application via Brun: Let g be a given positive integer and $N \geqslant 3$ a natural number. Then

$$\left|\left\{n\colon 1 \leqslant n \leqslant X, \left(\prod_{i=1}^{g} (in + 1), N\right) = 1\right\}\right|$$

is positive with $X = (\log N)^{20g}$. According to a note added at proof stage, Wang can prove this even with $X = C(g)\,(\log N)^{4g+3}$.

2.7. Theorem 2.4: cf. Klimov [2]. The case $g = 2$ is used by Barban–Vinogradov–Levin [1] to deduce that

$$\sum_{x^{1/4} \leqslant p \leqslant x} p\,\pi^2(x; p, l) \ll \frac{x^2}{\log^2 x}.$$

Corollary 2.4.1: cf. Ricci [10] ($k = 1$), Prachar [6], Sätze 4.5, 4.6. Applications in f.e. Erdös [2], Klimov [4].

Corollary 2.4.2: f.e. Erdös [4], [6] ($g = 3$).

Corollary 2.4.3: see Prachar [6], Satz 4.2; [5]. Applications in Rieger [6, 8]; Knödel [2] (Satz 2), [4]; Erdös–Rényi [1] (Lemma 1), corrected in Prachar [4].

Many other striking applications of Brun's method are to be found in the literature. For an interesting lower bound application related to the problem of Theorem 2.4 see Ricci [9]; another application in Chowla–Erdös–Straus [1] is described in the Notes for Chapter 8 (Section 5).

2.8. A result of the fundamental lemma type is already implicit in Brun's own work; according to Erdös and Kac [1] the following result follows from Brun [7], p.21: "If $\{m_n\}$, $\{s_n\}$ are two integer sequences tending to ∞ with n, in such a way that $m_n \to \infty$ more rapidly than any fixed power of s_n, then (with $\mathfrak{P} = \mathfrak{P}_1$)

$$|\{r: 1 \leqslant r \leqslant m_n, (r, P(s_n)) = 1\}| = e^{-\gamma}\frac{m_n}{\log s_n}\{1 + o(1)\}.$$

They use this to derive their well-known Central Limit Theorem for strongly additive arithmetic functions.

Theorem 2.5: cf. Halberstam–Richert [3, 4]. Kubilius [2] (Lemma 1.4) gives a general fundamental lemma in the context of additive functions, as does Uždavinis [1, 2, 3]. See also Buchstab [7, 8]; Barban [5]; Le Veque [1] (for a statement of a Rosser fundamental lemma); Iwaniec [1] (a Rosser fundamental lemma). Most of these also give applications, as do Barban [3, 9]; Barban–Vinogradov [1]; Chowla–Briggs [1] (by analytic methods), Eda–Yamano [1].

Theorem 2.5′: Halberstam–Richert [5]; also Lavrik [3], Levin [9]. Cf. Section 7.3 for a general fundamental lemma in Selberg theory.

Theorems 2.6, 2.6′: cf. Halberstam–Richert [3, 4, 5]. For special results of this kind see Buchstab [4, 6]; Lavrik [3], who discuss quasi-prime analogues of Goldbach.

For a discussion of quasi-primes see Linnik [3], Chapter 1.5.

2.9. An account of Rosser's sieve was to have appeared as a book on sieves by Harrington and Rosser, but this book was never published. We have been told that a manuscript exists, but we have been unable to gain access to it. There is reference to a special case of Rosser's sieve (of the fundamental lemma type) in Le Veque [1] and there is a brief description in Selberg [5]. For full details of important special cases (corresponding to $\kappa = 1$, $\kappa = \frac{1}{2}$) see Iwaniec [1, 2].

Chapter 3

The Simplest Selberg Upper Bound Method

1. The Method

In this first section we shall explain the simplest form of Selberg's upper sieve. We shall deal only with $S(\mathscr{A};\mathfrak{P},z)$.

Let $\lambda_1 = 1$ and let λ_d $(d \geqslant 2)$ be arbitrary real numbers. Selberg's sieve derives from the inequality

$$S(\mathscr{A};\mathfrak{P},z) \leqslant \sum_{a\in\mathscr{A}} \left(\sum_{\substack{d|a\\ d|P(z)}} \lambda_d \right)^2,$$

which is true without any further conditions on the numbers λ_d; for if $a \in \mathscr{A}$ and $(a, P(z)) = 1$, $d = 1$ is the only divisor appearing on the right and it makes a contribution 1 since $\lambda_1 = 1$; whilst all the other terms on the right, namely those associated with $a \in \mathscr{A}$, $(a, P(z)) > 1$, are non-negative because the λ_d's are real. If we square and interchange the order of summation, the inequality takes the form

$$S(\mathscr{A};\mathfrak{P},z) \leqslant \sum_{\substack{d_\nu|P(z)\\ \nu=1,2}} \lambda_{d_1}\lambda_{d_2} \sum_{\substack{a\in\mathscr{A}\\ a\equiv 0 \bmod D}} 1, \qquad D = [d_1, d_2],$$

where $[d_1, d_2]$ denotes the least common multiple of d_1 and d_2; and with the notation introduced in (1.4.14) it becomes

$$S(\mathscr{A};\mathfrak{P},z) \leqslant X \sum_{\substack{d_\nu|P(z)\\ \nu=1,2}} \lambda_{d_1}\lambda_{d_2} \frac{\omega(D)}{D} + \sum_{\substack{d_\nu|P(z)\\ \nu=1,2}} |\lambda_{d_1}\lambda_{d_2} R_D| = X\Sigma_1 + \Sigma_2, \quad (1.1)$$

say.

Selberg's idea, in principle, was to choose the numbers λ_d $(d \geqslant 2)$ in such a way that the expression on the right of (1.1) becomes a minimum. However, even for "well-behaved", simple sequences $\mathscr{A}$ this seems to be an extremely difficult problem, and so the idea had to be modified. One way of approximating to the ideal situation is to choose

$$\lambda_d = 0 \quad \text{for} \quad d \geqslant z, \quad (1.2)$$

and then to choose the remaining numbers λ_d $(2 \leqslant d < z)$ in such a way that Σ_1, which is a quadratic form in the λ_d's, becomes a minimum. Here the underlying idea is that Σ_2 may be regarded as a remainder term; in support of this we note that Σ_2 is composed of the remainder terms stemming from (1.4.14), and—which is even more important—in view of (1 2), Σ_2 does not contain too many terms.

Let us assume that (Ω_1) holds. With the aid of the function g defined in (1.4.17) we then introduce, for positive real x, the sum

$$G_k(x) = \sum_{\substack{d<x \\ (d,k)=1}} \mu^2(d)g(d). \tag{1.3}$$

Since

$$g(d) \geqslant 0$$

and g is multiplicative, we know that $g(1) = 1$ and hence that $G(z) \geqslant 1$. ($G(z) = G_1(z)$ is defined in (1.4.20)).

We now define also

$$\lambda_d = \frac{\mu(d)}{\prod_{p|d}(1 - \omega(p)/p)} \frac{G_d(z/d)}{G(z)}, \tag{1.4}$$

and observe immediately that this choice is consistent with the condition $\lambda_1 = 1$ and also with (1.2). Indeed, one can prove that the numbers (1.4) actually minimize Σ_1 subject to these conditions; but as this result is not needed here we omit the proof and instead refer the reader to the Notes.

Although it seems rather complicated to evaluate Σ_1 at (1.4), the calculation can in fact be accomplished in two fairly simple stages; and it will turn out that the value of Σ_1 at (1.4) is $1/G(z)$.

As a first step we observe that we may exclude from Σ_1 all terms for which $\omega([d_1, d_2]) = 0$, so that $\omega(p)$ is then non-zero for all prime divisors of both d_1 and d_2. Therefore we may assume in Σ_1 that

$$\omega(d_1) \neq 0, \qquad \omega(d_2) \neq 0, \qquad \omega([d_1, d_2]) \neq 0$$

(remember that ω is multiplicative), and we may therefore write, in Σ_1,

$$\frac{\omega(D)}{D} = \frac{\omega(d_1)\omega(d_2)}{d_1 d_2} \frac{(d_1, d_2)}{\omega((d_1, d_2))}.$$

Moreover, if follows from (1.4.17), if $\omega(p) \neq 0$, that

$$\frac{p}{\omega(p)} = 1 + \frac{p - \omega(p)}{\omega(p)} = 1 + \frac{1}{g(p)},$$

so that if we write $(d_1, d_2) = d$ we have

$$\frac{d}{\omega(d)} = \prod_{p|d}\left(1 + \frac{1}{g(p)}\right) = \prod_{l|d}\frac{1}{g(l)} \qquad (\mu(d) \neq 0).$$

Putting these remarks together with (1.2) and (1.4.19) we obtain

$$\begin{aligned}\Sigma_1 &= \sum_{d_1|P(z)}' \sum_{d_2|P(z)}' \lambda_{d_1}\lambda_{d_2}\frac{\omega(d_1)}{d_1}\frac{\omega(d_2)}{d_2}\sum_{\substack{l|d_1\\ l|d_2}}\frac{1}{g(l)} \\ &= \sum_{\substack{l<z\\ l|P(z)}}' \frac{1}{g(l)}\left(\sum_{\substack{d|P(z)\\ d\equiv 0 \bmod l}} \lambda_d \frac{\omega(d)}{d}\right)^2, \qquad (1.5)\end{aligned}$$

where the dash indicates that summation is over those numbers only for which $\omega \neq 0$.

So far we have used only the fact that the numbers λ_d given by (1.4) satisfy (1.2). We shall now use all the information contained in (1.4) to evaluate

$$\sum_{\substack{d|P(z)\\ d\equiv 0 \bmod l}} \lambda_d \frac{\omega(d)}{d}$$

when

$$1 \leqslant l < z \qquad \text{and} \qquad l|P(z).$$

Since g is multiplicative, the sum is equal to

$$\begin{aligned}&\sum_{\substack{m|P(z)\\ (m,\, l)=1}} \mu(lm)g(lm)\frac{G_{lm}(z/lm)}{G(z)} \\ &\qquad = \frac{\mu(l)g(l)}{G(z)}\sum_{\substack{m|P(z)\\ (m,\, l)=1}} \mu(m)g(m) \sum_{\substack{d<z/lm\\ (d,\, lm)=1}} \mu^2(d)g(d) \\ &\qquad = \frac{\mu(l)g(l)}{G(z)}\sum_{\substack{n<z/l\\ (n,\, l)=1}} \mu^2(n)g(n)\sum_{m|n}\mu(m) = \frac{\mu(l)g(l)}{G(z)},\end{aligned}$$

so that

$$\sum_{\substack{d|P(z)\\ d\equiv 0 \bmod l}} \lambda_d \frac{\omega(d)}{d} = \frac{\mu(l)g(l)}{G(z)} \quad \text{if} \quad 1 \leqslant l < z \quad \text{and} \quad l|P(z). \qquad (1.6)$$

If now we combine (1.5) and (1.6) and bear in mind (1.4.19) and (1.4.24), we arrive at

$$\Sigma_1 = \frac{1}{G^2(z)}\sum_{\substack{l<z\\ l|P(z)}}' g(l) = \frac{1}{G(z)}. \qquad (1.7)$$

The remainder term Σ_2 can be simplified. In fact, we shall prove that the numbers λ_d defined in (1.4) satisfy

$$|\lambda_d| \leqslant 1. \tag{1.8}$$

Because g is non-negative we have that

$$G(z) = \sum_{l|d} \sum_{\substack{m<z \\ (m,\,d)=l}} \mu^2(m)g(m) = \sum_{l|d} \sum_{\substack{h<z/l \\ (h,\,d/l)=1 \\ (h,\,l)=1}} \mu^2(lh)g(lh)$$

$$= \sum_{l|d} \mu^2(l)g(l)G_d(z/l) \geqslant \left(\sum_{l|d} \mu^2(l)g(l)\right) G_d(z/d); \tag{1.9}$$

and, by (1.4.17),

$$\sum_{l|d} \mu^2(l)g(l) = \prod_{p|d} (1 + g(p)) = \prod_{p|d} \frac{p}{p - \omega(p)} = 1\Big/\prod_{p|d}\left(1 - \frac{\omega(p)}{p}\right).$$

Hence

$$G_d(z/d) \leqslant \prod_{p|d} \left(1 - \frac{\omega(p)}{p}\right) G(z),$$

and (1.8) follows at once in view of (1.4). As a result we obtain

$$\Sigma_2 \leqslant \sum_{\substack{d_\nu < z \\ d_\nu | P(z) \\ \nu=1,2}} |R_{[d_1,d_2]}|. \tag{1.10}$$

Now in the sum on the right the numbers $d = [d_1, d_2]$ are necessarily less than z^2 and divide $P(z)$. Since d is squarefree, the number of terms arising from the same d is at most

$$|\{d_1, d_2\colon\ [d_1, d_2] = d\}| = 3^{\nu(d)}. \tag{1.11}$$

Therefore we are able to deduce from (1.10) that

$$\Sigma_2 \leqslant \sum_{\substack{d_\nu < z \\ d_\nu | P(z) \\ \nu=1,2}} |R_{[d_1,d_2]}| \leqslant \sum_{\substack{d<z^2 \\ d|P(z)}} 3^{\nu(d)}|R_d| \leqslant \sum_{\substack{d<z^2 \\ (d,\,\overline{\mathfrak{P}})=1}} \mu^2(d)\,3^{\nu(d)}|R_d|. \tag{1.12}$$

We often find that our condition

(R) $$|R_d| \leqslant \omega(d) \quad \text{if} \quad \mu(d) \neq 0, \qquad (d, \overline{\mathfrak{P}}) = 1,$$

is satisfied, and we may then take the simplification of Σ_2 still further. We have that

$$(R):\quad \sum_{\substack{d<z^2\\ d|P(z)}} 3^{\nu(d)}|R_d| \leqslant \sum_{\substack{d<z^2\\ d|P(z)}} 3^{\nu(d)}\omega(d)$$

$$\leqslant z^2 \sum_{d|P(z)} \frac{3^{\nu(d)}\omega(d)}{d} = z^2 \prod_{\substack{p<z\\ p\in\mathfrak{P}}} \left(1 + \frac{3\omega(p)}{p}\right)$$

$$\leqslant z^2 \prod_{p<z} \left(1 + \frac{\omega(p)}{p}\right)^3 \leqslant \frac{z^2}{W^3(z)} \tag{1.13}$$

by (1.4.12) and (1.4.16).

A combination of (1.1), (1.7), (1.12) and (1.13) leads to the following result.

THEOREM 3.1. (Ω_1), (R):

$$S(\mathscr{A}; \mathfrak{P}, z) \leqslant \frac{X}{G(z)} + \frac{z^2}{W^3(z)}.$$

If, on the other hand, we do not impose condition (R), then (1.1), (1.7) and (1.12) yield the more general

THEOREM 3.2. (Ω_1):

$$S(\mathscr{A}; \mathfrak{P}, z) \leqslant \frac{X}{G(z)} + \Sigma_2,$$

where

$$\Sigma_2 \leqslant \sum_{\substack{d<z^2\\ (d,\,\overline{\mathfrak{P}})=1}} \mu^2(d)\, 3^{\nu(d)}|R_d|, \tag{1.14}$$

or, equally well,

$$\Sigma_2 \leqslant \sum_{\substack{d_\nu<z\\ d_\nu|P(z)\\ \nu=1,2}} |R_{[d_1,d_2]}| \leqslant \sum_{\substack{d<z^2\\ d|P(z)}} 3^{\nu(d)}|R_d|. \tag{1.15}$$

We note that the estimates for Σ_2 in (1.15) are slightly sharper than that in (1.14), but (1.14) is in a more convenient form which suffices for most of our applications of Theorem 3.2.

2. THE CASE $\omega(d) = 1$, $|R_d| \leqslant 1$

The estimates in Theorems 3.1 and 3.2 are in as simple a form as the general circumstances allow. However, both involve the function $G(z)$, and it is a characteristic feature of the Selberg method that, in order to apply these basic estimates to any particular problem or class of problems, one has first to study the function G in the special circumstances of the problem (or class of problems).

In this section we shall deal with the very special but important case (conforming to (Ω_1) and (R)) when

$$\omega(d) = 1 \quad \text{and} \quad |R_d| \leqslant 1 \quad \text{whenever} \quad \mu(d) \neq 0 \quad \text{and} \quad (d, \mathfrak{P}) = 1.$$

Subject to this condition on ω, the behaviour of the corresponding function G is particularly easy to determine.

We require the following result.

LEMMA 3.1. *For integral k define*

$$H_k(x) = \sum_{\substack{d<x \\ (d,k)=1}} \frac{\mu^2(d)}{\phi(d)}.$$

Then

$$H_k(x) \geqslant \prod_{p|k} \left(1 - \frac{1}{p}\right) \log x. \tag{2.1}$$

Proof. If $k \neq 0$ (for $k = 0$ the result is trivially true) we have, as in the argument of (1.9), that

$$H_1(x) = \sum_{l|k} \sum_{\substack{d<x \\ (d,k)=l}} \frac{\mu^2(d)}{\phi(d)} = \sum_{l|k} \sum_{\substack{h<x/l \\ (h,k/l)=1 \\ (h,l)=1}} \frac{\mu^2(lh)}{\phi(lh)}$$

$$= \sum_{l|k} \frac{\mu^2(l)}{\phi(l)} H_k\left(\frac{x}{l}\right) \leqslant \left(\sum_{l|k} \frac{\mu^2(l)}{\phi(l)}\right) H_k(x);$$

and

$$\sum_{l|k} \frac{\mu^2(l)}{\phi(l)} = \prod_{p|k} \left(1 + \frac{1}{p-1}\right) = 1 \Big/ \prod_{p|k} \left(1 - \frac{1}{p}\right).$$

Hence it suffices to show that $H_1(x) \geqslant \log x$.

Let $k(n)$ denote the "kernel" (i.e. the largest squarefree divisor) of n. Then

$$H_1(x) = \sum_{d<x} \frac{\mu^2(d)}{d} \prod_{p|d} \left(1 - \frac{1}{p}\right)^{-1} = \sum_{k(n)<x} \frac{1}{n}.$$

Since $n < x$ implies that $k(n) < x$, we therefore find that

$$H_1(x) \geqslant \sum_{n<x} \frac{1}{n} \geqslant \sum_{n<x} \int_n^{n+1} \frac{dt}{t} \geqslant \int_1^x \frac{dt}{t} = \log x,$$

since 1 plus the greatest integer less than x is at least x. This completes the proof of (2.1).

We are now able easily to deduce

THEOREM 3.3. *Suppose that*

$$\omega(d) = 1 \quad \textit{and} \quad |R_d| \leqslant 1 \quad \textit{if} \quad \mu(d) \neq 0 \quad \textit{and} \quad (d, \bar{\mathfrak{P}}) = 1. \tag{2.2}$$

Then

$$S(\mathscr{A}; \mathfrak{P}, z) \leqslant \frac{X}{\prod_{\substack{p<z \\ p \notin \mathfrak{P}}} (1 - 1/p) \log z} + \Sigma_2$$

and

$$\Sigma_2 \leqslant \left(\sum_{\substack{d<z \\ (d, \bar{\mathfrak{P}})=1}} \mu^2(d) \right)^2 < z^2. \tag{2.3}$$

Proof. The given condition on ω (in (2.2)) implies, in view of (1.4.18), that

$$g(d) = \begin{cases} 1/\phi(d) & \text{if} \quad (d, \bar{\mathfrak{P}}) = 1, \\ 0 & \text{if} \quad (d, \bar{\mathfrak{P}}) > 1. \end{cases} \qquad (\mu(d) \neq 0)$$

Hence, by the definition of $H_k(z)$,

$$G(z) = H_k(z)$$

if

$$k = \prod_{p<z, p \notin \mathfrak{P}} p.$$

It follows from (2.1) that we may replace $G(z)$ in Theorem 3.2 by

$$\prod_{\substack{p<z \\ p \notin \mathfrak{P}}} \left(1 - \frac{1}{p}\right) \log z.$$

If now we replace $|R_{[d_1,d_2]}|$ in the first inequality of (1.15) by 1, as we may do in view of (2.2), inequality (2.3) follows at once.

3. Application to $\sum_{\substack{n \leqslant x \\ (n,k)=1}} 1$

Let us now consider a typical application of Theorem 3.3. Suppose that

$$1 \leqslant k < y \leqslant x,$$

and that k is coprime with $\mathfrak{P}$. With any integer l form the sequence

$$\mathscr{A} = \{n:\ x - y < n \leqslant x,\ \ n \equiv l \bmod k\}.$$

This sequence was analysed in Example 2. If d is a squarefree number satisfying $(d, \mathfrak{P}) = 1$ then, since $(k, \mathfrak{P}) = 1$, d and k are coprime. Hence, choosing $X = y/k$ as in (1.3.8), it is clear from (1.4.15), (1.3.7) and (1.3.9) that the conditions of Theorem 3.3 are satisfied. Thus Theorem 3.3 leads at once to

THEOREM 3.4. *If*

$$1 \leqslant k < y \leqslant x$$

and if $\mathfrak{P}$ *is any set of primes that such*

$$(k, \mathfrak{P}) = 1,$$

we have for any z $(\geqslant 2)$ *that*

$$|\{n:\ x - y < n \leqslant x,\ n \equiv l \bmod k,\ (n, P(z)) = 1\}|$$

$$\leqslant \frac{1}{\prod_{\substack{p<z \\ p \notin \mathfrak{P}}} (1 - 1/p)} \frac{y}{k \log z} + \Sigma_2,$$

and

$$\Sigma_2 \leqslant \left(\sum_{\substack{d<z \\ (d,\mathfrak{P})=1}} \mu^2(d) \right)^2 < z^2.$$

As a special case of Theorem 3.4 we can deal with the function

$$\Phi_k(x) := \sum_{\substack{n \leqslant x \\ (n,\ k)=1}} 1 \tag{3.1}$$

where k is a positive integer. Let $p(k)$ denote the largest prime factor of k and let $p(1) = 1$. When studying $\Phi_k(x)$ it is clearly no restriction to assume that $p(k) \leqslant x$. We shall prove the following result.

THEOREM 3.5. *If $x \geqslant e^6$ and $p(k) \leqslant x$ we have*

$$\sum_{\substack{n \leqslant x \\ (n,k)=1}} 1 < 7 \frac{\phi(k)}{k} x.$$

Proof. In Theorem 3.4 we take k (of Theorem 3.4) as 1, $y = x$ and

$$\mathfrak{P} = \{p\colon p|k\};$$

and we derive at once, for any $z \leqslant x$,

$$\Phi_k(x) \leqslant \left|\left\{n\colon n \leqslant x, \left(n, \prod_{\substack{p<z \\ p|k}} p\right) = 1\right\}\right| \leqslant \frac{x}{\prod\limits_{\substack{p<z \\ p\nmid k}} (1 - 1/p) \log z} + z^2,$$

or, since $p(k) \leqslant x$,

$$\frac{k}{\phi(k)} \frac{\Phi_k(x)}{x} \leqslant \frac{1}{\prod\limits_{p \leqslant x} (1 - 1/p)} \left(\frac{1}{\log z} + \frac{z^2}{x}\right).$$

It is known that

$$\prod_{p \leqslant x} \left(1 - \frac{1}{p}\right)^{-1} \leqslant e^{\gamma} \log x \left(1 + \frac{1}{2 \log^2 x}\right),$$

and if we now choose

$$z = x^{1/3}$$

we obtain

$$\frac{k}{\phi(k)} \frac{\Phi_k(x)}{x} \leqslant e^{\gamma} \log x \left(1 + \frac{1}{2 \log^2 x}\right) \left(\frac{3}{\log x} + \frac{1}{x^{1/3}}\right).$$

The expression on the right is decreasing, and for $x = e^6$ it is less than 7.

4. THE BRUN–TITCHMARSH INEQUALITY

In this section we shall apply our results to primes of an arithmetic progression lying in an interval. For this purpose we shall work with

$$\mathfrak{P}_K = \{p\colon p \nmid K\}$$

where K is an even natural number.

We prepare the ground with the following result.

THEOREM 3.6. *Let K be an even natural number, and suppose that*

$$K < y \leqslant x.$$

For $\mathfrak{P}$ *take the set* $\mathfrak{P}_K$. *Then, for any* $z(\geqslant 2)$, *we have*

$$|\{n: x - y < n \leqslant x,\ n \equiv l_1 \bmod K,\ (n, P(z)) = 1\}|$$

$$\leqslant \frac{y}{\phi(K)\log z} + \Sigma_2, \tag{4.1}$$

where

$$\Sigma_2 \leqslant \left(\frac{z-1}{2}\right)^2 \quad \textit{if} \quad z \geqslant 9, \tag{4.2}$$

and

$$\Sigma_2 \leqslant 24\frac{K}{\phi(K)}\frac{z^2}{\log^2 z} \quad \textit{if} \quad z \geqslant e^6. \tag{4.3}$$

Proof. With our set $\mathfrak{P}_K$ the form of the leading term (on the right of (4.1)) is a direct consequence of Theorem 3.4 which also gives, since K is even, that

$$\Sigma_2 \leqslant \left(\sum_{\substack{d<z\\(d,K)=1}} \mu^2(d)\right)^2 \leqslant \left(\sum_{\substack{d<z\\(d,2)=1}} \mu^2(d)\right)^2 \leqslant \left(\frac{z-1}{2}\right), \quad \text{if} \quad z \geqslant 9.$$

It remains to prove (4.3). In the given circumstances it is clear that $G(z) = H_K(z)$ and that $|R_d| \leqslant 1$ in Σ_2. Hence, by (1.1) and (1.4),†

$$\Sigma_2 \leqslant \left(\sum_{\substack{d<z\\d|P(z)}} |\lambda_d|\right)^2 = \left(\sum_{\substack{d<z\\(d,K)=1}} \mu^2(d)\frac{d}{\phi(d)}\frac{1}{H_K(z)}\sum_{\substack{m<z/d\\(m,Kd)=1}} \frac{\mu^2(m)}{\phi(m)}\right)^2$$

$$= \frac{1}{H_K{}^2(z)}\left(\sum_{\substack{n<z\\(n,K)=1}} \mu^2(n)\frac{\sigma(n)}{\phi(n)}\right)^2,$$

$$\leqslant \frac{\Phi_K(z)}{H_K{}^2(z)}\sum_{\substack{n<z\\(n,K)=1}} \mu^2(n)\frac{\sigma^2(n)}{\phi^2(n)}$$

† $\sigma(n) = \sum_{d|n} d.$

by an application of Cauchy's inequality. Since K is even,

$$\Sigma_2 \leqslant \frac{\Phi_K(z)}{H_K{}^2(z)} \sum_{\substack{n<z\\(n,\,2)=1}} \mu^2(n)\frac{\sigma^2(n)}{\phi^2(n)}.$$

Now

$$\frac{\sigma^2(n)}{\phi^2(n)} = \prod_{p|n}\left(\frac{p+1}{p-1}\right)^2 = \prod_{p|n}\left(1+\frac{4p}{(p-1)^2}\right) = \sum_{d|n}\frac{4^{\nu(d)}d}{\phi^2(d)} \quad (\mu(n)\neq 0),$$

so that†

$$\sum_{\substack{n<z\\(n,\,2)=1}} \mu^2(n)\frac{\sigma^2(n)}{\phi^2(n)} = \sum_{\substack{d<z\\(d,\,2)=1}} \mu^2(d)\frac{4^{\nu(d)}d}{\phi^2(d)} \sum_{\substack{m<z/d\\(m,\,2d)=1}} \mu^2(m)$$

$$\leqslant z\prod_{p>2}\left(1+\frac{4}{(p-1)^2}\right) < \frac{16}{5}z.$$

Hence by (2.1) and Theorem 3.5

$$\Sigma_2 < \frac{7(\phi(K)/K)\,z}{(\phi^2(K)/K^2)\log^2 z}\,\frac{16}{5}z = \frac{112}{5}\,\frac{K}{\phi(K)}\,\frac{z^2}{\log^2 z},$$

provided that $p(K) \leqslant z$. Finally, we may remove this last restriction on K. For on the one hand Σ_2 is independent of those prime factors of K that exceed z; and on the other the ratio $K/\phi(K)$ increases as new prime factors are included in K. With this remark the proof of (4.3), and so of Theorem 3.6, is complete.

We come now to the principal result of this section.

THEOREM 3.7. *Let x and y be real numbers and k and l integers, satisfying*

$$1 \leqslant k < y \leqslant x, \qquad (k, l) = 1.$$

Then

$$\pi(x;\,k, l) - \pi(x-y;\,k, l) < \frac{y}{\phi(k)\log\sqrt{(y/k)}}\left(1+\frac{4}{\log\sqrt{(y/k)}}\right) \tag{4.4}$$

and

$$\pi(x;\,k, l) - \pi(x-y;\,k, l) < \frac{3y}{\phi(k)\log(y/k)}. \tag{4.5}$$

† Here we use $\prod_p\left(1+\frac{4}{(p-1)^2}\right) = 15{\cdot}9396\ldots.$

Proof. Put

$$\Delta(x, y;\ k, l) = \pi(x;\ k, l) - \pi(x - y;\ k, l)$$

and

$$K = \begin{cases} k & \text{if } 2 \mid k \\ 2k & \text{if } 2 \nmid k. \end{cases}$$

Then, for a suitable integer l_1,

$$\Delta(x, y;\ k, l) \leqslant \Delta(x, y;\ K, l_1) + 1,$$

because if k is odd, every other term of the progression $l + mk$ is even, and at most one of the even terms is prime.

A trivial estimation is provided by

$$\Delta(x, y;\ K, l_1) \leqslant \sum_{\substack{x-y<n\leqslant x \\ n\equiv l_1 \bmod K}} 1 \leqslant \frac{y}{K} + 1,$$

so that

$$\Delta(x, y;\ k, l) \leqslant \frac{y}{K} + 2.$$

If we put

$$t = \sqrt{\frac{y}{k}}$$

and note that, whether k is even or odd,

$$\phi(k) = \phi(K) \leqslant \tfrac{1}{2} K,$$

we obtain

$$D: = \frac{\phi(k) \log \sqrt{(y/k)}}{y} \Delta(x, y;\ k, l) \leqslant \log t \left(\frac{1}{2} + \frac{2}{t^2}\right)$$

$$< \tfrac{3}{2} \quad \text{if} \quad 1 < t < e^{2 \cdot 9};$$

this proves (4.5) for the range $1 < t < e^{2 \cdot 9}$.

Now take $\mathscr{A}$ to be the sequence of Theorem 3.6. The expression on the left of (4.1) counts at least all those primes p in the interval $x - y < n \leqslant x$ which satisfy $p \geqslant z$ and $p \equiv l_1 \bmod K$. Hence, by (4.1),

$$\Delta(x, y;\ k, l) \leqslant \frac{y}{\phi(k) \log z} + \Sigma_2 + 1 + \pi(z; K, l_1). \tag{4.6}$$

It is easy to check that

$$1 + \pi(z; K, l_1) \leqslant \sum_{\substack{1 \leqslant d \leqslant z \\ (d,\,2)=1}} \mu^2(d)$$

$$\leqslant \frac{z-1}{2} \quad \text{if} \quad z \geqslant 9;$$

hence, by (4.6) and (4.2),

$$D \leqslant \log t \left\{ \frac{1}{\log z} + \frac{1}{t^2}\left(\left(\frac{z-1}{2}\right)^2 + \frac{z-1}{2}\right)\right\} < \log t \left\{\frac{1}{\log z} + \frac{z^2}{4t^2}\right\}$$

$$\text{if} \quad z \geqslant 9.$$

We now define u by

$$t = \frac{u}{\sqrt{2}}\, e^u$$

and choose

$$z = e^u.$$

Then $t \geqslant e^{2\cdot 9}$ implies that $z \geqslant 9$, and therefore

$$D \leqslant \log t \left\{\frac{1}{u} + \frac{1}{2u^2}\right\}. \tag{4.7}$$

If $u \geqslant \sqrt{(2e)}$ the function on the right is decreasing, and at $u = \sqrt{(2e)}$ it is less than 3/2. Hence the proof of (4.5) is complete.

There remains (4.4), and we begin with the observation that (4.4) is a consequence of (4.5) if $t \leqslant e^8$. Suppose then that $t > e^8$, so that the corresponding u exceeds 25/4. Then (4.7) shows that

$$D < 1{\cdot}4 < 1 + \frac{4}{\log t} \quad \text{if} \quad t < e^{10}.$$

For $t \geqslant e^{10}$ we return to (4.6) and apply (4.3); if $z \geqslant e^6$, we obtain

$$(\log t)(D - 1) \leqslant \log t \left\{\frac{\log t}{\log z} - 1 + 48\,\frac{\log t}{t^2}\,\frac{z^2}{\log^2 z} + \frac{\log t}{t^2}\, z\right\}. \tag{4.8}$$

We choose

$$\log z = \log t - 2$$

(so that even $z \geqslant e^8$), and the right side of (4.8) becomes

$$\log t \left\{\frac{2}{\log t - 2} + \frac{48}{e^4}\,\frac{\log t}{(\log t - 2)^2} + \frac{\log t}{e^2 t}\right\}.$$

This function is decreasing as t increases, and when $t = e^{10}$ it is less than 4. This completes the proof of (4.4).

If we specialise to $y = x$ in Theorem 3.7, we obtain at once a type of result that occurs in the literature in a large variety of forms and is usually referred to as the Brun–Titchmarsh inequality.

THEOREM 3.8. *If*

$$1 \leqslant k < x \quad \textit{and} \quad (k, l) = 1,$$

then

$$\pi(x; k, l) < \frac{x}{\phi(k) \log \sqrt{(x/k)}} \left(1 + \frac{4}{\log \sqrt{(x/k)}}\right) \tag{4.9}$$

and

$$\pi(x; k, l) < \frac{3x}{\phi(k) \log (x/k)} \, . \tag{4.10}$$

5. THE TITCHMARSH DIVISOR PROBLEM

The sieve results we have proved so far, and also many of the results we shall prove in the sequel, are often needed in arithmetical investigations in an auxiliary capacity. This is especially true of Theorem 3.8. The proof of the principal result of this section (Theorem 3.9 below), which was once regarded as an exceptionally difficult problem, illustrates this aspect of the sieve, and also serves to illustrate another use of Bombieri's theorem. Indeed, this latter result (Lemma 3.3) plays an important part in many of the most significant applications of sieve theory.

Our objective will be to derive an asymptotic formula for the sum

$$\sum_{l<p\leqslant x} \tau(p - l)$$

where l is a constant positive integer. All O-constants occurring below (in Section 5) may depend on l. This problem used to be known as the Titchmarsh divisor problem.

We shall prepare the ground.

LEMMA 3.2. *We have*

$$\sum_{\substack{n\leqslant x \\ (n,\, l)=1}} \frac{1}{\phi(n)} = \prod_{p\nmid l} \left(1 + \frac{1}{p(p-1)}\right) \frac{\phi(l)}{l} \log x + O_l(1).$$

Proof. The proof depends on the formula

$$\frac{1}{\phi(n)} = \frac{1}{n} \sum_{d|n} \frac{\mu^2(d)}{\phi(d)}.$$

Then

$$\sum_{\substack{n \leqslant x \\ (n,\, l)=1}} \frac{1}{\phi(n)} = \sum_{\substack{n \leqslant x \\ (n,\, l)=1}} \frac{1}{n} \sum_{d|n} \frac{\mu^2(d)}{\phi(d)} = \sum_{\substack{d \leqslant x \\ (d,\, l)=1}} \frac{\mu^2(d)}{d\phi(d)} \sum_{\substack{m \leqslant x/d \\ (m,\, l)=1}} \frac{1}{m},$$

and the inner sum on the right equals

$$\sum_{h|l} \frac{\mu(h)}{h} \sum_{n \leqslant x/(dh)} \frac{1}{n} = \sum_{h|l} \frac{\mu(h)}{h} \left\{ \log \frac{x}{d} - \log h + O(1) \right\}$$

$$= \frac{\phi(l)}{l} \log \frac{x}{d} + O_l(1).$$

Hence

$$\sum_{\substack{n \leqslant x \\ (n,\, l)=1}} \frac{1}{\phi(l)} = \frac{\phi(l)}{l} \log x \sum_{\substack{d \leqslant x \\ (d,\, l)=1}} \frac{\mu^2(d)}{d\phi(d)} + O_l(1)$$

$$= \frac{\phi(l)}{l} \log x \prod_{p \nmid l} \left(1 + \frac{1}{p(p-1)}\right) + O_l(1),$$

as required.

Our main tool is the following well-known theorem of Bombieri.

LEMMA 3.3. *Let*

$$E(x; d) = \max_{2 \leqslant y \leqslant x} \max_{(l,d)=1} |E(y; d, l)|. \tag{5.1}$$

Then, given any positive constant U, there exists a positive constant C such that

$$\sum_{d < \sqrt{(x)}/\log^C x} E(x; d) = O_U \left(\frac{x}{\log^U x} \right). \tag{5.2}$$

We are now in a position to prove

THEOREM 3.9. *We have, as $x \to \infty$,*

$$\sum_{l < p \leqslant x} \tau(p - l) = \prod_{p \nmid l} \left(1 + \frac{1}{p(p-1)}\right) \frac{\phi(l)}{l} x + O_l \left(\frac{x \log \log x}{\log x} \right).$$

Proof. Since

$$\tau(n) = 2 \sum_{\substack{d|n \\ d<\sqrt{n}}} 1 + \delta(n)$$

where $\delta(n) = 1$ if n is a perfect square and is otherwise 0, it follows, using the notation (5.1), and using also Theorem 3.8, that

$$\sum_{<p\leqslant x} \tau(p-l) = 2 \sum_{l<p\leqslant x} \sum_{\substack{d|p-l \\ d<(p-l)^{1/2}}} 1 + O(x^{1/2})$$

$$= 2 \sum_{d<(x-l)^{1/2}} \sum_{\substack{l+d^2<p\leqslant x \\ p\equiv l \bmod d}} 1 + O(x^{1/2})$$

$$= 2 \sum_{\substack{d<(x-l)^{1/2} \\ (d,\,l)=1}} \{\pi(x;d,l) - \pi(l+d^2;d,l)\} + O_l(x^{1/2})$$

$$= 2\,li\,x \sum_{\substack{d\leqslant x^{1/2}/\log^C x \\ (d,\,l)=1}} \frac{1}{\phi(d)} + O\Bigg(\sum_{d\leqslant x^{1/2}/\log^C x} E(x,d)\Bigg)$$

$$+\,O\left(\sum_{d<x^{1/2}/\log^C x} \frac{l+d^2}{\phi(d)\log\{(l+d^2)/d\}}\right)$$

$$+\,O\left(\sum_{x^{1/2}/\log^C x\leqslant d\leqslant x^{1/2}} \frac{x}{\phi(d)\log(x/d)}\right) + O_l(x^{1/2})$$

Therefore by Lemma 3.3 (actually it suffices to take $U = 1$)

$$\sum_{l<p\leqslant x} \tau(p-l) = 2\,li\,x \sum_{\substack{d\leqslant x^{1/2} \\ (d,l=1)}} \frac{1}{\phi(d)}$$

$$+\,O\left(\frac{x}{\log x}\left\{1 + \sum_{x^{1/2}/\log^C x\leqslant d\leqslant x^{1/2}} \frac{1}{\phi(d)}\right\}\right)$$

$$+\,O_l\left(\frac{x}{\log^{2C} x} \sum_{d\leqslant x^{1/2}/\log^C x} \frac{1}{\phi(d)}\right) + O_l(x^{1/2}),$$

and an appeal to Lemma 3.2 now completes the proof, since we may assume that $C \geqslant 1$.

6. The Case $\omega(p) = p/(p-1)$

We shall now derive from the basic results of Chapter 3 another class of applications all of which correspond to $\omega(p) = p/(p-1)$. Because of the nature of these applications it is not appropriate to impose at this stage also a common condition on R_d; but we shall work throughout with the sifting set $\mathfrak{P}_K$, where K is a (not necessarily positive) non-zero integer. In these circumstances Theorem 3.2 can be brought into the following form.

Theorem 3.10. *Let $\mathfrak{P} = \mathfrak{P}_K$ where $K \neq 0$ is an even integer. Suppose that*

$$\omega(p) = \frac{p}{p-1} \quad \textit{if} \quad p \in \mathfrak{P}_K \, (\textit{i.e. if } p \nmid K). \tag{6.1}$$

Then we have

$$S(\mathscr{A};\mathfrak{P}_K, z) \leqslant 2 \prod_{p>2}\left(1 - \frac{1}{(p-1)^2}\right) \prod_{2<p|K} \frac{p-1}{p-2} \frac{X}{\log z}\left\{1 + O\left(\frac{1}{\log z}\right)\right\}$$

$$+ \sum_{\substack{d<z^2\\(d,K)=1}} \mu^2(d) 3^{\nu(d)} |R_d|. \tag{6.2}$$

Proof. We apply Theorem 3.2; we may do so because the validity of (Ω_1) is ensured by the evenness of K. It obviously suffices to show that under our conditions

$$\frac{1}{G(z)} \leqslant 2 \prod_{p>2}\left(1 - \frac{1}{(p-1)^2}\right) \prod_{2<p|K} \frac{p-1}{p-2} \frac{1}{\log z}\left\{1 + O\left(\frac{1}{\log z}\right)\right\}. \tag{6.3}$$

We begin with the remark that

$$g(p) = \frac{\omega(p)/p}{1 - \omega(p)/p} = \frac{1}{p-2} = \frac{1}{p-1}\left\{1 + \frac{1}{p-2}\right\}$$

$$= \frac{1}{\phi(p)}\{1 + g(p)\} \qquad (p \nmid K),$$

so that

$$g(d) = \frac{1}{\phi(d)} \sum_{l|d} \mu^2(l) g(l) \quad \text{if} \quad (d, K) = 1.$$

With this representation of g we have that

$$G(z) = \sum_{\substack{d<z \\ (d,K)=1}} \frac{\mu^2(d)}{\phi(d)} \sum_{l|d} \mu^2(l)g(l)$$

$$= \sum_{\substack{l<z \\ (l,K)=1}} \frac{\mu^2(l)g(l)}{\phi(l)} \sum_{\substack{m<z/l \\ (m,K)=1 \\ (m,l)=1}} \frac{\mu^2(m)}{\phi(m)}$$

$$= \sum_{\substack{l<z \\ (l,K)=1}} \frac{\mu^2(l)g(l)}{\phi(l)} H_{Kl}\left(\frac{z}{l}\right),$$

and an application of (2.1) leads at once to

$$G(z) \geqslant \prod_{p|K}\left(1 - \frac{1}{p}\right) \sum_{\substack{l<z \\ (l,K)=1}} \frac{\mu^2(l)g(l)}{l} \log\frac{z}{l}$$

$$\geqslant \prod_{p|K}\left(1 - \frac{1}{p}\right) \sum_{\substack{l=1 \\ (l,K)=1}}^{\infty} \frac{\mu^2(l)g(l)}{l} \log\frac{z}{l}. \tag{6.4}$$

Since $g(p) = 1/(p-2)$ when $p \nmid K$,

$$\sum_{\substack{l=1 \\ (l,K)=1}}^{\infty} \frac{\mu^2(l)g(l)}{l} = \prod_{p \nmid K}\left(1 + \frac{1}{p(p-2)}\right),$$

and

$$\sum_{\substack{l=1 \\ (l,K)=1}}^{\infty} \frac{\mu^2(l)g(l)}{l} \log l = \sum_{\substack{l=1 \\ (l,K)=1}}^{\infty} \frac{\mu^2(l)g(l)}{l} \sum_{p|l} \log p$$

$$= \sum_{p \nmid K} \frac{\log p}{p(p-2)} \sum_{\substack{m=1 \\ (m,K)=1 \\ (m,p)=1}}^{\infty} \frac{\mu^2(m)g(m)}{m}$$

$$= \sum_{p \nmid K} \frac{\log p}{p(p-2)} \frac{1}{1 + \{p(p-2)\}^{-1}} \prod_{p' \nmid K}\left(1 + \frac{1}{p'(p'-2)}\right).$$

If now we substitute these relations in (6.4) we obtain

$$G(z) \geqslant \prod_{p|K}\left(1 - \frac{1}{p}\right) \prod_{p \nmid K}\left\{1 + \frac{1}{p(p-2)}\right\} \left(\log z - \sum_{p} \frac{\log p}{(p-1)^2}\right). \tag{6.5}$$

If z bounded by some absolute constant, (6.3) is true trivially, because then the factor depending on K on the right of (6.3) is at least 1. Otherwise (6.5) implies (6.3); for $2 \mid K$ implies that

$$\prod_{p \mid K}\left(1 - \frac{1}{p}\right)^{-1} \prod_{p \nmid K}\left\{1 + \frac{1}{p(p-2)}\right\}^{-1} = 2 \prod_{2<p \mid K} \frac{p}{p-1} \prod_{p \nmid K} \frac{p(p-2)}{p(p-2)+1}$$

$$= 2 \prod_{p>2}\left(1 - \frac{1}{(p-1)^2}\right) \prod_{2<p \mid K} \frac{p}{p-1} \times \frac{(p-1)^2}{p(p-2)},$$

and this completes the proof.

We shall need to make use of the following simple result.

LEMMA 3.4. *For any natural number h and for $x \geqslant 1$ we have*

$$\sum_{d<x} \mu^2(d) h^{\nu(d)} \leqslant x(\log x + 1)^h. \tag{6.6}$$

Proof. The sum on the left is at most

$$x \sum_{d<x} \frac{\mu^2(d) h^{\nu(d)}}{d},$$

and

$$\sum_{d<x} \frac{\mu^2(d) h^{\nu(d)}}{d} = \sum_{d_1 \ldots d_h < x} \frac{\mu^2(d_1 \ldots d_h)}{d_1 \ldots d_h}$$

$$\leqslant \left(\sum_{n<x} \frac{1}{n}\right)^h \leqslant (\log x + 1)^h. \tag{6.7}$$

Remark. A more careful argument shows that our sum is of order

$$x(\log x + 1)^{h-1} \qquad (x \geqslant 1);$$

but for our purpose the weaker result is sufficient.

To estimate our remainder term the following result will be required.

LEMMA 3.5. *Let h and k be positive integers, suppose that*

$$k \leqslant \log^A x,$$

and recall the definition

$$E(x, d) = \max_{2 \leqslant y \leqslant x} \max_{(l,d)=1} \left| \pi(y; d, l) - \frac{\operatorname{li} y}{\phi(d)} \right|. \tag{6.8}$$

Then, given any positive constant U_1, there exists a positive constant $C_1 = C_1(U_1, h, A)$ such that

$$\sum_{d < x^{1/2}/k \log^{C_1} x} \mu^2(d) h^{\nu(d)} E(x, dk) = O_{U_1,h,A}\left(\frac{x}{\phi(k) \log^{U_1} x}\right). \tag{6.9}$$

Proof. It is clear that we must make (6.9) depend on the important Lemma 3.3. We begin by noting the trivial estimate

$$E(x, dk) \ll \frac{x}{dk} \quad \text{if} \quad d \leqslant x/k.$$

Then, by Cauchy's inequality,

$$\sum_{d < x^{1/2}/(k \log^{C_1} x)} \mu^2(d) h^{\nu(d)} E(x, dk)$$

$$\ll \left(\frac{x}{k}\right)^{1/2} \sum_{d < x^{1/2}/(k \log^{C_1} x)} \frac{\mu^2(d) h^{\nu(d)}}{d^{1/2}} E^{1/2}(x, dk)$$

$$\ll \left(\frac{x}{k}\right)^{1/2} \left(\sum_{d < x^{1/2}} \frac{\mu^2(d) h^{2\nu(d)}}{d}\right)^{1/2} \left(\sum_{q < x^{1/2}/\log^{C_1} x} E(x, q)\right)^{1/2};$$

we apply (6.7) (with h^2 in place of h) to the second expression on the right, Lemma 3.3 with

$$U = 2U_1 + h^2 + A, \qquad C_1 = C(U)$$

to the third expression, and we use

$$\frac{1}{k} \leqslant \frac{1}{\phi^2(k)} \log^A x$$

in the first. Then the entire product on the right is

$$O_U\left(\frac{1}{\phi(k)} \left\{x \log^A x \log^{h^2} x \frac{x}{\log^{2U_1 + h^2 + A} x}\right\}^{1/2}\right) = O_U\left(\frac{x}{\phi(k) \log^{U_1} x}\right),$$

and (6.9) is proved.

7. The Prime Twins and Goldbach Problems: Explicit Upper Bounds

We shall now apply Theorem 3.10 to obtain upper estimates in the famous twin prime problem and its "conjugate", the Goldbach problem. We introduce the integers

$$N > 2 \quad \text{and} \quad h \neq 0,$$

and set out to derive good upper bounds for the numbers

$$|\{p\colon p \leqslant N,\ p + h = p'\}|,$$

which counts the number of "prime twins" p, $p' = p + h$; and

$$|\{p\colon p \leqslant N,\ N - p = p'\}|,$$

which counts the number of representations of N in the form $N = p + p'$. We shall prove the following result.

THEOREM 3.11. *As*† $N \to \infty$,

$$|\{p\colon p \leqslant N,\ p + h = p'\}|$$

$$\leqslant 8\prod_{p>2}\left(1 - \frac{1}{(p-1)^2}\right)\prod_{2<p|h}\frac{p-1}{p-2}\frac{N}{\log^2 N}\left\{1 + O\left(\frac{\log\log N}{\log N}\right)\right\} \quad \textit{if } 2\,|\,h \tag{7.1}$$

uniformly in h; and

$$|\{p\colon p \leqslant N,\ N - p = p'\}|$$

$$\leqslant 8\prod_{p>2}\left(1 - \frac{1}{(p-1)^2}\right)\prod_{2<p|N}\frac{p-1}{p-2}\frac{N}{\log^2 N}\left\{1 + O\left(\frac{\log\log N}{\log N}\right)\right\} \quad \textit{if } 2\,|\,N. \tag{7.2}$$

Remark. The range of magnitude of the second product in each of these inequalities can be seen from

$$1 \leqslant \prod_{2<p|k}\frac{p-1}{p-2} \ll \log\log 3\,|k|\,,$$

valid for any integer $k \neq 0$.

Proof. We define the sequences

$$\mathscr{X} = \{p + h\colon p \leqslant N\} \quad \text{and} \quad \mathscr{G} = \{N - p\colon p \leqslant N\},$$

and choose $K = h$ in the case of $\mathscr{X}$ and $K = N$ in the case of $\mathscr{G}$. Then

$$S(\mathscr{X}; \mathfrak{P}_h, z) = |\{p + h\colon p \leqslant N,\ (p + h, P(z)) = 1\}|$$

† For *odd* values of N and h the numbers being estimated in (7.1) and (7.2) are at most 1 and at most 2 respectively.

counts (at least) all the primes $p' \geqslant z$ among the numbers $p + h$, and

$$S(\mathcal{G}; \mathfrak{P}_N, z) = |\{N - p: p \leqslant N,\ (N - p, P(z)) = 1\}|$$

counts (at least) all the primes $p' \geqslant z$ among the numbers $N - p$. Hence

$$\begin{aligned} &|\{p: p \leqslant N,\ p + h = p'\}| \leqslant S(\mathcal{L}; \mathfrak{P}_h, z) + z \\ \text{and} \\ &|\{p: p \leqslant N,\ N - p = p'\}| \leqslant S(\mathcal{G}; \mathfrak{P}_N, z) + z, \end{aligned} \tag{7.3}$$

Both $\mathcal{L}$ and $\mathcal{G}$ are special cases of the sequence analysed in Example 5 (taking $x = N$ and $k = 1$ in Example 5). Accordingly we may take in each case (see (1.3.36))

$$X = li\,N \quad \text{and} \quad \omega(p) = \frac{p}{p-1} \quad \text{if} \quad p \nmid K,$$

and (see (1.3.37) and (6.8))

$$|R_d| \leqslant E(N, d).$$

Hence, by (7.3) and Theorem 3.10,

$$\begin{aligned} &|\{p: p \leqslant N,\ p + h = p'\}| \\ &\quad \leqslant 2 \prod_{p>2} \left(1 - \frac{1}{(p-1)^2}\right) \prod_{2<p|h} \frac{p-1}{p-2} \frac{li\,N}{\log z} \left\{1 + O\left(\frac{1}{\log z}\right)\right\} \\ &\quad + \sum_{d<z^2} \mu^2(d) 3^{\nu(d)} E(N, d) + z \end{aligned}$$

and

$$\begin{aligned} &|\{p: p \leqslant N,\ N - p = p'\}| \\ &\quad \leqslant 2 \prod_{p>2} \left(1 - \frac{1}{(p-1)^2}\right) \prod_{2<p|N} \frac{p-1}{p-2} \frac{li\,N}{\log z} \left\{1 + O\left(\frac{1}{\log z}\right)\right\} \\ &\quad + \sum_{d<z^2} \mu^2(d) 3^{\nu(d)} E(N, d) + z. \end{aligned}$$

If we take C_1 to be the constant which, in Lemma 3.5 (with $k = 1$), corresponds to the choice

$$U_1 = h = 3,$$

and choose

$$z^2 = \frac{N^{1/2}}{\log^{C_1} N},$$

we see that Lemma 3.5 implies

$$\sum_{d<z^2} \mu^2(d)3^{\nu(d)}E(N,d) = O\left(\frac{N}{\log^3 N}\right). \tag{7.4}$$

Finally, with the above choice of z

$$\frac{li\,N}{\log z}\left\{1 + O\left(\frac{1}{\log z}\right)\right\} = \frac{4N}{\log^2 N}\left\{1 + O\left(\frac{\log\log N}{\log N}\right)\right\},$$

and this remark evidently completes the proof of Theorem 3.11.

One final word about Theorem 3.11. We could as easily have proved (7.1) with $|p + h|$ in place of $p + h$, still uniformly in h. If we had done this, (7.2) would have followed at once, as the special case $h = -N$ of the modified result (7.1), and would not have required separate proof. We shall follow (essentially) this procedure in Theorem 9.2 (of Chapter 9).

8. The Problem $ap + b = p'$: An Explicit Upper Bound

A more general application of Theorem 3.10, one that corresponds closely to the case discussed in Example 5, is contained in the following result.

THEOREM 3.12. *Let a, b, k, l be integers satisfying*

$$ab \neq 0,\ (a,b) = 1,\ 2 \mid ab$$

and

$$(k,l) = 1,\ 1 \leqslant k \leqslant \log^A x.$$

Then as $x \to \infty$ we have, uniformly in a, b, k and l, that

$$|\{p\colon\ p \leqslant x,\ p \equiv l \bmod k,\ ap + b = p'\}|$$

$$\leqslant 8\prod_{p>2}\left(1 - \frac{1}{(p-1)^2}\right)\prod_{2<p|kab}\frac{p-1}{p-2}\,\frac{x}{\phi(k)\log^2 x}\left\{1 + O_A\left(\frac{\log\log x}{\log x}\right)\right\}.$$

(We note that the result is trivial if $(al + b, k) > 1$.)

Proof. We shall consider the sequence

$$\mathscr{A} = \{ap + b\colon\ p \leqslant x,\ p \equiv l \bmod k\}$$

which was analysed in Example 5, and we take

$$K = kab.$$

According to Example 5 (see especially (1.3.36)) we may also take

$$X = \frac{li\,x}{\phi(k)} \quad \text{and} \quad \omega(p) = \frac{p}{p-1} \quad \text{if} \quad p \nmid K,$$

and as a consequence, by (1.3.42),

$$|R_d| \leqslant E(x, dk).$$

Then

$$S(\mathscr{A}; \mathfrak{P}_K, z) = |\{ap + b \colon p \leqslant x,\ p \equiv l \bmod k,\ (ap + b, P(z)) = 1\}|$$

counts (at least) all the primes $p' \geqslant z$ that occur among the numbers $ap + b$, so that

$$|\{p \colon p \leqslant x, p \equiv l \bmod k, ap + b = p'\}| \leqslant S(\mathscr{A}; \mathfrak{P}_K, z) + z.$$

Since $2 \mid K$, Theorem 3.10 applies and yields

$$\begin{aligned} &|\{p\colon\ p \leqslant x,\ p \equiv l \bmod k,\ ap + b = p'\}| \\ &\quad \leqslant 2 \prod_{p>2} \left(1 - \frac{1}{(p-1)^2}\right) \prod_{2<p\mid kab} \frac{p-1}{p-2} \frac{li\,x}{\phi(k)\log z} \left\{1 + O\left(\frac{1}{\log z}\right)\right\} \\ &\quad + \sum_{d<z^2} \mu^2(d) 3^{\nu(d)} E(x, dk) + z. \end{aligned} \tag{8.1}$$

We take x sufficiently large and apply Lemma 3.5 with

$$U_1 = h = 3;$$

there is then a corresponding number $C_1 = C_1(A)$, and if we choose

$$z^2 = \frac{x^{1/2}}{k \log^{C_1} x},$$

we obtain, since $k \leqslant \log^A x$,

$$\sum_{d<z^2} \mu^2(d) 3^{\nu(d)} E(x, dk) = O_A\left(\frac{x}{\phi(k)\log^3 x}\right).$$

Also

$$\log z = \tfrac{1}{4} \log x \left\{1 + O_A\left(\frac{\log\log x}{\log x}\right)\right\},$$

whence

$$\frac{li\,x}{\log z} = \frac{4x}{\log^2 x} \left\{1 + O_A\left(\frac{\log\log x}{\log x}\right)\right\}.$$

These remarks together with (8.1) conclude the proof of Theorem 3.12.

NOTES

3.1. This section is based on Selberg [1]. For an analysis of the structure of Selberg's upper sieve see Halberstam–Roth [1], Chapter 4. There are several ways of proving that the numbers λ_d given by (1.4) minimize Σ_1 subject to the constraints (1.2) and $\lambda_1 = 1$. Writing

$$y_l = \sum_{\substack{d<z,\, d|P(z)\\ d\equiv 0 \bmod l}} \lambda_d \frac{\omega(d)}{d}, \qquad l < z, \quad l \mid P(z),$$

we see from (1.5) that

$$\Sigma_1 = \sum_{l<z,\, l|P(z)} \frac{1}{g(l)} y_l^2,$$

and the condition $\lambda_1 = 1$ is equivalent (by inversion) to

$$1 = \sum_{l<z,\, l|P(z)} \mu(l) y_l.$$

Thus the problem is one of minimizing a positive definite quadratic form subject to a single linear side-condition. Selberg used the method of Lagrangian multipliers, and this procedure was followed by Prachar [6]. Gelfond–Linnik [1] and Halberstam–Roth [1] used an argument based on completion of squares. A third argument is due to Turán, cf. Fluch [1] and Levin [9]. In this argument one applies Cauchy's inequality to the side-condition: thus

$$1 = \sum_{l<z,\, l|P(z)} \mu(l)\sqrt{g(l)}\, \frac{y_l}{\sqrt{g(l)}}$$

implies

$$1 \leqslant \Sigma_1 \,.\, \sum_{l<z,\, l|P(z)} \mu^2(l) g(l) = \Sigma_1 \,.\, G(z),$$

with equality provided that the numbers $\mu(l)y_l/g(l)$ $(l < z,\ l \mid P(z))$ are equal (necessarily to $1/G(z)$). This proves (1.7), and (1.4) follows easily from (1.6) by inversion.

(1.8): This was first observed by van Lint–Richert [2]. Actually $|\lambda_d|$ is much smaller for large d, and this fact is exploited in the proof of Theorem 3.6, (4.3).

(1.11): Let $d = p_1 \ldots p_r$ be the canonical prime decomposition of d, and let $d_1 = p^{\alpha_1} \ldots p_r^{\alpha_r}$, $d_2 = p_1^{\beta_1} \ldots p_r^{\beta_r}$ where the α's and β's take the values 0, 1 only. Then $[d_1, d_2] = d$ if and only if max $(\alpha_i, \beta_i) = 1$ for each $i = 1, \ldots r$, and there are precisely 3^r pairs d_1, d_2 for which this is true, since max $(\alpha_i, \beta_i) = 1$ if and only if $\alpha_i = \beta_i = 1$; $\alpha_i = 1$, $\beta_i = 0$; or $\alpha_i = 0$, $\beta_i = 1$.

(R): first introduced in the context of the Selberg sieve by Čulanovskiĭ [1], who was the first to derive effective upper bounds, under general conditions, by means of Selberg's method; $\omega(p) < p$ (cf. our (Ω_1)) occurs for the first time here also. Prachar [6] follows Čulanovskiĭ.

Theorem 3.1: It is worth noting that one can extend this method to "weighted" sums of the type

$$\sum_{\substack{a \in \mathscr{A} \\ (a, P(z))=1}} w(a),$$

provided that one has sufficient information about the distribution of the weights $w(a)$ modulo d, $d \mid P(z)$. Such an extension occurs f.e. in Hooley [5] and there is another in Ramachandra [2].

Up to 1956 there were, apart from Selberg [1], only five papers that dealt with, or used, Selberg's method: Selberg's own lectures [3] and [4]—which show him to have arrived at results, for example about lower bound sieves and limitations of sieves, that others were not to reach for several years—Čulanovskiĭ [1], Shapiro–Warga [1] and Ožigova [1]. Ožigova [1] and Čulanovskiĭ [1] (the latter with much more justification) are frequently quoted in subsequent Russian sieve literature.

For other formulations or applications, see f.e. Rieger [10, 13]; Halberstam [1]; Hooley [2]; Barban [4]; Klimov [2, 3].

3.2. This and the next two sections are based on van Lint-Richert [2].

Lemma 3.1, (2.1): Actually (cf. van Lint–Richert [1])

$$\begin{aligned} H_1(x) &= \log x + \gamma + \sum_p \frac{\log p}{p(p-1)} + o(1), \qquad && x \to \infty, \\ &\geqslant \log x + 1{\cdot}33258 \qquad\qquad\qquad , && x \geqslant x_0. \end{aligned}$$

(For the latter inequality see Rosser–Schoenfeld [1], (2.11)).

3.3. Theorem 3.5: According to van Lint–Richert [2], p. 212, the constant 7 may be replaced by 5, and the inequality is even then true for all $x \geqslant 1$. This result is clear-cut from a numerical point of view; in fact, the argument gives that

$$\Phi_k(x) \leqslant 2e^{\gamma} \frac{\phi(k)}{k} x\left(1 + O\left\{\frac{\log\log x}{\log x}\right\}\right), \qquad p(k) \leqslant x.$$

Recently R. R. Hall (*Acta Arith.*, to appear) has proved, by different methods, that this estimate holds without the factor 2 on the right; it is then, by reference to the prime number theorem, best possible (apart from the quality of the error term). (See also Prachar [9].)

The upper bound for

$$\prod_{p \leqslant x} \left(1 - \frac{1}{p}\right)^{-1}$$

is (3.2.9) on p.70 of Rosser–Schoenfeld [1].

3.4. The sum to which the footnote is attached is at most $\frac{5}{3}z$ (in place of $\frac{16}{5}z$) according to van Lint–Richert [2], p.214.

Theorems 3.7, 3.8: any one of these results is nowadays referred to as a Brun–Titchmarsh inequality (the nomenclature appears to have been first used in Linnik [3]). The constant factor 4 in (4.4) could be replaced by 3, according to van Lint–Richert [2]. For a similar result via the large sieve see Bombieri [2]. Hardy–Littlewood [2] proved that

$$\pi(x + y) - \pi(x) \ll y/\log y$$

using Brun [7] (Erdös [12] remarks that this is the only occasion on which Hardy and Littlewood used Brun's sieve) and Titchmarsh [1] showed that

$$\pi(x; k, l) \ll \frac{x}{\phi(k) \log x}, \qquad (l, k) = 1, \quad k \leqslant x^{\alpha} \quad (0 < \alpha < 1).$$

For other early versions of one or more of these results see Romanoff [1], Heilbronn–Landau–Scherk [1]; Klimov [1]; Selberg [4]; Prachar [6]; Klimov [3], Selberg [3] came close in precision to Theorem 3.7, and Čulanovskiĭ [1], Uchiyama [1] to Theorem 3.8.

These theorems are important as the only results known at present that are uniform in k for all $k < y$ in Theorem 3.7, and all $k < x$ in Theorem 3.8. Both have recently been improved using deeper methods based on the large sieve. Thus Montgomery–Vaughan [1] have recently proved (4.5) with 2 in place of 3 on the right, for $1 \leqslant k < y \leqslant x$, a result which is superior to (4.4) (see Montgomery [2], Chapter 4 for an earlier version). A further improvement, from 2 to $2 - \delta$ for some $\delta > 0$ would have important implications for the location of Siegel zeros; this was pointed out by Rodosskiĭ (quoted in Klimov [3]), Erdös [12, 16], Barban [8], Bombieri–Davenport [2] and others since then. See also Turán [1] and Siebert [3]. Taking $k = 1$, the result of Montgomery–Vaughan implies that

$$\pi(x) - \pi(x - y) < 2\pi(y), \qquad 1 \leqslant y \leqslant x;$$

whatever improvements may lie ahead, the old conjecture $\pi(x) - \pi(x - y) \leqslant \pi(y)$ $(1 \leqslant y \leqslant x)$ is probably false; for Hensley–Richards [1] have shown recently that it is inconsistent with the prime k-tuplets conjecture, and the balance of evidence favours the latter.

Recently also Hooley [8, 9] has shown by means of a striking combination of Selberg's sieve and analytic methods that better results hold on average (in various senses); and Motohashi [4, 5] and Goldfeld [2] have used Hooley's ideas to improve Theorem 3.8 for $k \leqslant x^{\frac{1}{4}}$ (their improvements are slight but a new avenue with great possibilities has been opened up by these researches). A similar combination of Selberg's sieve and analytic methods occurs in Haneke [1] (but this paper contains inaccuracies); see also Chapter 11.

Given the elementary nature of the argument and the explicit character of the result, Theorem 3.7 is very satisfactory. Nevertheless, it is perhaps worth indicating that the beautiful inequality

$$\pi(x; k, l) - \pi(x - y; k, l) < \frac{2y}{\phi(k) \log (y/k)}, \qquad 1 \leqslant k < y \leqslant x, \tag{1}$$

of Montgomery–Vaughan [1] is accessible via Selberg's method (we understand that this fact is known to Selberg). There are, indeed, two rather different approaches.

The first argument is simpler and amounts only to introducing greater precision into the proof of Theorem 3.7; we refer from now on, for the sake of brevity, only to the case $k = 1$. First of all, we know (see an earlier Note) that

$$H_1(z) \geqslant \log z + 1{\cdot}332 \qquad (z \geqslant z_0).$$

Next, the estimation of Σ_2 for large z, leading to (4.3), can obviously be improved (even beyond the numerical refinement already mentioned) by classical procedures e.g. from the theory of mean values of multiplicative arithmetic functions (cf. Wirsing [2], Satz 1); in fact, we get (see proof of (4.3) in Theorem 3.6 with $K = 2$)

$$\begin{aligned}\Sigma_2 &\leqslant (H_2(z))^{-2} \Bigg(\sum_{n<z,(n,2)=1} \mu^2(n)\sigma(n)/\phi(n) \Bigg)^2 \\ &\leqslant (H_2(z))^{-2} \left(\frac{6}{\pi^2} + o(1) \right)^2 z^2 \leqslant \left(\frac{144}{\pi^4} + o(1) \right) z^2/H_1{}^2(z)\end{aligned}$$

since $H_2(z) \geqslant \frac{1}{2}H_1(z)$ by Lemma 3.1. Hence (cf. (4.6)).

$$\begin{aligned}\pi(x) - \pi(x-y) &\leqslant 1 + \pi(z) + y/H_1(z) + \tfrac{3}{2} z^2/H_1{}^2(z), && z \geqslant z_1, \\ &\leqslant y/H_1(z) + 2z^2/H_1{}^2(z), && z \geqslant z_2, \\ &< \frac{2y}{\log y}, && y_0 \leqslant y \leqslant x,\end{aligned}$$

after a simple computation, choosing $z = y^{\frac{1}{2}}/e$ (for the sake of convenience). To determine the lowest possible y_0, and to extend the result to arithmetic progressions, would require more elaborate asymptotics and calculations, such as are to be found in the work of Montgomery–Vaughan; but in principle there are no new difficulties.

The second argument utilizes the illuminating connection noticed by Kobayashi [1] (also by Huxley [2]) between the Selberg and large sieves. Taking the more general situation of $\mathscr{A}$ as in Example 3 (of Chapter 1) and $\mathfrak{P} = \mathfrak{P}_1$, there is the remarkable identity

$$G(z) \sum_{\substack{d|F(n)\\ d|P(z)}} \lambda_d = \sum_{q<z} \sum_{\substack{h=1\\ (h,q)=1}}^{q} b_{q,h} \exp(-2\pi i h n/q)$$

for the Selberg λ's (cf. (1.4)), where

$$b_{q,h} = \frac{1}{q} \prod_{p|q} \left(1 - \frac{\rho(p)}{p}\right)^{-1} \sum_{\substack{l=1\\ (F(l),q)=1}}^{q} \exp\left(\frac{2\pi i h l}{q}\right);$$

moreover,

$$\mu^2(q) g(q) = \sum_{\substack{h=1\\ (h,q)=1}}^{q} |b_{q,h}|^2,$$

where

$$g(q) = \prod_{p|q} \rho(p)/(p - \rho(p)) \quad \text{if} \quad \mu(q) \neq 0.$$

Hence

$$G^2(z)\, S(\mathscr{A}; \mathfrak{P}_1, z) \leqslant \sum_{x-y<n\leqslant x} \Bigg(G(z) \sum_{\substack{d|F(n)\\ d|P(z)}} \lambda_d\Bigg)^2$$

$$\leqslant \sum_{x-y<n\leqslant x} \Bigg| \sum_{q<z} \sum_{\substack{h=1\\ (h,q)=1}}^{q} b_{q,h} \exp(-2\pi i h n/q)\Bigg|^2$$

and the expression on the right is the *dual* form of the usual large sieve expression; from Montgomery–Vaughan [1] we now know that this expression is at most

$$(y + z^2) \sum_{q<z} \sum_{\substack{h=1\\ (h,q)=1}}^{q} |b_{q,h}|^2 = (y + z^2)\, G(z),$$

so that we obtain (cf. Theorem 3.1)

$$S(\mathscr{A}; \mathfrak{P}_1, z) \leqslant (y + z^2)/G(z),$$

i.e. the Montgomery sieve (see Montgomery [2]) but via Selberg's method! Note that this result is actually inferior, for the case $F(n) = n$ considered above, to the estimate

$$S(\mathscr{A}; \mathfrak{P}_1, z) \leqslant y/H_1(z) + \tfrac{3}{2}z^2/H_1{}^2(z) \qquad (z \geqslant z_1)$$

in Theorem 3.6 ($K = 2$), so that up to this point our second line of argument is of methodological interest only.

However, Montgomery saw (see Montgomery [2]) that a certain weighted form of the large sieve would lead to improvements, and Montgomery–Vaughan [2] established a very precise result of this kind; we now point out that Kobayashi's argument can be adapted to take account of the weighted large sieve, and this is especially interesting for us because a new choice of Selberg's λ's is required. Remember that the λ's given by (1.4) minimize only Σ_1, with the magnitude of Σ_2 controlled only by virtue of the fact that $\lambda_d = 0$ for $d \geqslant z$. By contrast, we now choose the λ's, in effect, with *both* Σ_1 *and* Σ_2 in mind:

$$\lambda_d = \mu(d)d\prod_{p|d}(p - \rho(p))^{-1}\, G_d{}^*(z/d)/G^*(z), \qquad \mu(d) \neq 0,$$

where

$$G_d{}^*(z) = \sum_{\substack{q<z\\(q,d)=1}} \frac{\mu^2(q)}{y + \tfrac{3}{2}zdq}\, g(q), \qquad G^*(z) = G_1{}^*(z).$$

As before, $\lambda_d = 0$ if $d \geqslant z$ and $\lambda_1 = 1$; and, while these λ's no longer minimize Σ_1, they are close to the optimal choice (1.4) as well as damp down the effect of Σ_2. To be precise, we can show by adapting Kobayashi's argument and using the weighted large sieve that

$$(G^*(z))^2\, S(\mathscr{A}; \mathfrak{P}_1, z) \leqslant \sum_{q<z} \sum_{\substack{h=1\\(h,q)=1}}^{q} (y + \tfrac{3}{2}zq)^{-1}\, |b_{q,h}|^2 = G^*(z),$$

so that we obtain the elegant upper estimate

$$S(\mathscr{A}; \mathfrak{P}_1, z) \leqslant 1/G^*(z)$$

of Montgomery–Vaughan [1] (cf. Theorem 5.3). As they go on to show, this leads to (1) on taking $\mathscr{A} = \{n\colon\ x - y < n \leqslant x,\ n \equiv l \bmod k\}$.

3.5. Lemma 3.2: see Titchmarsh [1] for a sketch of an analytic proof. For an elementary proof see Estermann (*J. London Math. Soc.* **6** (1931), 250–251). Note that although k may be negative, the sums in the proof extend only over positive divisors of k.

Lemma 3.3: see Bombieri [1], Gallagher [2], Montgomery [2], Huxley [2], Vaughan (*J. London Math. Soc.*, to appear). This important result makes it possible in many contexts to circumvent use of the *GRH*. The first to obtain information of this kind (actually in a slightly different, weighted, form) was Rényi [2]; he used the large sieve as have all mathematicians who followed him in this investigation. (For Rényi's formulation see also Pan [1, 2].) Inequalities of type

$$\sum_{d \leqslant x^\alpha} \mu^2(d) \max_{(l,d)=1} |E(x; d, l)| \ll_U \frac{x}{\log^U x}$$

occur in Barban [2] (with $\alpha = \frac{1}{6} - \varepsilon$), Pan [3] (with $\alpha = \frac{1}{3}$, $U = 6$), Barban [6, 7, 8] (with $\alpha = \frac{3}{8} - \varepsilon$ and arbitrary but fixed U), A. I. Vinogradov [7] (with $\alpha = \frac{1}{2} - \varepsilon$ and U as in Barban). It has been conjectured that there exist admissible exponents $\alpha \geqslant \frac{1}{2}$ (Buchstab [10]; Halberstam–Jurkat–Richert [1]), and even that $\alpha = 1 - \varepsilon$ is admissible (Elliott–Halberstam [1, 2]). Bombieri's theorem has the defect that the implied constants are not effectively computable. An effective version is desirable and would be useful.

Theorem 3.9: our account derives from Rodriquez [1] (see also Halberstam [2]). (Actually Rodriquez proves a result uniform in l for

$$0 < |l| \ll x^{\frac{1}{4}} \log^{-3} x;$$

he also conjectures a sharper formula having a second dominant term.) The problem was first posed by Titchmarsh [1] and solved by him subject to *GRH*. The first unconditional proof was given by Linnik [3]. For some early investigations see Barban [7]; Erdös [2]; Hazelgrove [1]; and see Prachar [6] for a discussion on pp. 164–168. For further and deeper investigations see Bredehin [1]. Perhaps it is worth remarking that our proof requires information about the distribution of primes $\leqslant x$ in arithmetic progressions modulo d for $d \leqslant x^{\frac{1}{2}}$; roughly speaking, Bombieri's theorem supplies this information for $d \leqslant x^{\frac{1}{2}-\varepsilon}$ and Brun–Titchmarsh for the rest.

The conjugate sum $\sum_{p<n} \tau(n-p)$ may be treated in the same way, as also may the sums $\sum_{l<p\leqslant x} \tau(p-l)$, $\sum_{p<n} r(n-p)$ (though there is an additional difficulty here) where $r(m)$ is the number of ways of expressing m as a sum of two integer squares; and others (see Hooley [1], Linnik [3], Elliott–Halberstam [1]).

If $\tau_k(m)$ denotes the number of ways of representing m as a product of k positive factors (so that $\tau_2 = \tau$), Porter [1] has shown for each $k = 3, 4, \ldots$, that an asymptotic formula for $\sum_{l<p\leqslant x} \tau_k(p-l)$ can be established provided that $\alpha = 1 - 1/k - \varepsilon$ (see above discussion of Lemma 3.3) is admissible; and Vaughan (*J. London Math. Soc.* (2), **6** (1972), 43–55) has proved that Porter's formulae hold for almost all l.

A last remark apropos of results of Bombieri type, due to Barban [7] (see also Barban [10], Theorem 3.3; Karšiev [1]).
Let

$$\pi_2(x; k, l) = \sum_{\substack{p_1, p_2 \leqslant x^{1/2} \\ p_1 p_2 \equiv l \bmod k}} 1, \qquad (l, k) = 1,$$

so that

$$\pi_2(x; k, l) = \sum_{\substack{h=1 \\ (h,k)=1}}^{k} \pi(x^{\frac{1}{2}}; k, h)\, \pi(x^{\frac{1}{2}}; k, lh'), \qquad hh' \equiv 1 \bmod k;$$

then, by an easy calculation,

$$\left| \pi_2(x; k, l) - \frac{1}{\phi(k)} (li\, x^{\frac{1}{2}})^2 \right| \leqslant \sum_{\substack{h=1 \\ (h,k)=1}}^{k} |E(x^{\frac{1}{2}}; k, h)\, E(x^{\frac{1}{2}}; k, lh')|$$

$$\leqslant \sum_{\substack{h=1 \\ (h,k)=1}}^{k} E^2(x^{\frac{1}{2}}; k, h),$$

whence

$$\sum_{k \leqslant Q} \left| \pi_2(x; k, l) - \frac{1}{\phi(k)} (li\, x^{\frac{1}{2}})^2 \right| \leqslant \sum_{k \leqslant Q} \sum_{\substack{h=1 \\ (h,k)=1}}^{k} E^2(x^{\frac{1}{2}}; k, h)$$

$$\ll x \log^{-U} x \quad \text{if} \quad Q \leqslant x^{\frac{1}{2}} \log^{-U-1} x \ (U > 0),$$

using at the last step another result of Barban [7], Davenport–Halberstam [1], Gallagher [1]. Here then is a Bombieri result for the sequence of numbers that are products of two primes. For an interesting application of this kind of argument, see Greaves [2]. For a similar but deeper result see Chapter 11.

3.6. Theorem 3.10: f.e. Wang [11]; Shapiro–Warga [1] established the behaviour of $G(z)$ in this case.

Lemma 3.5: This is the form of Bombieri's theorem most appropriate for sieve applications.

3.7. Theorem 3.11: (7.1) (with the factor 8) was proved by Wang [11] subject to *GRH* and, unconditionally, by Bombieri–Davenport [1], also Kondakova–Klimov [1]. Both estimates (7.1), (7.2) are too large by a factor 4 when compared with the conjectured asymptotic formulae of Hardy–Littlewood [2] (cf. Shah–Wilson [1]). (Note that

$$\prod_{p>2} (1 - (p-1)^{-2}) = 0{\cdot}6601 \ldots;$$

Ricci [11], p. 7). Without a Bombieri-type theorem the best available results had 16 in place of 8 (Selberg [3], Čulanovskiĭ [1]). Montgomery has pointed out in correspondence that an improvement from 8 to $8 - \delta$ here would have the same consequences as the improvement from 2 to $2 - \delta$ in the Brun–Titchmarsh theorem.

For obvious reasons, both problems considered here have attracted much attention over the years, and we shall not attempt to list all the estimates of this kind, mostly in weaker form, that occur in the literature. There have been also many applications, and we mention just two. An early form of (7.1) played an important part in Schnirelmann's proof that there exists an integer s such that every large enough integer is the sum of at most s primes. We know now from Vinogradov's three primes theorem that $s \leqslant 4$; but an approach via Schnirelmann's density theorems and the sieve is an attractive and, in some ways, more elementary alternative approach (see Landau [2]). Shapiro–Warga [1] obtained $s \leqslant 20$ in this way, and their claim of $s \leqslant 18$ was confirmed by Yin [1, 2]. The best result now known is $s \leqslant 10$, but uses Bombieri's theorem (Siebert [1], Kuzjašev–Čečuro [1]).

The second application occurs in Bombieri–Davenport [1] who use (7.1) (with emphasis on the uniformity in h) to prove that, if p_n is the nth prime,

$$\liminf_{n\to\infty} \frac{p_{n+1} - p_n}{\log p_n} \leqslant \tfrac{1}{8}(2 + \sqrt{3}) = 0{\cdot}46650\ldots.$$

(Actually, Bombieri's theorem alone leads to the upper bound $\frac{1}{2}$, and the further improvement follows from using (7.1) in the treatment of the singular series.) This result is connected with the prime twin conjecture since, if the latter is true, this limit is 0.

Remark following Theorem 3.11: This estimate can be proved as follows:

$$1 \leqslant \prod_{2<p|k} \left(1 + \frac{1}{p-2}\right) \leqslant \exp\left(\sum_{p|k} \frac{1}{p} + \frac{1}{2}\right),$$

and

$$\begin{aligned}\sum_{p|k} \frac{1}{p} &\leqslant \sum_{p \leqslant \log 3|k|} \frac{1}{p} + \sum_{\substack{p|k \\ p > \log 3|k|}} \frac{1}{p} \\ &\leqslant \log\log\log 3\,|k| + O(1) + \frac{\nu(k)}{\log 3\,|k|} \\ &\leqslant \log\log\log 3\,|k| + O(1)\end{aligned}$$

since, trivially, $\nu(k) \log 2 \leqslant \log |k|$.

3.8. Theorem 3.12: See f.e. Klimov [1, 2]; and cf. Corollary 2.4.1.

Chapter 4

The Selberg Upper Bound Method (continued): O-Results

1. A Lower Bound for $G(x, z)$

In the later parts of Chapter 3 we dealt with two special classes of problems (those corresponding to $\omega(p) = 1$ and $\omega(p) = p/(p-1)$ respectively), among them some of the best known applications of the sieve method; but we now turn to the general sieve problem and obtain a Selberg upper bound for $S(\mathscr{A}; \mathfrak{P}, z)$ subject only to very weak general conditions on $\omega((\Omega_1)$ and $\Omega_2(\kappa))$. This result admits a wide range of applications; for example, Theorem 2.2 and all results of Sections 2.6 and 2.7 could be derived equally well from Theorem 4.1 below, after applying condition (R). Section 2 of this chapter will deal with some applications for which Theorem 4.1 is particularly suitable.

Theorem 4.1 is a result of $\ll$-type; it will therefore again prove most valuable in those cases where either numerical accuracy (with respect to the constant factor in the estimate) is not required, or is unobtainable in the present state of knowledge.

Before stating (and proving) Theorem 4.1, we need to deal with one of the technical difficulties inherent in applications of Selberg's method—finding a lower estimate for $G(z)$. The question is of some interest in itself; with a later purpose in mind, we shall deal with the more general function $G(x, z)$ defined in Chapter 1 as

$$G(x,z) = \sum_{\substack{d<x \\ d|P(z)}} g(d),$$

and we shall do so with more precision than Theorem 4.1 alone would warrant. Recall that, subject to (Ω_1), g is a non-negative multiplicative function and that $G(z, z) = G(z)$.

By (2.3.10)

$$(\Omega_1): \qquad G(x, z) \leqslant \sum_{d|P(z)} g(d) = \prod_{p<z} (1 + g(p))$$

$$= 1/\prod_{p<z}\left(1 - \frac{\omega(p)}{p}\right) = 1/W(z), \tag{1.1}$$

so that if $z \leqslant x$, it is appropriate to compare $G(x, z)$ with $1/W(z)$. In fact, we shall now establish a (one-sided) correspondence between these two functions.

LEMMA 4.1. (Ω_1), $(\Omega_2(\kappa))$: *For any* $\lambda > 0$ *we have*

$$\frac{1}{G(x,z)} \leqslant W(z)\left(1 + O\left(\exp\left\{-\lambda\frac{\log x}{\log z} + \left(\frac{2\kappa}{\lambda} + \frac{A_2}{\log z}\right)e^{\lambda}\right\}\right)\right) \quad \textit{if} \quad z \leqslant x, \tag{1.2}$$

where the O-constant is independent of λ.

Proof. For any natural number $n \geqslant 2$, $(\Omega_2(\kappa))$ implies that

$$\sum_{p<z} \frac{\omega(p)}{p} \log^n p = \int_2^z \sum_{t \leqslant p<z} \frac{\omega(p)\log p}{p}\, d(\log^{n-1} t)$$

$$\leqslant \int_1^z \left(\kappa \log\frac{z}{t} + A_2\right) d(\log^{n-1} t)$$

$$= (\kappa \log z + A_2)\log^{n-1} z - \kappa\frac{n-1}{n}\log^n z$$

$$= \frac{\kappa}{n}\log^n z + A_2 \log^{n-1} z, \tag{1.3}$$

and in view of $(\Omega_2(\kappa))$ this estimate can also be used for $n = 1$.

For any $s < 1$ we have by (1.1) that

$$\frac{1}{W(z)} - G(x, z) = \sum_{\substack{d \geqslant x \\ d|P(z)}} g(d) \leqslant \sum_{\substack{d \geqslant x \\ d|P(z)}} g(d)\left(\frac{d}{x}\right)^{1-s}$$

$$\leqslant x^{s-1}\prod_{p<z}\left(1 + \frac{\omega(p)}{p^s\{1 - \omega(p)/p\}}\right),$$

whence

$$1 - W(z)G(x, z) \leqslant x^{s-1} \prod_{p<z} \left(1 - \frac{\omega(p)}{p} + \frac{\omega(p)}{p^s}\right)$$

$$\leqslant \exp\left\{-(1-s)\log x + \sum_{p<z} \omega(p)\left(\frac{1}{p^s} - \frac{1}{p}\right)\right\}.$$

Since

$$\frac{1}{p^s} - \frac{1}{p} = \frac{1}{p}\exp\{(1-s)\log p\} - \frac{1}{p} = \frac{1}{p}\sum_{n=1}^{\infty} \frac{(1-s)^n \log^n p}{n!},$$

it follows using (1.3) that

$$\sum_{p<z} \omega(p)\left(\frac{1}{p^s} - \frac{1}{p}\right) = \sum_{n=1}^{\infty} \frac{(1-s)^n}{n!} \sum_{p<z} \frac{\omega(p)}{p} \log^n p$$

$$\leqslant \sum_{n=1}^{\infty} \frac{(1-s)^n \log^n z}{n!}\left(\frac{\kappa}{n} + \frac{A_2}{\log z}\right). \tag{1.4}$$

If now we put

$$1 - s = \frac{\lambda}{\log z}, \qquad \lambda > 0,$$

and estimate $1/n$ by $2/(n+1)$, we obtain

$$1 - W(z)G(x, z) \leqslant \exp\left(-\lambda\frac{\log x}{\log z} + 2\kappa\frac{e^\lambda}{\lambda} + \frac{A_2}{\log z}e^\lambda\right) \quad \text{if} \quad \lambda > 0, \tag{1.5}$$

so that

$$\frac{1}{G(x, z)} \leqslant W(z)\left(1 + W(z)^{-1} G(x,z)^{-1} \exp\left\{-\lambda\frac{\log x}{\log z} + \left(\frac{2\kappa}{\lambda} + \frac{A_2}{\log z}\right)e^\lambda\right\}\right)$$
$$\text{if} \quad \lambda > 0. \tag{1.6}$$

Let

$$c_2 = 2\kappa e + \frac{A_2 e}{\log 2} + \log 2.$$

Then, by (1.5) with $\lambda = 1$, we find that

$$W(z^{1/c_2})G(z, z^{1/c_2}) \geqslant 1 - \exp\left(-c_2 + 2\kappa e + \frac{A_2}{\log 2} e\right) = \tfrac{1}{2} \tag{1.7}$$

(provided, of course, that $z^{1/c_2} \geqslant 2$; but if this is not so, then the left hand side of (1.7) actually equals 1 and the rest of the proof is trivial).

If $x \geqslant z$ we have

$$G(x, z) \geqslant G(z, z) \geqslant G(z, z^{1/c_2}),$$

whence, by (2.3.5) and (1.7),

$$\frac{1}{W(z)\,G(x, z)} \leqslant \frac{W(z^{1/c_2})}{W(z)} \cdot \frac{1}{W(z^{1/c_2})G(z, z^{1/c_2})} = O(1);$$

and (1.2) now follows from (1.6).

We are now in a position to prove the main theorem of this chapter.

THEOREM 4.1. (Ω_1), $(\Omega_2(\kappa))$:

$$S(\mathscr{A}; \mathfrak{P}, z) \leqslant B_7 X W(z) + \sum_{\substack{d < z^2 \\ (d, \overline{\mathfrak{P}}) = 1}} \mu^2(d) 3^{\nu(d)} |R_d| . \tag{1.8}$$

Proof. We take $\lambda = 1$ in Lemma 4.1 and obtain, in view of (1.4.24),

$$\frac{1}{G(z)} = \frac{1}{G(z, z)} \ll W(z). \tag{1.9}$$

Theorem 4.1 now follows at once from Theorem 3.2.

2. APPLICATIONS

In this section we shall give an effective application of Theorem 4.1, involving, as $\mathscr{A}$, the sequence of values taken by a polynomial at the primes. The result, Theorem 4.2 below, sometimes yields more accurate information than Theorem 2.2 used in the manner of Section 2.7; to be precise, this happens when the sifting set $\mathfrak{P}$ has density less than 1, cf. Corollaries 4.2.1 and 4.2.2. The reason is that, when sifting by a thin set $\mathfrak{P}$, the latter method involves the use of a sifting function S which is arithmetically inappropriate and, indeed, much bigger than it should be. When the density of $\mathfrak{P}$ is 1, both approaches lead to results of the same quality; even though Theorem 4.2 depends on the use of Bombieri's theorem while the approach via Theorem 2.2 is elementary, it is fair to say that the application of Theorem 4.1 is more natural and slightly more straightforward (however, for a fuller discussion of some limitations involved when using Bombieri's theorem, see Section 5.5). The last two corollaries in this section afford an illustration of this remark.

THEOREM 4.2. *Let $F(n)$ be a polynomial of degree $g(\geqslant 1)$ with integer coefficients. Then, for any set $\mathfrak{P}$ of primes,*

$$|\{p: p \leqslant x,\ (F(p), \mathfrak{P}) = 1\}| \ll \prod_{\substack{p<x \\ p\in\mathfrak{P}}} \left(1 - \frac{\rho(p)}{p}\right) \prod_{\substack{p<x \\ p\in\mathfrak{P} \\ p|F(0)}} \left(1 - \frac{1}{p}\right)^{-1} \frac{x}{\log x}, \tag{2.1}$$

where $\rho(p)$ denotes the number of solutions of

$$F(n) \equiv 0 \bmod p,$$

and where the constant implied by the $\ll$-symbol depends only on g.

Remark. We are concerned here with estimating the number of primes $p \leqslant x$ for which $F(p)$ is not divisible by any prime of $\mathfrak{P}$.

Proof. We take as the sequence to be sifted

$$\mathscr{A} = \{F(p): p \leqslant x\},$$

which is the special case $k = 1$ of the sequence analysed in Example 6.

We may assume at once that F has no fixed prime divisor in $\mathfrak{P}$ since otherwise the left side of (2.1) is zero; this means that we may assume that

$$\rho(p) < p \quad \text{if} \quad p \in \mathfrak{P}, \tag{2.2}$$

and it follows from (1.3.54) that

$$\rho(p) \leqslant g \quad \text{if} \quad p \in \mathfrak{P}. \tag{2.3}$$

In accordance with (1.3.48) we choose

$$X = li\, x \tag{2.4}$$

(so that we require that $li\, x > 1$), and by (1.4.12) and (1.3.51)

$$\omega(p) = \begin{cases} \dfrac{\rho_1(p)}{p-1} p & \text{if } p \in \mathfrak{P}, \\ 0 & \text{if } p \in \overline{\mathfrak{P}}, \end{cases} \tag{2.5}$$

where $\rho_1(p)$ is the number of solutions of

$$F(m) \equiv 0 \bmod p, \qquad p \nmid m.$$

By (1.3.53) we have

$$\rho_1(p) = \begin{cases} \rho(p) & \text{if } \; p \nmid F(0) \\ \rho(p) - 1 & \text{if } \; p \mid F(0) \end{cases}, \tag{2.6}$$

and by (2.3)

$$\rho_1(d) \leqslant \rho(d) \leqslant g^{\nu(d)} \quad \text{if} \quad \mu(d) \neq 0, \qquad (d, \mathfrak{P}) = 1.$$

Hence, by (1.4.15) and (1.3.52), it follows that

$$|R_d| \leqslant g^{\nu(d)}(E(x, d) + 1) \quad \text{if} \quad \mu(d) \neq 0, \qquad (d, \mathfrak{P}) = 1. \tag{2.7}$$

We may assume further that

$$\rho_1(p) < p - 1 \quad \text{if} \quad p \in \mathfrak{P}; \tag{2.8}$$

otherwise, if there were a prime $p' \in \mathfrak{P}$ with $\rho_1(p') = p' - 1$, it would follow that $F(p) \equiv 0 \bmod p'$ for all $p \neq p'$ and this in turn would make the left side of (2.1) at most 1, making our theorem trivially true.

From (2.8) we deduce that

$$\frac{\omega(p)}{p} \leqslant \frac{p-2}{p-1} \leqslant 1 - \frac{1}{g} \quad \text{if} \quad p \leqslant g + 1,$$

and from (2.3) that

$$\frac{\omega(p)}{p} \leqslant \frac{g}{p-1} \leqslant 1 - \frac{1}{g+1} \quad \text{if} \quad p \geqslant g + 2,$$

so that (Ω_1) holds with $A_1 = g + 1$. Also by (2.3)

$$\omega(p) \leqslant \frac{g}{p-1} p \leqslant 2g,$$

so that (Ω_0) is satisfied with $A_0 = 2g$ and so too, by Lemma 2.2, is $(\Omega_2(\kappa))$ with $\kappa = A_2 = 2g$; hence, by (2.3.5),

$$\frac{W(z)}{W(x)} \ll \left(\frac{\log x}{\log z}\right)^{2g} \quad \text{if} \quad 2 \leqslant z \leqslant x. \tag{2.9}$$

We now apply Theorem 4.1: with the aid of (2.4), (2.7) and (2.9) we obtain, for any z, that

$$|\{p: p \leqslant x,\ (F(p), \mathfrak{P}) = 1\}|$$

$$\leqslant |\{p: p \leqslant x,\ (F(p), P(z)) = 1\}| = S(\mathscr{A}; \mathfrak{P}, z)$$

$$\ll W(x)\left\{li\, x . \frac{W(z)}{W(x)} + \frac{1}{W(x)} \sum_{d < z^2} \mu^2(d)(3g)^{\nu(d)}(E(x, d) + 1)\right\}$$

$$\ll \prod_{\substack{p < x \\ p \in \mathfrak{P}}} \left(1 - \frac{\rho_1(p)}{p - 1}\right) \left\{\frac{x}{\log x} \left(\frac{\log x}{\log z}\right)^{2g}\right.$$

$$\left. + \log^{2g} x \sum_{d < z^2} \mu^2(d)(3g)^{\nu(d)}(E(x, d) + 1)\right\}, \tag{2.10}$$

where the $\ll$-constant depends on g only.

In Lemma 3.5, with $h = 3g$ and $k = 1$, we now choose

$$U_1 = 2g + 1;$$

and we determine z by

$$z^2 = \frac{x^{1/2}}{\log^{C_1} x},$$

where $C_1 = C_1(2g + 1,\ 3g,\ 1)$ is the constant of Lemma 3.5. Since (2.3) holds trivially as long as $x = O_g(1)$, we may also assume that the required conditions

$$li\, x > 1, \qquad 2 \leqslant z \leqslant x$$

are satisfied. Then Lemmas 3.5 and 3.4 imply that

$$\sum_{d < z^2} \mu^2(d)(3g)^{\nu(d)}(E(x, d) + 1) = O_g\left(\frac{x}{\log^{2g+1} x}\right), \tag{2.11}$$

so that (2.10) yields the inequality

$$|\{p: p \leqslant x,\ (F(p), \mathfrak{P}) = 1\}| \ll \prod_{\substack{p < x \\ p \in \mathfrak{P}}} \left(1 - \frac{\rho_1(p)}{p - 1}\right) \frac{x}{\log x},$$

with the $\ll$-constant depending only on g.

Finally we use (2.6) and the identity (cf.(2.8.6))

$$1 - \frac{\rho(p) - 1}{p - 1} = \frac{1 - \rho(p)/p}{1 - 1/p} \tag{2.12}$$

to obtain

$$\prod_{\substack{p<x\\ p\in\mathfrak{P}}}\left(1-\frac{\rho_1(p)}{p-1}\right)=\prod_{\substack{p<x\\ p\in\mathfrak{P}\\ p\nmid F(0)}}\left(1-\frac{\rho(p)}{p-1}\right)\prod_{\substack{p<x\\ p\in\mathfrak{P}\\ p\mid F(0)}}\left(1-\frac{\rho(p)}{p}\right)$$

$$\times\prod_{\substack{p<x\\ p\in\mathfrak{P}\\ p\mid F(0)}}\left(1-\frac{\rho(p)-1}{p-1}\right)\left(1-\frac{\rho(p)}{p}\right)^{-1}$$

$$\leqslant\prod_{\substack{p<x\\ p\in\mathfrak{P}}}\left(1-\frac{\rho(p)}{p}\right)\prod_{\substack{p<x\\ p\in\mathfrak{P}\\ p\mid F(0)}}\left(1-\frac{1}{p}\right)^{-1};$$

and this completes the proof of Theorem 4.2.

In the following two corollaries we shall again use as $\mathfrak{P}$ the sets of primes that were introduced in (2.6.9), namely

$$\mathfrak{P}_{l,k}=\{p\colon p\equiv l \bmod k\} \quad\text{with}\quad (l,k)=1.$$

COROLLARY 4.2.1. *Let N and k be natural numbers, and let l be an integer such that $(l, k) = 1$. Then*

$$|\{p\colon p\leqslant N,\ (N-p,\mathfrak{P}_{l,k})=1\}|\ll\prod_{\substack{p\mid N\\ p\equiv l \bmod k}}\left(1-\frac{1}{p}\right)^{-1}\frac{N}{(\log N)^{1+1/\phi(k)}},$$

where the constant implied by the $\ll$-symbol depends only on k.

Proof. We consider the polynomial

$$F(n)=N-n,$$

which satisfies the conditions of Theorem 4.2 with

$$g=1,\ \rho(p)=1 \text{ for all } p, \text{ and } F(0)=N.$$

We take the special case $x = N$ and $\mathfrak{P} = \mathfrak{P}_{l,k}$ and obtain by (2.1), with an absolute $\ll$-constant,

$$|\{p\colon p\leqslant N,\ (N-p,\mathfrak{P}_{l,k})=1\}|$$

$$\ll\prod_{\substack{p<N\\ p\equiv l \bmod k}}\left(1-\frac{1}{p}\right)\prod_{\substack{p\mid N\\ p\equiv l \bmod k}}\left(1-\frac{1}{p}\right)^{-1}\frac{N}{\log N}. \tag{2.13}$$

By (2.6.11) the first product on the right is at most

$$\exp\left\{-\sum_{\substack{p<N\\ p\equiv l \bmod k}}\frac{1}{p}\right\} = \exp\left\{-\frac{1}{\phi(k)}\log\log N + O_k(1)\right\}$$
$$= O_k((\log N)^{-1/\phi(k)}),$$

and the result now follows from (2.13).

Because of its connection with the number of representations by sums of two squares, the case $k = 4$ is of particular interest. Here $\phi(k) = 2$, and if we take $l = 3$, Corollary 4.2.1 implies

COROLLARY 4.2.2. *Let N be a positive integer. Then*

$$|\{p: p \leqslant N,\ (N - p, \mathfrak{P}_{3,4}) = 1\}| \ll \prod_{\substack{p|N\\ p\equiv 3 \bmod 4}}\left(1 - \frac{1}{p}\right)^{-1}\frac{N}{\log^{3/2} N},$$

where the constant implied by the $\ll$*-symbol is absolute.*

We shall close this section by touching on a rather interesting question which we consider here for reasons already given at the beginning of the section; namely, the estimation of the number of representations of a positive integer N in the form

$$N = F(p) + p',$$

where the problem for us here is to obtain an upper bound uniform with respect to N.

COROLLARY 4.2.3. *Let N be a positive integer, and let F(n) be a polynomial of degree* $g > 1$ *with integer coefficients such that* $N - F(n)$ *is irreducible. Let* $\rho_N(p)$ *denote the number of solutions of*

$$F(n) \equiv N \bmod p$$

and suppose that

$$\rho_N(p) < p \quad \textit{for all} \quad p. \tag{2.14}$$

Then

$$|\{p: p \leqslant x,\ N - F(p) = p'\}| \ll \prod_{p<x}\left(1 - \frac{\rho_N(p) - 1}{p - 1}\right)$$
$$\times \prod_{\substack{p<x\\ p|N-F(0)}}\left(1 - \frac{1}{p}\right)^{-1}\frac{x}{\log^2 x},$$

where the constant implied by the $\ll$*-symbol depends only on g.*

Remark. The linear case $g = 1$ has been excluded because it is covered by the more precise Theorem 3.12. For $g \geqslant 2$ we may be sure that

$$N - F(0) \neq 0,$$

because $N - F(n)$ is by hypothesis irreducible. It is clear that the irreducibility condition excludes only trivial cases, as does (2.14), which merely expresses the requirement that $N - F(n)$ should have no fixed prime divisor.

Proof. The polynomial

$$F_N(n) = N - F(n)$$

satisfies the conditions of Theorem 4.2 with

$$\rho(p) = \rho_N(p), \qquad F_N(0) = N - F(0).$$

We take

$$\mathfrak{P} = \{p \colon p < x^{1/2}\}.$$

Then, by (2.1),

$$\left|\left\{p \colon p \leqslant x, \left(N - F(p), \prod_{p_1 < x^{1/2}} p_1\right) = 1\right\}\right|$$

$$\ll \prod_{p < x^{1/2}} \left(1 - \frac{\rho_N(p)}{p}\right) \prod_{\substack{p < x^{1/2} \\ p \mid N - F(0)}} \left(1 - \frac{1}{p}\right)^{-1} \frac{x}{\log x} \tag{2.15}$$

where the $\ll$-constant depends only on g.

If $p \leqslant x$ and $N - F(p)$ is a prime $p' \geqslant x^{1/2}$, it is counted on the left of (2.15). Otherwise $(0 <) N - F(p) < x^{1/2}$, that is,

$$N - x^{1/2} < F(p) < N,$$

and since, for each n, $F(p) = n$ has at most g solutions, their number is less than $gx^{1/2}$. Thus we have

$$|\{p \colon p \leqslant x,\ N - F(p) = p'\}|$$

$$\ll \prod_{p < x^{1/2}} \left(1 - \frac{\rho_N(p)}{p}\right) \prod_{\substack{p < x^{1/2} \\ p \mid N - F(0)}} \left(1 - \frac{1}{p}\right)^{-1} \frac{x}{\log x} + x^{1/2}$$

$$\ll \prod_{p < x} \left(1 - \frac{\rho_N(p)}{p}\right) \prod_{\substack{p < x \\ p \mid N - F(0)}} \left(1 - \frac{1}{p}\right)^{-1} \frac{x}{\log x} \left\{ \prod_{x^{1/2} \leqslant p < x} \left(1 - \frac{\rho_N(p)}{p}\right)^{-1} \right.$$

$$\left. + \prod_{p < x} \left(1 - \frac{\rho_N(p)}{p}\right)^{-1} \frac{\log x}{x^{1/2}} \right\}. \tag{2.16}$$

By (2.14) Lagrange's Theorem implies that

$$\rho_N(p) \leqslant g \quad \text{for all} \quad p.$$

Using $\rho_N(p) \leqslant p - 1$ for $p \leqslant g$ and (2.6.11) with $k = 1$, we find that

$$\prod_{z \leqslant p < x} \left(1 - \frac{\rho_N(p)}{p}\right)^{-1} \leqslant \exp\left\{\sum_{z \leqslant p < x} \frac{\rho_N(p)}{p - \rho_N(p)}\right\}$$

$$\leqslant \exp\left\{\sum_{z \leqslant p < x} \frac{g}{p} + O_g(1)\right\}$$

$$= \exp\left\{g \log \frac{\log x}{\log z} + O_g(1)\right\} = O_g\left(\left(\frac{\log x}{\log z}\right)^g\right).$$

If we use this estimate with $z = x^{1/2}$ and $z = 2$ in turn, we see that the last factor in (2.16) is $O_g(1)$. Finally we use the identity (2.12) and obtain

$$\prod_{p < x} \left(1 - \frac{\rho_N(p)}{p}\right) = \prod_{p < x} \left(1 - \frac{\rho_N(p) - 1}{p - 1}\right) \prod_{p < x} \left(1 - \frac{1}{p}\right)$$

$$= \prod_{p < x} \left(1 - \frac{\rho_N(p) - 1}{p - 1}\right) O\left(\frac{1}{\log x}\right),$$

by (1.5.7). Hence Corollary 4.2.3 follows from (2.16).

Corollary 4.2.3 covers many interesting applications. Let us single out the case

$$F(n) = n^2,$$

which leads to an estimate for the number of solutions of

$$N = p^2 + p'.$$

We may clearly assume that N is an even number and not a perfect square. Then†

$$\rho_N(p) = \begin{cases} 1 & \text{if} \quad p \mid N \\ 1 + (N/p) & \text{if} \quad p \nmid N, \end{cases}$$

† (N/p) denotes the Legendre symbol.

and the conditions of Corollary 4.2.3 are satisfied. Since $N - p^2 = p'$ implies that $p \leqslant N^{1/2}$, we choose

$$x = N^{1/2}.$$

Then we derive from Corollary 4.2.3

COROLLARY 4.2.4. *Let $N(\neq m^2)$ be an even positive integer. Then*

$$|\{p\colon N - p^2 = p'\}| \ll \prod_{\substack{p<N^{1/2}\\ p\nmid N}} \left(1 - \frac{(N/p)}{p-1}\right) \prod_{p|N} \frac{p}{p-1} \cdot \frac{N^{1/2}}{\log^2 N},$$

where the constant implied by the $\ll$-symbol is absolute.

NOTES

Note that this Chapter leads only to $\ll$-results. For explicit upper bound results see Chapter 5.

4.1. Lemma 4.1: The method of proof derives from Rankin [1]; see also Prachar [6], Lemma 5.2.

4.2. Theorem 4.2: Note that the result is virtually uniform relative to the usual constants associated with F such as the coefficients, discriminant.

Cor. 4.2.1: Kátai [3] states (with misprint) the corresponding result for $p + 1$ in place of $N - p$ (cf. Note attached to Section 2.6 apropos of Corollaries 2.3.4, 2.3.5).

Cor. 4.2.3: cf. Schwarz [1], Satz 4, where a less precise estimate is proved, but incorrectly (for correction see Schwarz [4]). Several mistakes of this kind are to be found in the literature. A correct and detailed proof of a more general result is to be found in Russell [1], who established an explicit upper bound for the counting number corresponding to hypothesis H_N (See Notes for Introduction) analogous to our Theorem 5.3, the hypothesis H explicit upper bound (see next chapter).

Chapter 5

The Selberg Upper Bound Method: Explicit Estimates

1. A Two-Sided Ω_2-Condition

In Chapter 4 we were able to establish a form of Selberg's upper bound sieve (see Theorem 4.1) that is capable of *direct* application to a very general class of sieve problems. The main results of Chapter 4 were proved subject to the condition

$$(\Omega_2(\kappa)) \qquad \sum_{w \leqslant p < z} \frac{\omega(p)\log p}{p} \leqslant \kappa \log \frac{z}{w} + A_2 \quad \text{if} \quad 2 \leqslant w \leqslant z.$$

We shall now obtain more precise results by introducing the further condition

$$\sum_{w \leqslant p < z} \frac{\omega(p)\log p}{p} \geqslant \kappa \log \frac{z}{w} - L \quad \text{if} \quad 2 \leqslant w \leqslant z,$$

which is clearly a lower bound complementing $(\Omega_2(\kappa))$. We combine the two into the new two-sided condition

$$(\Omega_2(\kappa,L)) \qquad -L \leqslant \sum_{w \leqslant p < z} \frac{\omega(p)\log p}{p} - \kappa \log \frac{z}{w} \leqslant A_2 \quad \text{if} \quad 2 \leqslant w \leqslant z,$$

where $A_2(\geqslant 1)$ and $L(\geqslant 1)$ are independent of z and w, and A_2 is like another A-constant in the sense that the constants implied by the O- and $\ll$-notations may depend on it; but we shall regard L more like an error term, and keep all O- and $\ll$-constants independent of it.

Clearly $(\Omega_2(\kappa,L))$ is only one of many ways in which to express the fact that $\omega(p)$ is, on average, equal to κ, and sieve literature contains numerous variations of this condition. The number κ is an important parameter, and we shall refer to κ as the *dimension of the sieve.*

As was the case in Chapter 4, our main problem is to investigate the behaviour of $G(z)$, this time with the stronger condition $(\Omega_2(\kappa,L))$ at our disposal. A certain amount of preparation will be required.

From Mertens' prime number theory we shall need

$$\sum_{w \leqslant p < z} \frac{\log p}{p} = \log\left|\frac{z}{w}\right| + O(1) \quad \text{if} \quad 2 \leqslant w \leqslant z, \tag{1.1}$$

$$\sum_{w \leqslant p < z} \frac{1}{p} = \log\frac{\log z}{\log w} + O\left(\frac{1}{\log w}\right) \quad \text{if} \quad 2 \leqslant w \leqslant z \tag{1.2}$$

and (cf. (1.4.25))

$$V(z) = \prod_{p<z}\left(1 - \frac{1}{p}\right) = \frac{e^{-\gamma}}{\log z}\left\{1 + O\left(\frac{1}{\log z}\right)\right\}, \tag{1.3}$$

where γ denotes Euler's constant.

LEMMA 5.1. *We have*

$$\frac{V(w)}{V(z)} = \frac{\log z}{\log w}\left\{1 + O\left(\frac{1}{\log w}\right)\right\} \quad \textit{if} \quad 2 \leqslant w \leqslant z, \tag{1.4}$$

and, for any natural number k,

$$\sum_{p|k} \frac{\log p}{p} \ll \log\log 3k. \tag{1.5}$$

Proof. (1.4) is an immediate consequence of (1.3). As for (1.5), we note that

$$\sum_{p|k} \log p \leqslant \log k$$

so that, by (1.1),

$$\sum_{p|k} \frac{\log p}{p} \leqslant \sum_{p<\log 3k} \frac{\log p}{p} + \frac{1}{\log 3k}\sum_{p|k}\log p \leqslant \log\log 3k + O(1).$$

2. TECHNICAL PREPARATION

In this section we shall study some consequences of the condition $(\Omega_2(\kappa,L))$.

LEMMA 5.2. (Ω_1), $(\Omega_2(\kappa,L))$: *Then if*

$$2 \leqslant w \leqslant z,$$

we have

$$-\frac{L}{\log w} \leqslant \sum_{w \leqslant p < z} \frac{\omega(p)}{p} - \kappa \log \frac{\log z}{\log w} \leqslant \frac{A_2}{\log w} \tag{2.1}$$

and

$$O\left(\frac{L}{\log w}\right) \leqslant \sum_{w \leqslant p < z} \frac{g(p)}{p^s} - \kappa \sum_{w \leqslant p < z} \frac{1}{p^{s+1}} \leqslant O\left(\frac{1}{\log w}\right) \tag{2.2}$$

uniformly in $s \geqslant 0$.

Proof. As was shown in Lemma 2.3 ((2.3.3)), the right-hand side of (2.1) is a consequence of $(\Omega_2(\kappa))$, that is, of the right-hand part of $(\Omega_2(\kappa,L))$; and the other part of (2.1) follows in the same way (using partial summation) from the left-hand inequality of $(\Omega_2(\kappa,L))$.

By (1.2) it follows from (2.3.4) that

$$\sum_{w \leqslant p < z} g(p) - \kappa \sum_{w \leqslant p < z} \frac{1}{p} \leqslant O\left(\frac{1}{\log w}\right),$$

and since

$$g(p) \geqslant \frac{\omega(p)}{p},$$

the left half of (2.1) implies that

$$\sum_{w \leqslant p < z} g(p) - \kappa \sum_{w \leqslant p < z} \frac{1}{p} \geqslant O\left(\frac{L}{\log w}\right);$$

a straightforward partial summation now shows that (2.2) holds uniformly in $s \geqslant 0$.

We turn next to the product

$$W(z) = \prod_{p < z} \left(1 - \frac{\omega(p)}{p}\right).$$

LEMMA 5.3. (Ω_1),$(\Omega_2(\kappa,L))$: *If*

$$2 \leqslant w \leqslant z,$$

we have

$$\prod_{w \leqslant p < z} \left(1 + \frac{g(p)}{p^s}\right)\left(1 - \frac{1}{p^{s+1}}\right)^{\kappa} = 1 + O\left(\frac{L}{\log w}\right), \tag{2.3}$$

uniformly in $s \geqslant 0$,

$$\frac{W(w)}{W(z)} = \frac{\log^{\kappa} z}{\log^{\kappa} w}\left\{1 + O\left(\frac{L}{\log w}\right)\right\}, \tag{2.4}$$

and

$$W(z) = \prod_{p}\left(1 - \frac{\omega(p)}{p}\right)\left(1 - \frac{1}{p}\right)^{-\kappa}\frac{e^{-\gamma\kappa}}{\log^{\kappa} z}\left\{1 + O\left(\frac{L}{\log z}\right)\right\}; \tag{2.5}$$

in the case of (2.5) *the infinite product is convergent, uniformly positive and we have in fact the explicit lower estimate*

$$\prod_{p}\left(1 - \frac{\omega(p)}{p}\right)\left(1 - \frac{1}{p}\right)^{-\kappa} \geqslant \exp\{-A_1A_2(1 + \kappa + A_2)\} > 0. \tag{2.6}$$

Proof. We have by (Ω_1) and (2.3.9) that

$$\sum_{w \leqslant p < z} g^2(p) = O\left(\frac{1}{\log w}\right).$$

Using the fact that

$$\log(1 + x) = x + O(x^2) \qquad (x \geqslant -\tfrac{1}{2}),$$

we therefore obtain from (2.2), if $s \geqslant 0$ and $2 \leqslant w \leqslant z$, that

$$\prod_{w \leqslant p < z}\left(1 + \frac{g(p)}{p^s}\right)\left(1 - \frac{1}{p^{s+1}}\right)^{\kappa}$$

$$= \exp\left\{\sum_{w \leqslant p < z}\left(\frac{g(p)}{p^s} + O(g^2(p)) - \kappa\frac{1}{p^{s+1}} + O\left(\frac{1}{p^2}\right)\right)\right\}$$

$$= \exp\left\{O\left(\frac{L}{\log w}\right)\right\} = 1 + O\left(\frac{L}{\log w}\right)$$

if $L/\log w$ is sufficiently small. Otherwise it is sufficient to prove that the product in (2.3) is $O(1)$, and this can be done by using above merely the right hand inequality of (2.2).

Putting $s = 0$ (c.f. (2.3.10)) the product in (2.3) is precisely

$$\frac{W(w)}{W(z)} \cdot \frac{V^{\kappa}(z)}{V^{\kappa}(w)},$$

so that (2.3) yields

$$\frac{W(w)}{W(z)} = \frac{V^{\kappa}(w)}{V^{\kappa}(z)}\left\{1 + O\left(\frac{L}{\log w}\right)\right\},$$

and (2.4) follows at once from (1.4).

If we take $s = 0$ in (2.3), let $z \to \infty$, and then write z in place of w, we obtain

$$\prod_{p \geqslant z}\left(1 - \frac{\omega(p)}{p}\right)^{-1}\left(1 - \frac{1}{p}\right)^{\kappa} = 1 + O\left(\frac{L}{\log z}\right). \tag{2.7}$$

Hence

$$W(z) = \prod_{p<z}\left(1 - \frac{\omega(p)}{p}\right)\prod_{p \geqslant z}\left(1 - \frac{\omega(p)}{p}\right)\left(1 - \frac{1}{p}\right)^{-\kappa}\left\{1 + O\left(\frac{L}{\log z}\right)\right\}$$

$$= \prod_{p}\left(1 - \frac{\omega(p)}{p}\right)\left(1 - \frac{1}{p}\right)^{-\kappa} V^{\kappa}(z)\left\{1 + O\left(\frac{L}{\log z}\right)\right\},$$

and an appeal to (1.3) completes the proof of (2.5); moreover, we see from (2.7) that the infinite product in (2.5) is convergent.

Although it would suffice to be able to estimate this infinite product from below by some positive B-constant, it is more useful to have available an explicit (if crude) lower estimate. We base our argument on the obvious inequality

$$\prod_{p<z}(1 + g(p))\left(1 - \frac{1}{p}\right)^{\kappa} \leqslant \exp\left\{\sum_{p<z} g(p) - \kappa\sum_{p<z}\frac{1}{p}\right\}. \tag{2.8}$$

We estimate the first sum on the right by (2.3.11) with $w = e$; we have then that

$$\sum_{p<z} g(p) \leqslant g(2) + \kappa \log\log z + A_2 + A_1 A_2(\kappa + A_2),$$

and $g(2) \leqslant A_1 - 1$ by (2.3.10) and (Ω_1). Since

$$\sum_{p<z}\frac{1}{p} > \log\log z, \tag{2.9}$$

it follows that the product in (2.8) is at most

$$\exp\{A_1 - 1 + A_2 + A_1 A_2(\kappa + A_2)\} \leqslant \exp\{A_1 A_2(1 + \kappa + A_2)\}.$$

Letting $z \to \infty$, we arrive, in view of (2.3.10), at

$$\prod_p \left(1 - \frac{\omega(p)}{p}\right)^{-1} \left(1 - \frac{1}{p}\right)^{\kappa} \leqslant \exp\{A_1 A_2(1 + \kappa + A_2)\},$$

and this proves (2.6).

3. Asymptotic Formula for $G(z)$

We are now in a position to tackle the important function $G(z)$ under the new condition $(\Omega_2(\kappa,L))$.

Lemma 5.4. $(\Omega_1),(\Omega_2(\kappa,L))$: *We have*

$$\frac{1}{G(z)} = W(z)e^{\gamma\kappa}\Gamma(\kappa + 1)\left\{1 + O\left(\frac{L}{\log z}\right)\right\}; \tag{3.1}$$

more precisely,

$$\frac{1}{G(z)} = W(z)e^{\gamma\kappa}\Gamma(\kappa + 1)\left\{1 + O\left(\frac{\min(L, \log z)}{\log z}\right)\right\}. \tag{3.2}$$

Proof. By (4.1.9) we have

$$\frac{1}{G(z)} = O(W(z)).$$

Thus there is no loss of generality in assuming also that

$$L \leqslant \frac{1}{B_8} \log z \tag{3.3}$$

where $B_8(\geqslant 2)$ is sufficiently large; and this means that we have only to prove (3.1).

We can now embark on the main argument; we shall give the first part (leading up to relation (3.5)) in a more general form than is needed here, because we shall have to return to it in Chapter 6.

Let p be a prime divisor of $P(z)$. Then

$$G(x, z) = \sum_{\substack{d<x \\ d|P(z)}} g(d) = \sum_{\substack{d<x \\ d|P(z) \\ (d,p)=1}} g(d) + g(p) \sum_{\substack{m<x/p \\ m|P(z) \\ (m,p)=1}} g(m)$$

$$= G_p(x, z) + g(p)G_p(x/p, z)$$

where

$$G_p(x, z) = \sum_{\substack{d<x \\ d|P(z) \\ (d,p)=1}} g(d).$$

If we multiply by $\left(1 - \frac{\omega(p)}{p}\right)$ we see that

$$\left(1 - \frac{\omega(p)}{p}\right) G(x, z) = G_p(x, z) - \frac{\omega(p)}{p} \{G_p(x, z) - G_p(x/p, z)\},$$

and if we replace x by x/p in this formula it becomes, after rearrangement,

$$G_p(x/p, z) = \left(1 - \frac{\omega(p)}{p}\right) G(x/p, z) + \frac{\omega(p)}{p} \{G_p(x/p, z) - G_p(x/p^2, z)\}. \tag{3.4}$$

We shall need this identity almost at once, when we study the sum

$$\sum_{\substack{d<x \\ d|P(z)}} g(d) \log d.$$

This sum is equal to

$$\sum_{\substack{d<x \\ d|P(z)}} g(d) \sum_{p|d} \log p = \sum_{p<z} g(p) \log p \sum_{\substack{m<x/p \\ m|P(z) \\ (m,p)=1}} g(m)$$

$$= \sum_{p<z} g(p) \log p \, G_p\left(\frac{x}{p}, z\right);$$

and using (3.4) we deduce that

$$\sum_{\substack{d<x \\ d|P(z)}} g(d) \log d = \sum_{p<z} \frac{\omega(p)}{p} \log p \sum_{\substack{d<x/p \\ d|P(z)}} g(d)$$

$$+ \sum_{p<z} \frac{g(p)\omega(p)}{p} \log p \sum_{\substack{x/p^2 \leqslant d<x/p \\ d|P(z) \\ (d,p)=1}} g(d)$$

$$= \sum_{\substack{d<x \\ d|P(z)}} g(d) \sum_{p<\min(x/d,z)} \frac{\omega(p)}{p} \log p$$

$$+ \sum_{\substack{xz^{-2} \leqslant d<x \\ d|P(z)}} g(d) \sum_{\substack{\sqrt{(x/d)} \leqslant p<\min(x/d,z) \\ p \nmid d}} \frac{g(p)\omega(p)}{p} \log p.$$

For the first inner sum we use $(\Omega_2(\kappa,L))$ in the form

$$\sum_{p<y} \frac{\omega(p)}{p} \log p = \kappa \log y + O(L);$$

but for the second inner sum, since all its terms are non-negative, we are satisfied, using (2.2.8) and (2.3.4), to obtain the upper estimate

$$\sum_{\substack{\sqrt{(x/d)} \leqslant p < \min(x/d, z) \\ p \nmid d}} \frac{g(p)\omega(p)}{p} \log p = O\left(\sum_{\sqrt{(x/d)} \leqslant p < x/d} g(p)\right) = O(1).$$

With these estimates we now obtain

$$\sum_{\substack{d<x \\ d|P(z)}} g(d) \log d = \sum_{\substack{x/z \leqslant d < x \\ d|P(z)}} g(d) \left(\kappa \log \frac{x}{d} + O(L)\right)$$
$$+ \sum_{\substack{d<x/z \\ d|P(z)}} g(d) (\kappa \log z + O(L)) + O(G(x,z))$$
$$= \kappa \sum_{\substack{d<x \\ d|P(z)}} g(d) \log \frac{x}{d} - \kappa \sum_{\substack{d<x/z \\ d|P(z)}} g(d) \log \frac{x/z}{d} + O(LG(x,z));$$

if we add the first sum on the right to both sides, and introduce the function

$$T(x,z) := \int_1^x G(t,z) \frac{dt}{t} = \sum_{\substack{d<x \\ d|P(z)}} g(d) \log \frac{x}{d},$$

we arrive at

$$G(x,z) \log x = (\kappa+1) T(x,z) - \kappa T\left(\frac{x}{z}, z\right) + O(LG(x,z)). \tag{3.5}$$

This relation will be needed in the next chapter. Here it suffices to consider the special case $x = z$; we have, by (1.4.23), that

$$T(z) := T(z,z) = \int_1^z G(t) \frac{dt}{t}, \qquad G(z,z) = G(z), \tag{3.6}$$

and we therefore derive from (3.5) the special result

$$G(z) \log z = (\kappa+1) T(z) + G(z) \log z \,.\, r(z), \tag{3.7}$$

where

$$r(z) = O\left(\frac{L}{\log z}\right).$$

By (3.3) we may assume that

$$|r(y)| \leqslant \tfrac{1}{2} \qquad \text{if} \qquad y \geqslant z,$$

and write (3.7) as

$$G(z) = \frac{1}{1 - r(z)} \frac{\kappa + 1}{\log z} T(z). \tag{3.8}$$

Let

$$E(y) := \log \left\{ \frac{\kappa + 1}{\log^{\kappa+1} y} T(y) \right\};$$

then if $y \geqslant z$ we have by (3.8) that

$$E'(y) = -\frac{\kappa + 1}{y \log y} + \frac{G(y)}{y T(y)} = -\frac{\kappa + 1}{y \log y} + \frac{1}{1 - r(y)} \frac{\kappa + 1}{y \log y}$$

$$= \frac{\kappa + 1}{y \log y} \frac{r(y)}{1 - r(y)} = O\left(\frac{L}{y \log^2 y} \right),$$

so that the integral

$$\int_z^\infty E'(y) dy$$

is convergent. Hence we obtain; with some constant c_3,

$$\frac{\kappa + 1}{\log^{\kappa+1} z} T(z) = \exp E(z) = c_3 \exp \left\{ -\int_z^\infty E'(y) dy \right\} = c_3 \left\{ 1 + O\left(\frac{L}{\log z} \right) \right\};$$

in other words

$$T(z) = \frac{c_3}{\kappa + 1} \log^{\kappa+1} z \left\{ 1 + O\left(\frac{L}{\log z} \right) \right\}. \tag{3.9}$$

But

$$\frac{1}{1 - r(z)} = 1 + \frac{r(z)}{1 - r(z)} = 1 + O\left(\frac{L}{\log z} \right),$$

so that we conclude from (3.8) and (3.9) that

$$G(z) = c_3 \log^\kappa z \left\{ 1 + O\left(\frac{L}{\log z} \right) \right\}. \tag{3.10}$$

We still have to determine c_3. For this we shall need the formulae

$$\int_1^\infty \frac{\log^{\lambda-1} y}{y^{s+1}}\,dy = \frac{1}{s^\lambda}\int_0^\infty e^{-t}t^{\lambda-1}dt = \frac{\Gamma(\lambda)}{s^\lambda} \quad \text{if} \quad \lambda > 0,\ s > 0,$$

and

$$\prod_p \left(1 + \frac{g(p)}{p^s}\right) = \sum_{d=1}^\infty \frac{\mu^2(d)g(d)}{d^s} = s\int_0^\infty \frac{G(y)}{y^{s+1}}\,dy \quad \text{if} \quad s > 0.$$

Formula (3.10) has been proved subject to the condition (3.3). Taking $x = z = y$ in (4.1.1) and making use of (2.3.6), we see that

$$G(y) \ll \log^\kappa y. \tag{3.11}$$

It therefore follows that

$$G(y) = c_3 \log^\kappa y + O(L \log^{\kappa-1} y) \quad \text{if} \quad y > 1; \tag{3.12}$$

and if we substitute in the above integral we obtain, for $s > 0$,

$$\prod_p \left(1 + \frac{g(p)}{p^s}\right) = s\int_1^\infty \frac{c_3 \log^\kappa y + O(L \log^{\kappa-1} y)}{y^{s+1}}\,dy$$

$$= c_3 \frac{\Gamma(\kappa+1)}{s^\kappa} + O\left(\frac{L}{s^{\kappa-1}}\right),$$

whence

$$c_3 = \frac{1}{\Gamma(\kappa+1)} \lim_{s\to+0} s^\kappa \prod_p \left(1 + \frac{g(p)}{p^s}\right).$$

On the other hand, it is a well-known property of the Riemann ζ-function

$$\zeta(s+1) = \prod_p (1 - p^{-s-1})^{-1} \qquad (s > 0)$$

that

$$\lim_{s\to+0} s\zeta(s+1) = 1.$$

Hence

$$c_3 = \frac{1}{\Gamma(\kappa+1)} \lim_{s\to+0} \prod_p \left(1 + \frac{g(p)}{p^s}\right)\left(1 - \frac{1}{p^{s+1}}\right)^\kappa,$$

so that, by (2.3) and (2.3.10),

$$c_3 = \frac{1}{\Gamma(\kappa+1)} \prod_p \left(1 - \frac{\omega(p)}{p}\right)^{-1}\left(1 - \frac{1}{p}\right)^\kappa. \tag{3.13}$$

In view of (3.3), we are now able to deduce from (3.10) that

$$\frac{1}{G(z)} = \Gamma(\kappa + 1)\prod_p \left(1 - \frac{\omega(p)}{p}\right)\left(1 - \frac{1}{p}\right)^{-\kappa} \frac{1}{\log^\kappa z}\left\{1 + O\left(\frac{L}{\log z}\right)\right\};$$

and (3.1) follows at once from this and (2.5). The proof of Lemma 5.4 is therefore complete.

4. The Main Theorems

We are now in a position to prove the first main result of this chapter.

Theorem 5.1. (Ω_1), $(\Omega_2(\kappa,L))$, (R): *We have*

$$S(\mathscr{A};\mathfrak{P},z) \leqslant \Gamma(\kappa + 1)\prod_p \left(1 - \frac{\omega(p)}{p}\right)\left(1 - \frac{1}{p}\right)^{-\kappa}$$

$$\times \frac{X}{\log^\kappa z}\left\{1 + O\left(\frac{\log\log 3z + L}{\log z}\right)\right\} \text{ if } z \leqslant X^{\frac{1}{2}}, \tag{4.1}$$

where the infinite product on the right is convergent. In particular,

$$S(\mathscr{A};\mathfrak{P},X^{\frac{1}{2}}) \leqslant 2^\kappa\Gamma(\kappa + 1)\prod_p \left(1 - \frac{\omega(p)}{p}\right)\left(1 - \frac{1}{p}\right)^{-\kappa}$$

$$\times \frac{X}{\log^\kappa X}\left\{1 + O\left(\frac{\log\log 3X + L}{\log X}\right)\right\}. \tag{4.2}$$

Remark. We shall see from the proof that (4.1) holds without the factor $\log\log 3z$ in the error term if

$$z \leqslant \frac{X^{\frac{1}{2}}}{\log^{2\kappa+\frac{1}{2}} X}.$$

Proof. It follows by (2.3.6) that

$$\frac{z^2}{W^3(z)} = XW(z)\frac{z^2}{XW^4(z)} = XW(z)\,O\left(\frac{z^2\log^{4\kappa} z}{X}\right).$$

Hence Theorem 3.1 and (3.1) yield

$$S(\mathscr{A};\mathfrak{P},z) \leqslant XW(z)e^{\gamma\kappa}\Gamma(\kappa + 1)\left\{1 + O\left(\frac{L}{\log z}\right) + O\left(\frac{z^2\log^{4\kappa} z}{X}\right)\right\},$$

so that, by (2.5),

$$S(\mathscr{A};\mathfrak{P},z) \leqslant \Gamma(\kappa+1)\prod_{p}\left(1-\frac{\omega(p)}{p}\right)\left(1-\frac{1}{p}\right)^{-\kappa}$$

$$\times \frac{X}{\log^{\kappa} z}\left\{1+O\left(\frac{L}{\log z}\right)+O\left(\frac{z^2\log^{4\kappa} z}{X}\right)\right\}, \tag{4.3}$$

where the infinite product is convergent.

Since our theorem is trivial for $X = O(1)$ we may assume that X is greater than a sufficiently large number which may depend on κ. Defining z_0 by

$$z_0{}^2 = \frac{X}{\log^{4\kappa+1} X},$$

we see that (4.1) follows from (4.3) if $z \leqslant z_0$. The remaining case $z_0 \leqslant z \leqslant X^{\frac{1}{2}}$ follows easily from the inequality

$$S(\mathscr{A};\mathfrak{P},z) \leqslant S(\mathscr{A};\mathfrak{P},z_0);$$

for the sum $S(\mathscr{A};\mathfrak{P},z_0)$ can be estimated by (4.1) (with $z \leqslant z_0$) and

$$\frac{1}{\log z_0} = \frac{1}{\log z}\left\{1+\frac{\log(z/z_0)}{\log z_0}\right\} = \frac{1}{\log z}\left\{1+O\left(\frac{\log\log z}{\log z}\right)\right\}.$$

We may omit (R), and obtain by Theorem 3.2 and (3.1) the following more general result:

THEOREM 5.2. (Ω_1), $(\Omega_2(\kappa,L))$: *We have*

$$S(\mathscr{A},\mathfrak{P},z) \leqslant XW(z)\,e^{\gamma\kappa}\,\Gamma(\kappa+1)\left\{1+O\left(\frac{L}{\log z}\right)\right\} \tag{4.4}$$

$$+ \sum_{\substack{d<z^2\\(d,\bar{\mathfrak{P}})=1}} \mu^2(d)\,3^{\nu(d)}\,|R_d|\,.$$

We remark that in both Theorems, 5.1 and 5.2, we could replace L by $\min(L, \log z)$ (cf. Lemma 5.4). These results then include Theorem 4.1 (and also Theorem 2.2) a circumstance that is hardly surprising since the conditions used here include the conditions under which Theorems 4.1 (and 2.2) were proved: namely, (Ω_1), $(\Omega_2(\kappa))$ (and (R)).

5. Two ways of Dealing with Polynomial Sequences $\{F(p)\}$: Discussion

We follow here the pattern of earlier chapters: having established some general theorems, in this case Theorems 5.1 and 5.2, we shall now derive a variety of applications. In fact, we shall devote the next three sections to applications of Theorem 5.1, and Section 5.9 (the last section) to applications of Theorem 5.2.

Before we begin, we shall discuss briefly the difference between the two main theorems, and the general principles which guide our decisions to classify particular problems as applications of one rather than the other theorem.

Although for practical reasons the two theorems have been put in somewhat different form, the essential difference between them is that Theorem 5.2 still has a remainder term (the sum on the right of (4.4)), whereas Theorem 5.1 has not; this amounts to saying that the additional condition (R) present in Theorem 5.1 has had the effect of keeping the remainder term fairly small. We may therefore infer that not only will Theorem 5.1 always be simpler to apply, but any (genuine) application of Theorem 5.2 will require deeper information, because we may then assume that condition (R) is violated (in this application), and the remainder sum is then that much harder to estimate.

We shall see presently that there are some problems—indeed many problems—which can be formulated in different ways: either in terms of a sequence satisfying (R), or in terms of another sequence which does not. We should not suppose that in such cases one ought automatically to choose the technically simpler procedure (via Theorem 5.1); sometimes the other way leads to the better result†, and we shall now discuss when to use one way and when the other. In the process of doing this, we shall also come up against various limitations, resulting usually from deficiencies in the present state of knowledge, which restrict the scope of application of one method or another, sometimes for reasons extraneous to the method.

Let us be more explicit and consider the following problem: find an upper estimate for the number

$$|\{p\colon\ x - y < p \leqslant x,\ p \equiv l \bmod k,\ F(p) = p'\}| \tag{5.1}$$

of primes p in the interval $x - y < p \leqslant x$, lying in the arithmetic progression $l \bmod k$, such that at these primes the polynomial F is also a prime number. (Of course, we ought also to specify here some natural conditions on the problem, such as that $1 < y \leqslant x$, $(l, k) = 1$, F has integer coefficients, is irreducible, etc; these, however, are not relevant to our present discussion, but will be formulated precisely when we come to the actual applications.) We have two ways of attacking this problem (c.f. Chapter 2): we might consider it in terms of the sequence

$$\mathscr{A} = \{nF(n)\colon\ x - y < n \leqslant x,\ n \equiv l \bmod k\}, \tag{5.2}$$

† We explained this in Section 4.2 when sifting with thin $\mathfrak{P}$.

and apply a "double" sieve (as was done in the proof of Theorem 2.4), estimating the number of integers n such that *both* n *and* $F(n)$ are prime (namely $n = p$, $F(n) = p'$); or we might deal with the sequence

$$\mathscr{A} = \{F(p): \ x - y < p \leqslant x, \ p \equiv l \bmod k\} \tag{5.3}$$

and use a direct sifting procedure to count the number of primes p such that $F(p) = p'$ (as was done, for example, in Theorem 2.8′). A useful way of speaking of these two approaches is to say that the second is a "*linearized*" form of the first. The "non-linearized" approach (in terms of the sequence (5.2)) is technically the simpler because, as a moment's reflection shows, condition (R) is fulfilled and therefore Theorem 5.1 can be applied. The "linearized" approach, on the other hand, does not have (R) available and so must operate through Theorem 5.2. Indeed, it is only in recent years that this method has become effective; although it has always been the natural attack, it used to require unproved hypotheses such as the extended Riemann conjecture, or something only slightly weaker, in order to be at all successful. By contrast, we now have Bombieri's deep theorem (Lemma 3.3; Bombieri's result was preceded by a number of results almost as good), and we have already seen (e.g. in Theorems 3.11, 3.12) how this theorem of Bombieri's can sometimes be made to show that the awkward looking remainder term in Theorem 5.2 is small enough.

There are many important problems of type (5.1) in which linearizing improves the constant in the sieve estimate by a factor 2; as can be seen, for example, by comparing Theorems 5.9 and 5.6 for $y = x$. At the same time, the more delicate remainder term imposes certain restrictions (which have more to do with our ignorance about the distribution of prime numbers than with any weakness inherent in the sieve method). The restrictions are mainly the following two: (i) having linearized, we cannot obtain results for an arbitrary interval $(x - y, x]$; we need to impose on y the condition

$$y \geqslant \frac{x}{\log^A x}. \tag{5.4}$$

(ii) We cannot deal with an arbitrary arithmetic progression $l \bmod k$; we have to require that

$$k \leqslant \log^A x. \tag{5.5}$$

(In each case A is some arbitrary but fixed constant.) These technical limitations do give some of our results a somewhat artificial look.

To explain a third limitation, let us turn finally to the extension of problem (5.1) to g polynomials. We ask (under appropriate conditions) for an upper estimate of the number

$$|\{p: \ x - y < p \leqslant x, \ p \equiv l \bmod k, \ F_i(p) \text{ prime} \quad \text{for} \quad i = 1, \ldots, g\}|; \tag{5.6}$$

clearly (5.1) is the special case $g = 1$. Writing

$$F(n) = F_1(n) \dots F_g(n), \tag{5.7}$$

we may, as before, work either in terms of the sequence $\{nF(n)\}$ (as in the proof of Theorem 2.4) with a $(g+1)$-dimensional sieve, or in terms of $\{F(p)\}$ with a g-dimensional sieve. Our comments above still apply, in particular those concerning the restrictions (5.4) and (5.5). However, we encounter a new feature which affects, or may affect, the decision whether or not to linearize. Whereas in the most important case $g = 1$ linearizing definitely leads to an improvement in the estimate by a factor of 2, if $g > 1$ the magnitude of g becomes highly relevant: if $g = 2$ the improvement is only by a factor 3/2, while if $g \geqslant 3$, *linearizing does not lead to any improvement.* When $g \geqslant 3$, there is therefore no point in choosing the deeper and sometimes more restrictive method; when $g = 3$ we actually get the same constant by either method, but if $g > 3$ we obtain a better constant if we do *not* linearize!

Looking back on what we did in earlier chapters in the light of these remarks, it is clear that in Theorem 3.12 (also its special case Theorem 3.11) we linearized in order to get a better numerical constant. We find condition (5.5) on k there, and although we did not consider it worthwhile to generalize to an "interval" case, we could have done so subject to a restriction of form (5.4).

In Chapter 2 we dealt with a problem of type (5.6) in Theorem 2.4 and its corollaries. There we were looking for $\ll$-estimates only, so that linearizing would have no advantage; whereas our non-linearized approach allowed us to use Theorem 2.2 and to avoid restrictions such as (5.4) and (5.5). (Note that here, as in Theorem 2.4, we have a sifting set $\mathfrak{P}$ of density 1; cf. the discussion at the beginning of Section 4.2.)

In this chapter we shall deal mainly with problems of type (5.6), sometimes by the linearized approach (as in Section 5.9), sometimes by the non-linearized approach (as we do e.g. in Theorem 5.3). The results of this chapter have the advantage over those of Chapters 2 and 4 in that they give explicit numerical constants; however, they lack for the most part the uniformity properties of the results in Sections 4.2 (and 2.7)—by this we mean that an application of Theorems 5.1 and 5.2 to an (upper) estimate for the number of p's in (5.6) will involve an O-constant that may depend not only on g but on the polynomial F in (5.7), i.e. on the coefficients and on the degrees of the polynomials F_i, as well as on their number g. We express this, as usual, by writing O_F whenever this is appropriate; in Sections 4.2 (and 2.7), as we recall, the O-constants depend at most on g.

There is just one exception (to be found in Section 5.8) to the preceding remarks regarding O-constants; there, because we know enough about the function $\omega(p)$, we can make the O-constants depend, as in Sections 4.2 (and 2.7), at most on g.

We conclude the discussion with a few remarks about the contents of the rest of this chapter.

In the Sections 6, 7 and 8 we take the non-linearized approach whenever a problem of type (5.6) is studied; for example, we apply Theorem 5.1 to lead us via Theorem 5.3 to Theorem 5.5. As has been pointed out, we are able in these circumstances to admit an arbitrary interval.

In Section 5.9 we linearize. Consequently we have to use Theorem 5.2, and this leads us to Theorem 5.9. We could extend the latter to an interval subject to a condition (5.4). Moreover, whereas Theorem 5.5 covers the general case of g polynomials, Theorem 5.9 deals with the case $g = 1$ only. In this, Theorem 5.9 gives the better result, and we might have extended Theorem 5.9 to the case $g = 2$ so as to improve Theorem 5.5 in this case also. For $g > 2$, however, such an extension of Theorem 5.9 would be useless, since the result is already covered by Theorem 5.5. which, in this context, is both technically simpler to obtain and independent of the restrictions induced by linearizing.

In Theorem 5.6 we shall extend the most important special case $g = 1$ of Theorem 5.5 to an arithmetic progression, and there will be no restriction of type (5.5) such as is present in Theorem 5.9. However, Theorem 5.6 has a numerical constant worse than that of Theorem 5.9 by a factor of 2, in agreement with earlier comments.

6. Primes Representable by Polynomials

We now proceed with the actual applications.

Theorem 5.3. *Let $F_1(n), \dots, F_g(n)$ be distinct irreducible polynomials with integral coefficients, and positive leading coefficients. Write*

$$F(n) = F_1(n) \dots F_g(n),$$

let $\rho(p)$ denote the number of solutions of

$$F(n) \equiv 0 \bmod p,$$

and suppose that

$$\rho(p) < p \quad \textit{for all} \quad p. \tag{6.1}$$

Let y and x be real numbers satisfying

$$1 < y \leqslant x. \tag{6.2}$$

Then

$$|\{n\colon x - y < n \leqslant x,\ F_i(n)\ \textit{prime for}\ i = 1, \ldots, g\}|$$
$$\leqslant 2^g g! \prod_p \left(1 - \frac{\rho(p) - 1}{p - 1}\right)\left(1 - \frac{1}{p}\right)^{-g+1} \tag{6.3}$$
$$\times \frac{y}{\log^g y}\left\{1 + O_F\left(\frac{\log\log 3y}{\log y}\right)\right\}.$$

Remark 1. The O-constant in (6.3) is independent of y and x; it may, however, depend on F, that is, on the coefficients and degrees of the F_i ($i = 1, \ldots, g$), and on g.

Remark 2. Condition (6.1) ensures that F has no fixed prime divisor. It is not really a restriction at all; whilst it excludes only a trivial case, it is convenient both for the statement and proof of the theorem.

Remark 3. If one polynomial F_i has zero degree, then (6.1) implies that F_i must be identically 1; in that event the left-hand side of (6.3) is zero and the result trivial. We may therefore assume that all the polynomials ($1 \leqslant i \leqslant g$) are of degree $\geqslant 1$.

Proof. We take as the sequence to be sifted

$$\mathscr{A} = \{F(n)\colon x - y < n \leqslant x\},$$

and as $\mathfrak{P}$ the set $\mathfrak{P}_1$ of all primes. $\mathscr{A}$ has been analysed in Example 3. We choose

$$X = y \quad \text{and} \quad \omega(d) = \rho(d), \tag{6.4}$$

where $\rho(d)$ denotes the number of solutions of

$$F(n) \equiv 0 \bmod d,$$

and we obtain, by (1.4.15) and (1.3.14), that

$$|R_d| \leqslant \omega(d),$$

so that condition (R) of Theorem 5.1 is satisfied. Condition (6.1) and Lagrange's theorem (see (1.3.16)) imply (Ω_1) with some $A_1 = O_F(1)$.

Let $\rho_i(p)$ denote the number of solutions of

$$F_i(n) \equiv 0 \bmod p,$$

$i = 1, \ldots, g$; then (*cf* (1.3.17))

$$\sum_{p<w} \frac{\rho_i(p)}{p} \log p = \log w + O_F(1) \tag{6.5}$$

and for all but at most $O_F(1)$ primes we have

$$\rho(p) = \rho_1(p) + \ldots + \rho_g(p).$$

Hence condition $(\Omega_2(\kappa,L))$ is satisfied with

$$\kappa = g,\ A_2 = O_F(1) \quad \text{and} \quad L = O_F(1).$$

We are now able to apply Theorem 5.1. Noting (6.4) and that $\kappa = g$, we obtain by (4.2)

$$S(\mathscr{A}; \mathfrak{P}_1, y^{1/2}) \leqslant 2^g g! \prod_p \left(1 - \frac{\rho(p)}{p}\right)\left(1 - \frac{1}{p}\right)^{-g} \times \frac{y}{\log^g y}\left\{1 + O_F\left(\frac{\log\log 3y}{\log y}\right)\right\}. \tag{6.6}$$

For each n counted on the left of (6.3) but not in $S(\mathscr{A}; \mathfrak{P}_1, y^{1/2})$ we have

$$F_i(n) < y^{1/2}$$

for at least one i. However, we know that

$$F_i(n) > \tfrac{1}{2}n \qquad (i = 1, \ldots, g)$$

for all but at most $O_F(1)$ values of n, and it follows that

$$|\{n\colon x - y < n \leqslant x,\ F_i(n) \text{ prime for } i = 1, \ldots, g\}| \leqslant S(\mathscr{A}; \mathfrak{P}_1, y^{1/2}) + O_F(y^{1/2}).$$

In view of the identity (2.8.6) and (2.6) we now see (6.3) to be a consequence of (6.6).

We go on to derive two special cases of Theorem 5.3. Both correspond to the case

$$g = 2,$$

and both can be viewed as generalizations of the prime twins problem.

Corollary 5.3.1. *If* $1 < y \leqslant x$ *we have*

$$|\{n\colon x - y < n \leqslant x,\ n^2 + 1 = p,\ n^2 + 3 = p'\}| \leqslant 32 \prod_{p>2}\left(1 - \frac{1}{(p-1)^2}\right) \prod_{p\equiv \pm 1 \bmod 12}\left(1 - \frac{2\chi(p)}{p-2}\right) \times \frac{y}{\log^2 y}\left\{1 + O\left(\frac{\log\log 3y}{\log y}\right)\right\},$$

where

$$\chi(p) = \begin{cases} 1 & \text{if} \quad p \equiv 1 \bmod 12 \\ -1 & \text{if} \quad p \equiv -1 \bmod 12, \end{cases}$$

and the constant implied by the O-symbol is absolute.

Proof. We take

$$F_1(n) = n^2 + 1, \qquad F_2(n) = n^2 + 3$$

in Theorem 5.3 and obtain

$$|\{n: x - y < n \leqslant x,\ n^2 + 1 = p,\ n^2 + 3 = p'\}|$$

$$\leqslant 8 \prod_p \left(1 - \frac{\rho(p) - 1}{p - 1}\right)\left(1 - \frac{1}{p}\right)^{-1} \frac{y}{\log^2 y}\left(1 + O\left(\frac{\log\log 3y}{\log y}\right)\right)$$

if $\rho(p) < p$ for all p, where $\rho(p)$ denotes the number of solutions of

$$(n^2 + 1)(n^2 + 3) \equiv 0 \bmod p.$$

Clearly

$$\rho(2) = 1, \qquad \rho(3) = 1,$$

and for $p > 2$ the two factors cannot have a common zero mod p. Hence if $p > 3$,

$$\rho(p) = 1 + \left(\frac{-1}{p}\right) + 1 + \left(\frac{-3}{p}\right) = \begin{cases} 4 & \text{if} \quad p \equiv 1 \bmod 12 \\ 0 & \text{if} \quad p \equiv -1 \bmod 12 \\ 2 & \text{if} \quad p \equiv \pm 5 \bmod 12; \end{cases} \tag{6.7}$$

and we see incidentally that $\rho(p) < p$.

To deal with the product we use the fact that $\rho(2) = \rho(3) = 1$ and the identity (for arbitrary ξ)

$$\frac{1 - \xi/(p-1)}{1 - 1/p} = \left(1 - \frac{1}{(p-1)^2}\right)\left(1 - \frac{\xi - 1}{p - 2}\right) \quad \text{if} \quad p > 2 \tag{6.8}$$

to obtain

$$\prod_p \left(1 - \frac{\rho(p) - 1}{p - 1}\right)\left(1 - \frac{1}{p}\right)^{-1} = 2 \prod_{p>2} \left(1 - \frac{1}{(p-1)^2}\right)$$

$$\times\, 2 \prod_{p>3} \left(1 - \frac{\rho(p) - 2}{p - 2}\right);$$

this together with (6.7) proves the result.

COROLLARY 5.3.2. *If* $1 < y \leqslant x$, *we have*

$$|\{n:\ x - y < n \leqslant x,\ n^2 - 2n + 2 = p,\ n^2 + 2n + 2 = p'\}|$$
$$\leqslant 16 \prod_{p>2}\left(1 - \frac{1}{(p-1)^2}\right)\prod_{p>2}\left(1 - \frac{2(-1)^{(p-1)/2}}{p-2}\right)\frac{y}{\log^2 y}$$
$$\times \left\{1 + O\left(\frac{\log\log 3y}{\log y}\right)\right\},$$

where the constant implied by the O-symbol is absolute.

Proof. The proof is very similar to that of Corollary 5.3.1. We take

$$F_1(n) = n^2 - 2n + 2, \qquad F_2(n) = n^2 + 2n + 2,$$

and Theorem 5.3 yields

$$|\{n:\ x - y < n \leqslant x,\ n^2 - 2n + 2 = p,\ n^2 + 2n + 2 = p'\}|$$
$$\leqslant 8 \prod_{p}\left(1 - \frac{\rho(p) - 1}{p - 1}\right)\left(1 - \frac{1}{p}\right)^{-1}\frac{y}{\log^2 y}\left\{1 + O\left(\frac{\log\log 3y}{\log y}\right)\right\}$$

if $\rho(p) < p$ for all p, where $\rho(p)$ is now the number of solutions of

$$\{(n-1)^2 + 1\}\{(n+1)^2 + 1\} \equiv 0 \bmod p.$$

Here

$$\rho(2) = 1$$

and

$$\rho(p) = 2(1 + (-1/p)) = 2 + 2(-1)^{(p-1)/2} \quad \text{if} \quad p > 2.$$

Hence $\rho(p) < p$ for all p, and (6.8) implies that

$$\prod_{p}\left(1 - \frac{\rho(p) - 1}{p - 1}\right)\left(1 - \frac{1}{p}\right)^{-1} = 2\prod_{p>2}\left(1 - \frac{1}{(p-1)^2}\right)$$
$$\times \prod_{p>2}\left(1 - \frac{2(-1)^{(p-1)/2}}{p-2}\right).$$

Next we shall state the important special case $g = 1$ of Theorem 5.3 as a separate theorem. We are concerned here with the problem of primes represented by a polynomial.

THEOREM 5.4. *Let $F(n)$ be an irreducible polynomial with integral coefficients and positive leading coefficient Let $\rho(p)$ denote the number of solutions of*

$$F(n) \equiv 0 \bmod p,$$

and suppose that

$$\rho(p) < p \quad \textit{for all } p. \tag{6.9}$$

If y and x are real numbers satisfying

$$1 < y \leqslant x,$$

then

$$\begin{aligned} &|\{n: x - y < n \leqslant x, F(n) = p\}| \\ &\leqslant 2\prod_{p}\left(1 - \frac{\rho(p) - 1}{p - 1}\right)\frac{y}{\log y}\left\{1 + O_F\left(\frac{\log\log 3y}{\log y}\right)\right\}. \end{aligned} \tag{6.10}$$

We might stress again that the O_F-constant is independent of y and x; it may, of course, depend on the coefficients and degree of F.

Again, we shall derive as corollaries two special applications of this result.

COROLLARY 5.4.1. *Let $a(\neq -b^2)$ be an integer. Then if $1 < y \leqslant x$, we have*

$$\begin{aligned} &|\{n: x - y < n \leqslant x, n^2 + a = p\}| \\ &\leqslant 2\prod_{2 < p \nmid a}\left(1 - \frac{(-a/p)}{p - 1}\right)\frac{y}{\log y}\left\{1 + O_a\left(\frac{\log\log 3y}{\log y}\right)\right\}. \end{aligned} \tag{6.11}$$

Remark. Since the O-constant in Theorem 5.4 may depend on F, the O-constant in (6.11), while independent of y and x, may depend on a.

Proof. We take

$$F(n) = n^2 + a.$$

Obviously

$$\rho(2) = 1 \quad \text{and} \quad \rho(p) = 1 \quad \text{if} \quad p \mid a,$$

whereas

$$\rho(p) = 1 + (-a/p) \quad \text{if} \quad p \nmid a.$$

Since $a \neq -b^2$, all the conditions of Theorem 5.4 are satisfied and (6.10) at once gives our result.

Next consider

$$F(n) = n^4 + 1.$$

If $\rho(p)$ is the number of solutions of

$$n^4 + 1 \equiv 0 \bmod p,$$

$\rho(2) = 1$ and, if $p > 2$,

$$\rho(p) = \begin{cases} 4 & \text{if} \quad p \equiv 1 \bmod 8 \\ 0 & \text{if} \quad p \not\equiv 1 \bmod 8. \end{cases}$$

Hence Theorem 5.4 yields

COROLLARY 5.4.2. *If* $1 < y \leqslant x$ *we have*

$$|\{n\colon\ x - y < n \leqslant x,\ n^4 + 1 = p\}|$$
$$\leqslant 2 \prod_{p>2} \left(1 - \frac{1 + 2\varepsilon(p)}{p - 1}\right) \frac{y}{\log y} \left\{1 + O\left(\frac{\log\log 3y}{\log y}\right)\right\},$$

where

$$\varepsilon(p) = \begin{cases} 1 & \text{if} \quad p \equiv 1 \bmod 8 \\ -1 & \text{if} \quad p \not\equiv 1 \bmod 8, \end{cases}$$

and the constant implied by the O-symbol is absolute.

We now turn to problem (5.6) and derive from Theorem 5.3 a general result (Theorem 5.5 below) whose quality reflects the delinearized approach of Theorem 5.3. Thus, for example, where Theorem 5.5 compares with Theorem 5.9 (that is, in the special case $g = 1, y = x$ of Theorem 5.5), it does not quite reach the quality of the latter.

THEOREM 5.5. *Let* $F_1(n), \ldots, F_g(n)$ *be* g *distinct irreducible polynomials with integral coefficients and positive leading coefficients, and suppose that*

$$F_i(n) \neq n \qquad (i = 1, \ldots, g). \tag{6.12}$$

Write

$$F(n) = F_1(n) \ldots F_g(n)$$

and let $\rho(p)$ *denote the number of solutions of*

$$F(n) \equiv 0 \bmod p. \tag{6.13}$$

Suppose that

$$\rho(p) < p \quad \textit{for all } p \tag{6.14}$$

and that

$$\rho(p) < p - 1 \quad \textit{if} \quad p \nmid F(0). \tag{6.15}$$

Let y and x be real numbers satisfying

$$1 < y \leqslant x. \tag{6.16}$$

Then we have

$$\begin{aligned} |\{p:\ x - y < p \leqslant x,\ F_i(p)\ \textit{prime for}\ i = 1, \ldots, g\}| \\ \leqslant 2^{2g+1}(g+1)! \prod_{p>2}\left(1 - \frac{1}{(p-1)^2}\right) \prod_{2<p\nmid F(0)}\left(1 - \frac{\rho(p)-1}{p-2}\right) \\ \times \left(1 - \frac{1}{p}\right)^{-g+1} \prod_{2<p\mid F(0)}\left(1 - \frac{\rho(p)-2}{p-2}\right)\left(1 - \frac{1}{p}\right)^{-g+1} \frac{y}{\log^{g+1} y} \\ \times \left\{1 + O_F\left(\frac{\log\log 3y}{\log y}\right)\right\}. \end{aligned} \tag{6.17}$$

Remark 1. By virtue of (6.12) the possibility of $F(0)$ being zero has been excluded, so that

$$F(0) \neq 0, \tag{6.18}$$

and the last product on the right of (6.17) is always finite.

Remark 2. Condition (6.15) does not constitute a restriction, but serves merely to exclude a trivial case; for if (6.15) were false, and there existed a prime p_0 such that $\rho(p_0) = p_0 - 1, p_0 \nmid F(0)$, this would imply that $F(n) \equiv 0 \bmod p_0$ for all $n \not\equiv 0 \bmod p_0$ and in particular that $p_0 \mid F(p)$ for all $p \neq p_0$.

Remark 3. The remarks following Theorem 5.3 apply here too.

Proof. We shall apply Theorem 5.3 with $g + 1$ in place of g, and in order to avoid confusion we shall mark all the data in Theorem 5.3 with a dash ($'$). With this convention we choose

$$F_i'(n) = F_i(n) \quad (i = 1, \ldots, g), \qquad F'_{g+1}(n) = n, \quad y' = y, \quad x' = x, \tag{6.19}$$

and see at once that, in view of (6.12) and (6.16), all the conditions of Theorem 5.3 are satisfied, with the possible exception of (6.1). As to (6.1), we note that

$$F'(n) = nF(n),$$

so that $\rho'(p)$ is the number of solutions of

$$nF(n) \equiv 0 \bmod p.$$

Comparing with (6.13), we see that $\rho'(p) = \rho(p) + 1$ if $n \equiv 0 \bmod p$ is not a solution of (6.13), i.e. if $p \nmid F(0)$; otherwise $\rho'(p) = \rho(p)$ and so

$$\rho'(p) = \begin{cases} \rho(p) + 1 & \text{if } p \nmid F(0), \\ \rho(p) & \text{if } p \mid F(0). \end{cases} \tag{6.20}$$

Hence (6.1) is a consequence of (6.15) and (6.14), and we may apply Theorem 5.3 (with $g + 1$ in place of g). Since, by (6.19),

$$|\{n\colon x' - y' < n \leqslant x',\ F_i'(n) \text{ prime for } i = 1, \ldots, g + 1\}|$$
$$= |\{p\colon x - y < p \leqslant x,\ F_i(p) \text{ prime for } i = 1, \ldots, g\}|,$$

we derive

$$|\{p\colon x - y < p \leqslant x,\ F_i(p) \text{ prime for } i = 1, \ldots, g\}|$$
$$\leqslant 2^{g+1}(g+1)!\prod_p \left(1 - \frac{\rho'(p) - 1}{p - 1}\right)\left(1 - \frac{1}{p}\right)^{-g} \frac{y}{\log^{g+1} y}$$
$$\times \left\{1 + O_F\left(\frac{\log\log 3y}{\log y}\right)\right\}. \tag{6.21}$$

As usual, it remains to elucidate the product. We note from (6.14), (6.15) and (6.20) that $\rho'(2) = 1$, and by means of the identity (6.8) we obtain

$$\prod_p \left(1 - \frac{\rho'(p) - 1}{p - 1}\right)\left(1 - \frac{1}{p}\right)^{-g} = 2^g \prod_{p>2}\left(1 - \frac{1}{(p-1)^2}\right)$$
$$\times \prod_{p>2}\left(1 - \frac{\rho'(p) - 2}{p - 2}\right)\left(1 - \frac{1}{p}\right)^{-g+1}.$$

Hence, using (6.20), our result follows from (6.21).

Theorem 5.5 applies to the problem dealt with in Theorem 2.4 (for $k = 1$) with

$$F_i(n) = a_i n + b_i, \qquad a_i \geqslant 1, \qquad i = 1, \ldots, g,$$

if we impose the natural conditions (6.14) and (6.15). This time we obtain explicit numerical estimates, but have to sacrifice uniformity in the coefficients a_i, b_i (whereas this uniformity was present in Theorem 2.4). We therefore give here an application only to a special case of Theorem 2.4, or rather of Corollary 2.4.2, namely to the problem of "prime quadruplets", where this lack of uniformity is not relevant.

COROLLARY 5.5.1. *If* $1 < y \leqslant x$, *we have*

$$|\{p\colon x - y < p \leqslant x,\ p - 2 = p',\ p + 4 = p'',\ p + 6 = p'''\}|$$
$$\leqslant 2^{10}3^2 \prod_{p>2}\left(1 - \frac{1}{(p-1)^2}\right)\prod_{p>3}\left(1 - \frac{4}{(p-2)^2}\right)\frac{y}{\log^4 y}$$
$$\times \left\{1 + O\left(\frac{\log\log 3y}{\log y}\right)\right\},$$

where the constant implied by the O-symbol is absolute.

Proof. We apply Theorem 5.5 with $g = 3$ to

$$F_1(n) = n - 2, \qquad F_2(n) = n + 4, \qquad F_3(n) = n + 6.$$

Here

$$\rho(2) = 1, \qquad \rho(3) = 2, \qquad \rho(p) = 3 \qquad \text{for} \quad p > 3,$$

so that the conditions of Theorem 5.5 are satisfied and we obtain for the number being estimated the upper bound

$$2^7.4! \prod_{p>2}\left(1 - \frac{1}{(p-1)^2}\right)\prod_{p>3}\left(1 - \frac{2}{p-2}\right)\left(1 - \frac{1}{p}\right)^{-2}$$
$$\times \left(1 - \frac{1}{3}\right)^{-2}\frac{y}{\log^4 y}\left\{1 + O\left(\frac{\log\log 3y}{\log y}\right)\right\}.$$

In the second product we use (6.8) (with $\xi = (p + 2)/(p - 2)$) and the result follows.

We add yet another illustration of the way in which Theorem 5.5 can be applied.

COROLLARY 5.5.2. *We have*

$$|\{p\colon x - y < p \leqslant x,\ p + 2 = p',\ p^2 + 4 = p''\}|$$

$$\leqslant 2^7 3 \prod_{p>2}\left(1 - \frac{1}{(p-)^2}\right)^2\prod_{p>3}\left(1 - \frac{1}{(p-2)^2}\right)\prod_{p>3}\left(1 - \frac{(-1)^{(p-1)/2}}{p-3}\right)$$
$$\times \frac{y}{\log^3 y}\left\{1 + O\left(\frac{\log\log 3y}{\log y}\right)\right\},$$

where the constant implied by the O-symbol is absolute.

Proof. In Theorem 5.5 we take

$$F_1(n) = n + 2, \qquad F_2(n) = n^2 + 4.$$

If $\rho(p)$ is the number of solutions of

$$(n + 2)(n^2 + 4) \equiv 0 \bmod p,$$

then

$$\rho(2) = 1, \qquad \rho(p) = 2 + \left(\frac{-4}{p}\right) = 2 + (-1)^{(p-1)/2} \quad \text{for} \quad p > 2,$$

and the conditions of Theorem 5.5 are evidently satisfied. Hence the counting number being estimated is, by (6.17), at most

$$2^5 3! \prod_{p>2}\left(1 - \frac{1}{(p-2)^2}\right) \prod_{p>2}\left(1 - \frac{1 + (-1)^{(p-1)/2}}{p-2}\right)\left(1 - \frac{1}{p}\right)^{-1}$$

$$\times \frac{y}{\log^3 y}\left\{1 + O\left(\frac{\log\log 3y}{\log y}\right)\right\},$$

and since the second product can be written in the form

$$\prod_{p>2}\left(1 - \frac{1 + (-1)^{(p-1)/2}}{p-2}\right)\left(1 - \frac{1}{p}\right)^{-1}$$

$$= \frac{3}{2}\prod_{p>3}\left(1 - \frac{1}{(p-1)^2}\right)\left(1 - \frac{1}{(p-2)^2}\right)\left(1 - \frac{(-1)^{(p-1)/2}}{p-3}\right),$$

the stated result follows.

7. Primes Representable by Polynomials $F(p)$: The Non-linearized Approach

As we indicated in Section 5, we shall show, only in the simplest case $g = 1$, how Theorem 5.5 can be extended to an arithmetic progression modulo k.

THEOREM 5.6. *Let $F(n)$ ($\neq n$) be an irreducible polynomial with integral coefficients and with a positive leading coefficient. Let $\rho(p)$ denote the number of solutions of*

$$F(n) \equiv 0 \bmod p,$$

and suppose both that

$$\rho(p) < p \quad \textit{for all} \quad p \tag{7.1}$$

and that

$$\rho(p) < p - 1 \quad \text{if} \quad p \nmid F(0). \tag{7.2}$$

Let k and l be integers, and let y and x be real numbers with

$$1 \leqslant k < y \leqslant x.$$

Then

$$|\{p\colon x - y < p \leqslant x,\ p \equiv l \bmod k,\ F(p) = p'\}|$$

$$\leqslant 16 \prod_{p>2} \left(1 - \frac{1}{(p-1)^2}\right) \prod_{\substack{2<p\nmid F(0)\\ p\nmid k}} \left(1 - \frac{\rho(p)-1}{p-2}\right) \prod_{\substack{2<p\mid F(0)\\ p\nmid k}} \left(1 - \frac{\rho(p)-2}{p-2}\right)$$
$$\times \prod_{2<p\mid k} \frac{p-1}{p-2} \frac{y}{\phi(k)\log^2(y/k)} \left\{1 + O_F\left(\frac{\log\log 3y}{\log(y/k)}\right)\right\}. \tag{7.3}$$

Remark 1. The O-constant is independent of x, k and l, but may well depend on the coefficients and degree g of F.

Remark 2. We have excluded the case $F(n) = n$ because it does not interest us here; by doing so we have ruled out the possibility of $F(0)$ vanishing, so that the conditions imply that

$$F(0) \neq 0. \tag{7.4}$$

Remark 3. It follows from (7.1) (see (1.3.16)) that

$$\rho(p) \leqslant g \quad \text{for all } p; \tag{7.5}$$

hence (7.2) can be replaced by the condition

$$\rho(p) < p - 1 \quad \text{if} \quad p \nmid F(0) \quad \text{and} \quad p \leqslant g + 1. \tag{7.6}$$

(See also the second remark following Theorem 5.5.)

Remark 4. The case $g = 0$ is not excluded. However, if $g = 0$ our conditions imply that $F(n) = 1$ so that (7.3) is trivial. We may therefore assume that

$$g \geqslant 1. \tag{7.7}$$

Remark 5. The theorem is trivial if $(l, k) > 1$ or if $(F(l), k) > 1$, since the conditions defining the sequence being sifted (see the left side of (7.3)) imply that $(F(l), k) \mid F(p)$ for all p that are counted. Hence we may assume that

$$(l, k) = 1 \quad \text{and} \quad (F(l), k) = 1. \tag{7.8}$$

Proof. We take

$$\mathscr{A} = \{nF(n)\colon\ x - y < n \leqslant x,\ n \equiv l \bmod k\}$$

and

$$\mathfrak{P} = \mathfrak{P}_k = \{p\colon\ p \nmid k\}.$$

Then if $(d,k) = 1$ (that is, if $(d, \overline{\mathfrak{P}}) = 1$),

$$|\mathscr{A}_d| = |\{n\colon\ x - y < n \leqslant x,\ n \equiv l \bmod k,\ nF(n) \equiv 0 \bmod d\}|$$

$$= \sum_{\substack{m=1 \\ mF(m) \equiv 0 \bmod d}}^{d} |\{n\colon\ x - y < n \leqslant x,\ n \equiv l \bmod k,\ n \equiv m \bmod d\}|,$$

so that

$$|\mathscr{A}_d| = \rho'(d)\left\{\frac{y}{kd} + \theta\right\} \quad \text{if} \quad (d,k) = 1 \qquad (|\theta| \leqslant 1);$$

here $\rho'(d)$ denotes the number of solutions of

$$mF(m) \equiv 0 \bmod d,$$

and we see that

$$\rho'(p) = \begin{cases} \rho(p) + 1, & p \nmid F(0), \\ \rho(p), & p \mid F(0). \end{cases} \tag{7.9}$$

Accordingly we choose

$$X = \frac{y}{k},\ \omega(p) = \rho'(p) \quad \text{if} \quad p \nmid k,$$

and we find that

$$|R_d| \leqslant \omega(d) \quad \text{if} \quad (d,k) = 1;$$

thus (R) is satisfied. From (7.2), (7.1) and (7.5) it follows that (Ω_1) holds with $A_1 = g + 2$. By (7.9) and (cf. (1.3.17))

$$\sum_{p<z} \frac{\rho(p)}{p} \log p = \log z + O_F(1), \tag{7.10}$$

noting also that $F(0) \neq 0$, we see that $(\Omega_2(\kappa, L))$ is satisfied with

$$\kappa = 2,\ A_2 = O_F(1)$$

and, by (1.5),

$$L \leqslant O_F(1) + O_g(1) \sum_{p|k} \frac{\log p}{p} = O_F(\log\log 3k).$$

We can now apply Theorem 5.1 and we obtain, by (4.2),

$$S(\mathscr{A}; \mathfrak{P}_k, \sqrt{(y/k)})$$
$$\leqslant 8\prod_{p\nmid k}\left(1-\frac{\rho'(p)}{p}\right)\left(1-\frac{1}{p}\right)^{-2}\prod_{p\mid k}\frac{p}{p-1}\frac{y}{\phi(k)\log^2(y/k)}$$
$$\times\left\{1+O_F\left(\frac{\log\log(3y/k)+\log\log 3k}{\log(y/k)}\right)\right\}. \tag{7.11}$$

Consider the numbers n satisfying $x-y<n\leqslant x, n\equiv l \bmod k$, for which both n and $F(n)$ are prime numbers. If both are primes $\geqslant (y/k)^{1/2}$, then n is counted in $S(\mathscr{A}; \mathfrak{P}_k, \sqrt{(y/k)})$; otherwise $n<(y/k)^{1/2}$ or $F(n)<(y/k)^{1/2}$ (or both). Therefore

$$|\{p:\ x-y<p\leqslant x,\ p\equiv l \bmod k, F(p)=p'\}|$$
$$\leqslant S(\mathscr{A}; \mathfrak{P}_k, \sqrt{(y/k)})+O_F(\sqrt{(y/k)}), \tag{7.12}$$

and it is a simple matter to check, by (2.6), that $O_F(\sqrt{(y/k)})$ can be absorbed in the error term on the right of (7.11).

It remains to deal with the product. By (7.2) and (7.1) we have that $\rho'(2)=1$, so that by (4.2.12) and (6.8)

$$8\prod_{p\nmid k}\left(1-\frac{\rho'(p)}{p}\right)\left(1-\frac{1}{p}\right)^{-2}\prod_{p\mid k}\frac{p}{p-1}$$
$$=16\prod_{2<p\nmid k}\left(1-\frac{\rho'(p)-1}{p-1}\right)\left(1-\frac{1}{p}\right)^{-1}\prod_{2<p\mid k}\frac{p}{p-1}$$
$$=16\prod_{2<p\nmid k}\left(1-\frac{1}{(p-1)^2}\right)\prod_{2<p\nmid k}\left(1-\frac{\rho'(p)-2}{p-2}\right)\prod_{2<p\mid k}\frac{p}{p-1}$$
$$=16\prod_{p>2}\left(1-\frac{1}{(p-1)^2}\right)\prod_{\substack{2<p\nmid F(0)\\ p\nmid k}}\left(1-\frac{\rho(p)-1}{p-2}\right)$$
$$\times\prod_{\substack{2<p\mid F(0)\\ p\nmid k}}\left(1-\frac{\rho(p)-2}{p-2}\right)\prod_{2<p\mid k}\frac{p-1}{p-2}.$$

We have now only to observe that

$$\log\log(3y/k)+\log\log 3k\leqslant 2\log\log 3y$$

in order to see that (7.3) follows from (7.11) and (7.12).

We give two applications of Theorem 5.6.

COROLLARY 5.6.1. *Let a, k and l be integers, and let y and x be real numbers, satisfying*

$$a \neq -b^2,\ 2 \mid a, \quad a \not\equiv -1 \bmod 3, \quad 1 \leqslant k < y \leqslant x.$$

Then

$$|\{p\colon x - y < p \leqslant x,\ p \equiv l \bmod k,\ p^2 + a = p'\}|$$
$$\leqslant 16 \prod_{p>2}\left(1 - \frac{1}{(p-1)^2}\right) \prod_{2<p\nmid ak}\left(1 - \frac{(-a/p)}{p-2}\right) \prod_{2<p\mid ak} \frac{p-1}{p-2}$$
$$\times \frac{y}{\phi(k)\log^2(y/k)}\left\{1 + O_a\left(\frac{\log\log 3y}{\log(y/k)}\right)\right\}.$$

Proof. We recall the Remark 5 following Theorem 5.6 to the effect that the result is trivial unless $(l, k) = 1$ and $(l^2 + a, k) = 1$.

The number $\rho(p)$ of solutions of

$$n^2 + a \equiv 0 \bmod p$$

is given by

$$\rho(p) = \begin{cases} 1 & \text{if } p \mid a \\ 1 + (-a/p) & \text{if } p \nmid a. \end{cases}$$

The conditions here imply that $F(n) = n^2 + a$ satisfies the conditions of Theorem 5.6, and the result now follows at once.

COROLLARY 5.6.2. *Let k and l be integers, and let y and x be real numbers satisfying*

$$1 \leqslant k < y \leqslant x.$$

Then

$$|\{p\colon x - y < p \leqslant x,\ p \equiv l \bmod k,\ p^2 + p + 1 = p'\}|$$
$$\leqslant 16 \prod_{p>2}\left(1 - \frac{1}{(p-1)^2}\right) \prod_{3<p\nmid k}\left(1 - \frac{\chi(p)}{p-2}\right) \prod_{2<p\mid k} \frac{p-1}{p-2}$$
$$\times \frac{y}{\phi(k)\log^2(y/k)}\left\{1 + O\left(\frac{\log\log 3y}{\log(y/k)}\right)\right\},$$

where

$$\chi(p) = \begin{cases} 1 & \text{if } p \equiv 1 \bmod 3, \\ -1 & \text{if } p \equiv -1 \bmod 3. \end{cases}$$

Proof. If $F(n) = n^2 + n + 1$ and ρ has its usual meaning, then

$$\rho(2) = 0, \qquad \rho(3) = 1;$$

and if we note that, for $p > 3$, $n^2 + n + 1 \equiv 0 \bmod p$ is equivalent to

$$(2n + 1)^2 \equiv -3 \bmod p,$$

we see that

$$\rho(p) = 1 + (-3/p) = 1 + (p/3) = 1 + \chi(p) \quad \text{if} \quad p > 3. \tag{7.13}$$

Thus the conditions of Theorem 5.6 are satisfied, and (7.3) gives the result.

8. PRIME k-TUPLETS

In this section we return to the problem considered in Sections 2.6 and 2.7. The underlying sequence is that analysed in Example 4. We shall be able in this problem to obtain results with O-constants depending at most on the degree g of the polynomial—as, indeed, was the case in Chapter 2; but now we shall be able also to make the numerical constants in the dominant terms explicit.

THEOREM 5.7. *Let g be natural number, and let a_i, b_i $(i = 1, \ldots, g)$ be integers satisfying*

$$E := \prod_{i=1}^{g} a_i \prod_{1 \leqslant r < s \leqslant g} (a_r b_s - a_s b_r) \neq 0. \tag{8.1}$$

Let $\rho(p)$ denote the number of solutions of

$$\prod_{i=1}^{g} (a_i n + b_i) \equiv 0 \bmod p,$$

and suppose that

$$\rho(p) < p \quad \textit{for all} \quad p. \tag{8.2}$$

Let y and x be real numbers satisfying

$$1 < y \leqslant x.$$

Then

$$|\{n:\ x - y < n \leqslant x,\ a_i n + b_i \textit{ prime for } i = 1, \ldots, g\}|$$

$$\leqslant 2^g g! \prod_p \left(1 - \frac{\rho(p) - 1}{p - 1}\right)\left(1 - \frac{1}{p}\right)^{-g+1} \frac{y}{\log^g y}$$

$$\times \left\{1 + O\left(\frac{\log\log 3y + \log\log 3|E|}{\log y}\right)\right\}, \tag{8.3}$$

where the constant implied by the O-symbol depends at most on g.

Remark. We observe that (8.2) implies

$$(a_i, b_i) = 1 \quad \text{for} \quad i = 1, \ldots, g;$$

otherwise $p_0 \mid (a_j, b_j)$ for some j would imply that $\rho(p_0) = p_0$. Hence, by (1.3.27),

$$\rho(p) \leqslant g \quad \text{for all } p, \tag{8.4}$$

and we recall from (1.3.28) that

$$\rho(p) = g \quad \text{if} \quad p \nmid E. \tag{8.5}$$

Proof. We proceed, as in the proof of Theorem 2.3, to work with

$$\mathscr{A} = \left\{ \prod_{i=1}^{g} (a_i n + b_i) \colon \; x - y < n \leqslant x \right\},$$

but this time we take for $\mathfrak{P}$ the set $\mathfrak{P}_1$ of all primes. Now as then we choose

$$X = y \quad \text{and} \quad \omega(p) = \rho(p),$$

so that (R) is satisfied and (Ω_1) holds with $A_1 = g + 1$. So far as $(\Omega_2(\kappa, L))$ is concerned, (8.5) implies that (remember that $\mathfrak{P}$ is now the set of *all* primes)

$$\sum_{w \leqslant p < z} \frac{\rho(p) \log p}{p} = g \sum_{w \leqslant p < z} \frac{\log p}{p} - \sum_{w \leqslant p < z} \frac{g - \rho(p)}{p} \log p,$$

and we see that, by virtue of (1.1) and (1.5), $(\Omega_2(\kappa, L))$ is satisfied with

$$\kappa = g, \qquad A_2 = O_g(1) \quad \text{and} \quad L = O_g(\log\log 3|E|).$$

We can now apply Theorem 5.1. Choosing

$$z = y^{1/2},$$

(4.2) yields

$$S(\mathscr{A}; \mathfrak{P}_1, y^{1/2}) \leqslant 2^g g! \prod_p \left(1 - \frac{\rho(p)}{p}\right)\left(1 - \frac{1}{p}\right)^{-g} \frac{y}{\log^g y}$$

$$\times \left\{1 + O\left(\frac{\log\log 3y + \log\log 3|E|}{\log y}\right)\right\}, \tag{8.6}$$

where the O-constant depends at most on g.

If an integer n from $(x - y, x]$ has the property that each factor $a_i n + b_i$ is a prime $\geqslant y^{1/2}$, then n is counted in $S(\mathscr{A}; \mathfrak{P}_1, y^{1/2})$. So far as the remaining n's counted on the left of (8.3) are concerned, there is at least one i such that

$$(0 <)\, a_i n + b_i < y^{1/2},$$

and since, by (8.1), $a_i \neq 0$, their number is $O_g(y^{1/2})$; this latter quantity can be absorbed into the error term (in (8.6)) by (2.6) and so, by (8.6), the number that was to be estimated is at most

$$2^g g! \prod_p \left(1 - \frac{\rho(p)}{p}\right)\left(1 - \frac{1}{p}\right)^{-g} \frac{y}{\log^g y}\left\{1 + O\left(\frac{\log\log 3y + \log\log 3|E|}{\log y}\right)\right\}.$$

The result now follows in view of the identity (4.2.12).

We shall show next that one can deduce from Theorem 5.7 the result corresponding to Theorem 2.4.

THEOREM 5.8. *Let g be a natural number and let a_i, b_i $(i = 1, \ldots, g)$ be integers satisfying*

$$E := \prod_{i=1}^{g} a_i \prod_{1 \leqslant r < s \leqslant g} (a_r b_s - a_s b_r) \neq 0, \qquad b_0 := b_1 \ldots b_g \neq 0. \tag{8.7}$$

Let $\rho(p)$ denote the number of solutions of

$$\prod_{i=1}^{g} (a_i n + b_i) \equiv 0 \bmod p, \tag{8.8}$$

and suppose that

$$\rho(p) < p \quad \textit{for all} \quad p \tag{8.9}$$

as well as that

$$\rho(p) < p - 1 \quad \textit{if} \quad p \nmid b_0. \tag{8.10}$$

Let k and l be integers and y and x real numbers, satisfying

$$1 \leqslant k < y \leqslant x, \tag{8.11}$$

and

$$(l, k) = 1, \qquad \left(\prod_{i=1}^{g} (a_i l + b_i), k\right) = 1. \tag{8.12}$$

Then

$$|\{p:\ x-y<p\leqslant x,\ p\equiv l \bmod k,\ a_i p+b_i \ \textit{prime for}\ \ i=1,\ldots,g\}|$$

$$\leqslant 2^{2g+1}(g+1)!\prod_{p>2}\left(1-\frac{1}{(p-1)^2}\right)\prod_{\substack{2<p\nmid b_0\\ p\nmid k}}\left(1-\frac{\rho(p)-1}{p-2}\right)\left(1-\frac{1}{p}\right)^{-g+1}$$

$$\times\prod_{\substack{2<p|b_0\\ p\nmid k}}\left(1-\frac{\rho(p)-2}{p-2}\right)\left(1-\frac{1}{p}\right)^{-g+1}$$

$$\times\prod_{2<p|k}\frac{p-1}{p-2}\left(\frac{p}{p-1}\right)^{g-1}\frac{y}{\phi(k)\log^{g+1}(y/k)}$$

$$\times\left\{1+O\left(\frac{\log\log 3|Eb_0|(y/k)}{\log(y/k)}\right)\right\}, \tag{8.13}$$

where the constant implied by the O-symbol depends at most on g.

Remark. We note that, by (1.3.28),

$$\rho(p)=g \quad \text{if} \quad p\nmid E, \tag{8.14}$$

a fact that will be useful when we come to apply Theorem 5.8.

Proof. It is convenient to assume that

$$l\leqslant 0, \tag{8.15}$$

which is clearly no restriction. We shall apply Theorem 5.7 with $g+1$ in place of g, and to avoid confusion we shall (as we did in the proof of Theorem 5.5) mark all the data of Theorem 5.7 with a dash (′).

With this convention we now select the parameters of Theorem 5.7 so as to apply Theorem 5.7 to the present situation. Accordingly we take

$$a_i'=a_i k,\ b_i'=a_i l+b_i \quad (i=1,\ldots,g), \qquad a_{g+1}'=k,\ b_{g+1}'=l. \tag{8.16}$$

Then

$$E'=\prod_{i=1}^{g+1}a_i'\prod_{1\leqslant r<s\leqslant g+1}(a_r'b_s'-a_s'b_r')$$

$$=\prod_{i=1}^{g}(a_i k)k\prod_{1\leqslant r<s\leqslant g}k(a_r b_s-a_s b_r)\prod_{1\leqslant r<g+1}(-kb_r),$$

or

$$E'=Eb_0(-1)^g k^{2g+1+g(g-1)/2},$$

so that, by (8.7), condition (8.1) is fulfilled.

By (8.16), $\rho'(p)$ is the number of n's in a complete system of residues modulo p for which

$$(kn + l) \prod_{i=1}^{g} (a_i(kn + l) + b_i) \equiv 0 \bmod p.$$

Hence, in view of (8.12),

$$\rho'(p) = 0 \quad \text{if} \quad p \mid k, \tag{8.17}$$

and if $p \nmid k$ we can simply replace $kn + l$ in the above congruence by n, so that $\rho'(p)$ is then also the number of solutions of

$$n \prod_{i=1}^{g} (a_i n + b_i) \equiv 0 \bmod p.$$

We see at once that $\rho'(p) = \rho(p) + 1$ if $n \equiv 0 \bmod p$ is not a solution of (8.8), that is, if $p \nmid b_0$, and that $\rho'(p) = \rho(p)$ otherwise. Thus

$$\rho'(p) = \begin{cases} \rho(p) + 1 & \text{if} \quad p \nmid b_0 \\ \rho(p) & \text{if} \quad p \mid b_0, \end{cases} \qquad (p \nmid k), \tag{8.18}$$

so that, by (8.10) and (8.9), $\rho'(p) < p$ for all p.

Furthermore, we choose

$$y' = \frac{y}{k} \quad \text{and} \quad x' = \frac{x - l}{k} \tag{8.19}$$

in Theorem 5.7, so that, by (8.11) and (8.15), condition $1 < y' \leqslant x'$ is satisfied.

We may therefore apply Theorem 5.7, and we obtain

$$\begin{aligned} &|\{n\colon\ x' - y' < n \leqslant x',\ a_i'n + b_i' \text{ prime for } i = 1, \ldots, g + 1\}| \\ &\leqslant 2^{g+1} (g + 1)! \prod_p \left(1 - \frac{\rho'(p) - 1}{p - 1}\right) \left(1 - \frac{1}{p}\right)^{-g} \\ &\times \frac{y'}{\log^{g+1} y'} \left\{1 + O\left(\frac{\log\log 3y' + \log\log 3|E'|}{\log y'}\right)\right\}, \end{aligned} \tag{8.20}$$

where the O-constant depends at most on g.

By (8.16) and (8.19) the conditions

$$x' - y' < n \leqslant x', \qquad a_{g+1}' n + b_{g+1}' \text{ prime}$$

and

$$x - y < p \leqslant x, \qquad p \equiv l \bmod k$$

are equivalent, and (in this notation) we have

$$a_i'n + b_i' = a_ip + b_i \quad \text{for} \quad i = 1, \ldots, g.$$

Hence the numbers on the left hand sides of (8.20) and (8.13) are the same.

By (6.8) and (8.17) the product on the right of (8.20) can be written as

$$\frac{1-(\rho'(2)-1)}{(1-1/2)^g}\prod_{p>2}\left(1-\frac{1}{(p-1)^2}\right)\prod_{\substack{p>2\\ p\nmid k}}\left(1-\frac{\rho'(p)-2}{p-2}\right)\left(1-\frac{1}{p}\right)^{-g+1}$$
$$\times \prod_{2<p\mid k}\left(1+\frac{2}{p-2}\right)\left(1-\frac{1}{p}\right)^{-g+1};$$

(8.17) also tells us that $\rho'(2) = 0$ if $2 \mid k$, and if $2 \nmid k$ then $\rho'(2) = 1$ by (8.10) and (8.9). Hence

$$2 - \rho'(2) = \frac{k}{\phi(k)}\prod_{2<p\mid k}\frac{p-1}{p},$$

and so, by (8.18), our product may be written in the form

$$2^g\prod_{p>2}\left(1-\frac{1}{(p-1)^2}\right)\prod_{\substack{2<p\nmid b_0\\ p\nmid k}}\left(1-\frac{\rho(p)-1}{p-2}\right)\left(1-\frac{1}{p}\right)^{-g+1}$$
$$\times \prod_{\substack{2<p\mid b_0\\ p\nmid k}}\left(1-\frac{\rho(p)-2}{p-2}\right)\left(1-\frac{1}{p}\right)^{-g+1}\prod_{2<p\mid k}\frac{p-1}{p-2}\left(\frac{p}{p-1}\right)^{g-1}\frac{k}{\phi(k)}.$$

If we substitute this on the right of (8.20) and take note of (8.19), we see that the proof is complete since

$$\log\log 3y' + \log\log 3\,|E'| = O_g(\log\log(3\,|Eb_0|\,y/k)).$$

Some of the most important special cases of Theorem 5.8 correspond to $g = 1$.

COROLLARY 5.8.1. *Let a, b, k and l be integers, and let y and x be real numbers, satisfying*

$$\left.\begin{aligned} &ab \neq 0,\ (a,b) = 1,\ 2 \mid ab,\\ &(l,k) = 1,\ (al+b,k) = 1,\ 1 \leqslant k < y \leqslant x.\end{aligned}\right\} \tag{8.21}$$

Then

$$|\{p:\ x-y<p\leqslant x,\ p\equiv l \bmod k,\ ap+b=p'\}|$$

$$\leqslant 16\prod_{p>2}\left(1-\frac{1}{(p-1)^2}\right)\prod_{2<p|kab}\frac{p-1}{p-2}\,\frac{y}{\phi(k)\log^2(y/k)}$$

$$\times\left\{1+O\left(\frac{\log\log 3|ab|(y/k)}{\log(y/k)}\right)\right\}, \tag{8.22}$$

where the constant implied by the O-symbol is absolute.

Proof. We take $g=1, a_1=a, b_1=b$ in Theorem 5.8, so that

$$E=a, \qquad b_0=b,$$

and (8.7) is satisfied. Since $(a,b)=1$ we have

$$\rho(p)=\begin{cases}1, & p\nmid a,\\ 0, & p\mid a,\end{cases}$$

so that (8.9) is true; moreover, so is (8.10) because for odd b the condition $2\mid ab$ implies that $2\mid a$. The remaining conditions required for an application of Theorem 5.8 to be valid, occur explicitly in (8.21). Since $(a,b)=1$, the counting number on the left of (8.22) is, by (8.13), at most

$$16\prod_{p>2}\left(1-\frac{1}{(p-1)^2}\right)\prod_{\substack{2<p\nmid b\\ p\nmid k\\ p|a}}\left(1+\frac{1}{p-2}\right)\prod_{\substack{2<p|b\\ p\nmid k}}\left(1+\frac{1}{p-2}\right)$$

$$\times\prod_{2<p|k}\frac{p-1}{p-2}\,\frac{y}{\phi(k)\log^2(y/k)}\left\{1+O\left(\frac{\log\log 3|ab|(y/k)}{\log(y/k)}\right)\right\},$$

and, if we invoke $(a,b)=1$ once more, the last three products together equal

$$\prod_{2<p|kab}\frac{p-1}{p-2}.$$

This result is a remarkable extension of Theorem 3.12 in that the result is valid for an arbitrary interval (as was also the case in Corollary 2.4.1) and for a virtually unrestricted range of values of k; however, the numerical constant is worse by a factor 2.

If we take $a=1$ and replace b by h in Corollary 5.8.1, we obtain

COROLLARY 5.8.2. *Let k, l and h be integers, and let y and x be real numbers, satisfying*

$$h \neq 0,\ 2 \mid h,\ (l, k) = 1,\ (l + h, k) = 1,\ 1 \leqslant k < y \leqslant x.$$

Then

$$|\{p\colon\ x - y < p \leqslant x,\ p \equiv l \bmod k,\ p + h = p'\}|$$
$$\leqslant 16 \prod_{p>2}\left(1 - \frac{1}{(p-1)^2}\right) \prod_{2<p|kh} \frac{p-1}{p-2} \frac{y}{\phi(k)\log^2(y/k)}$$
$$\times \left\{1 + O\left(\frac{\log\log 3|h|(y/k)}{\log(y/k)}\right)\right\}$$

where the constant implied by the O-symbol is absolute.

If we take $a = -1$ and $b = N$ (in Corollary 5.8.1) we obtain the "conjugate" of Corollary 5.8.2, namely

COROLLARY 5.8.3. *Let N, k and l be positive integers, and let y and x be real numbers, such that*

$$2 \mid N, \qquad (l, k) = 1, \qquad (N - l, k) = 1, \qquad k < y \leqslant x.$$

Then

$$|\{p\colon\ x - y < p \leqslant x,\ p \equiv l \bmod k,\ N - p = p'\}|$$
$$\leqslant 16 \prod_{p>2}\left(1 - \frac{1}{(p-1)^2}\right) \prod_{2<p|kN} \frac{p-1}{p-2} \frac{y}{\phi(k)\log^2(y/k)}$$
$$\times \left\{1 + O\left(\frac{\log\log 3N(y/k)}{\log(y/k)}\right)\right\}$$

where the constant implied by the O-symbol is absolute.

The cases $x = N$ and/or $k = 1$, and the case $y = x = N$, are of special interest. Corollaries 5.8.2. and 5.8.3 correspond to Theorem 3.11, and the remarks we made when comparing Corollary 5.8.1 and Theorem 3.12 apply equally well here.

We close the section with another interesting application of the case $g = 1$ of Theorem 5.8—that is, of Corollary 5.8.1.

COROLLARY 5.8.4. *Let q and h be integers, and let y and x be real numbers, satisfying*

$$h \neq 0,\ (q, h) = 1,\ 2 \mid qh, \qquad 1 \leqslant q < y \leqslant x.$$

Then

$$|\{p\colon x - y < p \leqslant x,\ (p-h)/q = p'\}|$$

$$\leqslant 16 \prod_{p>2}\left(1 - \frac{1}{(p-1)^2}\right) \prod_{2<p|qh} \frac{p-1}{p-2}\, \frac{y/q}{\log^2 (y/q)}$$

$$\times \left\{1 + O\left(\frac{\log\log 3|h|y}{\log (y/q)}\right)\right\}, \tag{8.23}$$

where the constant implied by the O-symbol is absolute.

Proof. If we distinguish the parameters of Corollary 5.8.1 by a dash ('), we choose

$$a' = q, \qquad b' = h, \qquad k' = l' = 1, \qquad y' = y/q, \qquad x' = (x-h)/q,$$

and obtain

$$\left|\left\{p'\colon \frac{x-h}{q} - \frac{y}{q} < p' \leqslant \frac{x-h}{q},\ qp' + h = p\right\}\right|$$

$$\leqslant 16 \prod_{p>2}\left(1 - \frac{1}{(p-1)^2}\right) \prod_{2<p|qh} \frac{p-1}{p-2}\, \frac{y/q}{\log^2 (y/q)}$$

$$\times \left\{1 + O\left(\frac{\log\log 3|h|y}{\log (y/q)}\right)\right\};$$

if we now rewrite the left hand side we obtain (8.23). (If h is positive, we have actually proved the result subject to $y \leqslant x - h$ only; but the extension of the range of y to $y \leqslant x$ is trivial.)

9. Primes Representable by Polynomials $F(p)$: The Linearized Approach

In this last Section we turn to applications of Theorem 5.2.

Theorem 5.9. *Let $F(n)$ $(\neq n)$ be an irreducible polynomial with integer coefficients and positive leading coefficient. Let $\rho(p)$ denote the number of solutions of*

$$F(n) \equiv 0 \bmod p,$$

and suppose that

$$\rho(p) < p \quad \textit{for all} \quad p \tag{9.1}$$

as well as that

$$\rho(p) < p - 1 \quad \textit{if} \quad p \nmid F(0). \tag{9.2}$$

Let k and l be integers, and x a real number, such that with some constant A

$$1 \leqslant k \leqslant \log^A x, \ (l, k) = 1. \tag{9.3}$$

Then, as $x \to \infty$,

$$\begin{aligned} &|\{p\colon p \leqslant x,\ p \equiv l \bmod k,\ F(p) = p'\}| \\ &\leqslant 8 \prod_{p>2} \left(1 - \frac{1}{(p-1)^2}\right) \prod_{\substack{2<p\nmid F(0) \\ p\nmid k}} \left(1 - \frac{\rho(p)-1}{p-2}\right) \\ &\times \prod_{\substack{2<p\mid F(0) \\ p\nmid k}} \left(1 - \frac{\rho(p)-2}{p-2}\right) \prod_{2<p\mid k} \frac{p-1}{p-2} \frac{x}{\phi(k)\log^2 x} \\ &\times \left\{1 + O_F\left(\frac{\log\log x}{\log x}\right)\right\}. \end{aligned} \tag{9.4}$$

Remark. All the remarks we made following the statement of Theorem 5.6 apply here also; in particular, we have that

$$\rho(p) \leqslant g \quad \text{for all } p. \tag{9.5}$$

Proof. We proceed as in the proof of Theorem 4.2, but this time with the sequence

$$\mathscr{A} = \{F(p)\colon p \leqslant x,\ p \equiv l \bmod k\}$$

which we studied in Example 6; and with

$$\mathfrak{P} = \mathfrak{P}_k = \{p\colon p \nmid k\}.$$

Then $(d, \mathfrak{P}) = 1$ means that $(d, k) = 1$, and in accordance with (1.3.48), (1.3.51) and (1.4.15) we choose

$$X = \frac{li\, x}{\phi(k)}, \quad \omega(p) = \frac{\rho_1(p)}{p-1} p \quad \text{for} \quad p \nmid k, \tag{9.6}$$

where

$$\rho_1(p) = \begin{cases} \rho(p) & \text{if} \quad p \nmid F(0) \\ \rho(p) - 1 & \text{if} \quad p \mid F(0). \end{cases} \tag{9.7}$$

With these choices we obtain, by (1.4.15), (1.3.52) and (1.3.55), that

$$|R_d| \leqslant g^{\nu(d)} (E(x, kd) + 1) \quad \text{if} \quad \mu(d) \neq 0, (d, k) = 1.$$

Let us apply Lemma 3.5 with

$$U_1 = 3, \ h = 3g$$

and with the corresponding number $C_1 = C_1(3, 3g, A)$. We choose z by

$$z^2 = \frac{x^{1/2}}{k \log^{C_1} x}$$

and obtain by Lemma 3.4, in view of (9.3),

$$\sum_{\substack{d < z^2 \\ (d,k)=1}} \mu^2(d)\, 3^{\nu(d)} |R_d| \ll \frac{x}{\phi(k) \log^3 x}, \tag{9.8}$$

where the $\ll$-constant depends only on g and A.

By (9.7) and (9.5) we have

$$\frac{\omega(p)}{p} \leqslant \frac{g}{p-1} \leqslant 1 - \frac{1}{g+1} \quad \text{if} \quad p \geqslant g + 2,$$

and by (9.2) and (9.1)

$$\frac{\omega(p)}{p} \leqslant \frac{p-2}{p-1} \leqslant 1 - \frac{1}{g} \quad \text{if} \quad p \leqslant g + 1,$$

so that (Ω_1) is satisfied with $A_1 = g + 1$. Also, condition $(\Omega_2(\kappa, L))$ is satisfied, in view of (7.10), with

$$\kappa = 1, \quad A_2 = O_F(1)$$

and (using (1.5))

$$L \leqslant O_F(1) + O_g\left(\sum_{p|k} \frac{\log p}{p}\right) = O_F(\log \log 3k).$$

Theorem 5.2 may now be applied, and we obtain by (9.8), using

$$li\, x = \frac{x}{\log x}\left\{1 + O\left(\frac{1}{\log x}\right)\right\},$$

that

$$S(\mathscr{A}; \mathfrak{P}_k, z) \leqslant \frac{x}{\phi(k) \log x} W(z) e^{\gamma}\left\{1 + O_F\left(\frac{\log\log 3k}{\log x}\right)\right\} + O_F\left(\frac{x}{\phi(k) \log^3 x}\right).$$

For each p counted on the left of (9.4) but not in $S(\mathscr{A}; \mathfrak{P}_k, z)$ we have necessarily that

$$F(p) < z,$$

and there are at most gz primes p satisfying that inequality. Hence

$$\begin{aligned}|\{p\colon p \leqslant x, p \equiv l \bmod k,\ F(p) = p'\}| &\leqslant S(\mathscr{A}; \mathfrak{P}_k, z) + gz\\ &\leqslant \frac{x}{\phi(k)\log x} W(z)e^{\gamma}\left\{1 + O_F\left(\frac{\log\log 3k}{\log x}\right) + O_F\left(\frac{1}{W(z)\log^2 x}\right)\right\}.\end{aligned}$$

By (2.5)

$$W(z)e^{\gamma} = \prod_{p\nmid k}\left(1 - \frac{\rho_1(p)}{p-1}\right)\left(1 - \frac{1}{p}\right)^{-1}\prod_{p|k}\frac{p}{p-1}\frac{1}{\log z}$$

$$\times \left\{1 + O_F\left(\frac{\log\log 3k}{\log z}\right)\right\},$$

and we note that

$$\frac{1}{\log z} = \frac{4}{\log x}\left\{1 + O_F\left(\frac{\log\log x}{\log x}\right)\right\};$$

taking account of (9.3) and (2.1.6) it follows that

$$\begin{aligned}&|\{p\colon p \leqslant x, p \equiv l \bmod k,\ F(p) = p'\}|\\ &\leqslant 4\prod_{p\nmid k}\left(1 - \frac{\rho_1(p)}{p-1}\right)\left(1 - \frac{1}{p}\right)^{-1}\prod_{p|k}\frac{p}{p-1}\frac{x}{\phi(k)\log^2 x}\\ &\times\left\{1 + O_F\left(\frac{\log\log x}{\log x}\right)\right\}.\end{aligned} \tag{9.9}$$

Finally there is, as usual the product. Since (9.7), (9.2) and (9.1) imply that $\rho_1(2) = 0$ always, the identity (6.8) shows that

$$\begin{aligned}&4\prod_{p\nmid k}\left(1 - \frac{\rho_1(p)}{p-1}\right)\left(1 - \frac{1}{p}\right)^{-1}\prod_{p|k}\frac{p}{p-1}\\ &= 8\prod_{p<2\nmid k}\left(1 - \frac{1}{(p-1)^2}\right)\prod_{2<p\nmid k}\left(1 - \frac{\rho_1(p)-1}{p-2}\right)\prod_{2<p|k}\frac{p}{p-1}\\ &= 8\prod_{p>2}\left(1 - \frac{1}{(p-1)^2}\right)\prod_{2<p\nmid k}\left(1 - \frac{\rho_1(p)-1}{p-2}\right)\prod_{2<p|k}\frac{p-1}{p-2};\end{aligned}$$

and Theorem 5.9 now follows from (9.9) and (9.7).

The simplest non-trivial case of Theorem 5.9 is the linear case

$$F(n) = an + b$$

where a and b are coprime integers, $a > 0$, $b \neq 0$ and $2 \mid ab$. However, here Theorem 5.9 gives only the same result as Theorem 3.12, and the latter is more general in that it holds uniformly in F. Thus Theorem 5.9 is of interest only for polynomials $F(n)$ of degree $g \geqslant 2$.

We shall consider, by way of illustration, special cases of the problems dealt with in Corollaries 5.6.1 and 5.6.2.

First consider

$$F(n) = n^2 + 4, \ k = 1.$$

Here

$$\rho(p) = 1 + (-1/p) = 1 + (-1)^{(p-1)/2} \quad \text{for} \quad p > 2,$$

so that the conditions of Theorem 5.9 are satisfied. We deduce

COROLLARY 5.9.1. *We have*

$$|\{p \colon p \leqslant x, \ p^2 + 4 = p'\}|$$

$$\leqslant 8 \prod_{p>2} \left(1 - \frac{1}{(p-1)^2}\right) \prod_{p>2} \left(1 - \frac{(-1)^{(p-1)/2}}{p-2}\right) \frac{x}{\log^2 x} \left\{1 + O\left(\frac{\log\log 3x}{\log x}\right)\right\}.$$

Next take

$$F(n) = n^2 + n + 1, \ k = 1.$$

Then by (7.13),

$$\rho(2) = 0, \ \rho(3) = 1, \ \rho(p) = 1 + \chi(p) \quad \text{for} \quad p > 3,$$

and Theorem 5.9 at once yields

COROLLARY 5.9.2. *We have*

$$|\{p \colon p \leqslant x, p^2 + p + 1 = p'\}|$$

$$\leqslant 8 \prod_{p>2} \left(1 - \frac{1}{(p-1)^2}\right) \prod_{p>3} \left(1 - \frac{\chi(p)}{p-2}\right) \frac{x}{\log^2 x} \left\{1 + O\left(\frac{\log\log 3x}{\log x}\right)\right\},$$

where

$$\chi(p) = \begin{cases} 1 \text{ if } p \equiv 1 \bmod 3, \\ -1 \text{ if } p \equiv -1 \bmod 3. \end{cases}$$

NOTES

5.1. Mertens prime number theory: for proofs of (1.1), (1.2), (1.3) see *H–W* Sections 22.6–8.

5.2. (2.9): see Rosser–Schoenfeld [1], Theorem 5 on p. 70.

5.3. Lemma 5.4: taken from Halberstam–Richert [1]; cf. Levin–Faĭnleĭb [2]. The basic approach derives from Wirsing [2]. (See also Levin–Tuljaganova [1]). A similar result is stated in Ankeny–Onishi [1], and is proved in Onishi [1]. Various results of this kind, corresponding to different levels of generality and/or precision, are scattered through literature; see, for example, Klimov [1], [2]; Levin [1, 6, 7]; Wang [8]; Vinogradov [2].

(3.10): the proof has been, essentially, using the method of variation of parameters.

The calculation of c_3 corresponds to an abelian method for Dirichlet series (cf. the tauberian theorem for Dirichlet series of Hardy–Littlewood, *Messenger of Math.* **42** (1913), 89–93). The corresponding argument in Levin–Faĭnleĭb [2] is faulty.

5.4. Theorem 5.1: cf. Ankeny–Onishi [1], Theorems 1.1, 3.1.

5.5. Bombieri's theorem and its precursors: see Notes attached to Lemma 3.3.

The term "linearized": not the most accurate description of what happens; linearizing, more precisely, means reducing the dimension of the sieve problem by 1.

5.6. Theorem 5.3: cf. Bateman–Stemmler [1], Lemma 3 (with $y = x$ and only $O(1)$ error term); Wang [3] ($y = x$, but for integer-valued polynomials; statement only). Bateman–Stemmler [1] comment that the estimate is $2^g g!\, h_1 \ldots h_g$ times the conjectured asymptotic value ($h_i = \deg F_i$; see Notes for Introduction). For the analogous explicit upper bound corresponding to to hypothesis H_N^* see Russell [1]; the dependence on N of the discriminant of the polynomial to which H_N^* refers, makes this a much more complicated calculation.

It seems likely that Theorem 5.3. can be sharpened so far as the error term is concerned by use of the the refinements sketched in connection with the Brun–Titchmarsh theorem (see Notes for Chapter 3).

Corollary 5.3.1: stated in Wang [3] (with $y = x$, $o(1)$ error only).

Theorem 5.4: see Ricci [1, 6, 7]; Wang [1], [3] (with $y = x$).

Corollary 5.4.1: see Selberg [3] (with $a = 1$); Klimov [1], Theorem 3, [2, 3] (with arithmetic progression).

Numerical data relating to various other special cases of Theorem 5.4 are to be found in several papers cited in Notes for Introduction.

Theorem 5.5: cf. the more general but less precise Theorem 4.2; also the less precise Ricci [6, 7].

5.7. Corollary 5.6.2: generalizes Bateman–Stemmler [1] (Lemma 5) to an arbitrary arithmetic progression and arbitrary interval; for $y = x$ and $k = 1$ the two results are the same except that we have sharper error term.

5.8. Theorem 5.7: see Čulanovskiĭ [1]; Klimov [2, 3] for somewhat weaker results. See Hardy–Littlewood [2], Hypothesis X_1 for conjectured asymptotic formula.

Theorem 5.8: cf. Klimov [1], Theorem 1; our Theorem 2.4.

Corollary 5.8.1: cf. Theorem 3.12. The estimate here is worse by a factor 2, but there is greater uniformity in k.

5.9 Theorem 5.9: Bateman–Stemmler [1], Lemma 4, consider the special case $F(p) = (p^r - 1)/(p^d - 1)$ where r is a prime power and d is the largest proper divisor of r. If r is a power of 2, the counting number being estimated in Theorem 5.9 is, trivially, $\leqslant 1$; hence one may take r odd. Our estimate is better by a factor 1/2.

Corollary 5.9.2: cf. our Corollary 5.6.2; Bateman–Stemmler [1], Lemma 5 (the special case $r = 3$ of above Note). Using their notation, our constant factor is $4c_3$ as compared with their $8c_3$, and since $c_3 = 1{\cdot}522\ldots$ (Bateman–Horn [1], p. 365), our estimate is

$$6.1 \frac{x}{\log^2 x}, \quad x \geqslant x_0.$$

On heuristic grounds one expects the correct numerical factor to be $\frac{1}{2}c_3$.

Chapter 6

An Extension of Selberg's Upper Bound Method

1. The Method

The last three chapters have been based on Selberg's upper sieve estimate as formulated in Theorems 3.1 and 3.2. Broadly speaking, we have developed from these results a theory which has made it possible to find good upper bounds for the counting numbers associated with an extensive class of representation problems in classical prime number theory. The nature of these problems has always required us to work with the least drastic truncation of the appropriate set $\mathfrak{P}$ permitted by our methods—in other words, when estimating $S(\mathscr{A}; \mathfrak{P}, z)$ from above we have always chosen z as *large* as possible.

From this chapter onward our point of view will gradually change. Of course, we are still interested in the same problems, but our (much more difficult) aim now will be to look for the corresponding *lower* bounds in these problems. Indeed, we know from the "prime twins" problem, or Goldbach's problem, or the problem of showing that $n^2 + 1$ represents a prime infinitely often, that the solution of these (and many other) problems depends on establishing (positive) lower bounds for the appropriate counting numbers. We shall not succeed in settling any of these famous questions, but it will be clear from the Introduction (where several of our final results are quoted), or by reference to the theorems of Chapters 9, 10 and 11 that we shall prove many results which are sufficiently good approximations to these old conjectures to be interesting in their own right.

In Selberg's lower sieve theory upper sieve estimates continue to play a vital part, but the question we put in relation to them has a somewhat different emphasis. We shall ask, not how large can we take z, but, *given* z, what is a good upper bound for $S(\mathscr{A}; \mathfrak{P}, z)$. To answer this question we shall need to extend the method of Section 3.1 by the introduction of a new parameter (to be denoted by ξ: see (1.3) below). Moreover, we shall work with the more general sifting function $S(\mathscr{A}_q; \mathfrak{P}, z)$ where q is subject to the conditions (1.4.5). Our motive is not just to generalize our results; when, in the next three chapters, we come to study the lower bound sieve we shall find many occasions when it is essential to be able to estimate sums over q (satisfying (1.4.5)) of functions of the type $S(\mathscr{A}_q; \mathfrak{P}, \ldots)$.

As was the case in Section 3.1, (Ω_1) is the only condition we impose in order to obtain the general upper bound. Indeed, in this section we follow closely the development in Section 3.1.

As before, suppose that $\lambda_1 = 1$ and that λ_d $(d \geqslant 2)$ are arbitrary real numbers. Then the starting point of Selberg's method is the inequality

$$S(\mathscr{A}_q; \mathfrak{P}, z) \leqslant \sum_{a \in \mathscr{A}_q} \Big(\sum_{\substack{d \mid a \\ d \mid P(z)}} \lambda_d \Big)^2 = \sum_{\substack{d_\nu \mid P(z) \\ \nu = 1,2}} \lambda_{d_1} \lambda_{d_2} \sum_{\substack{a \in \mathscr{A} \\ a \equiv 0 \bmod q \\ a \equiv 0 \bmod D}} 1$$

where D is used as shorthand for the least common multiple $[d_1, d_2]$. Since $(q, P(z)) = 1$, D and q are relatively prime and the two congruences $a \equiv 0 \bmod q$, $a \equiv 0 \bmod D$ are therefore equivalent to $a \equiv 0 \bmod qD$. Now q is squarefree, $(q, \overline{\mathfrak{P}}) = 1$ and we recall that ω is a multiplicative function. Hence, by (1.4.14),

$$\sum_{\substack{a \in \mathscr{A} \\ a \equiv 0 \bmod q \\ a \equiv 0 \bmod D}} 1 = \frac{\omega(q)}{q} \cdot \frac{\omega(D)}{D} X + R_{qD},$$

and if we insert this above we obtain

$$S(\mathscr{A}_q; \mathfrak{P}, z) \leqslant \frac{\omega(q)}{q} X \sum_{\substack{d_\nu \mid P(z) \\ \nu = 1,2}} \lambda_{d_1} \lambda_{d_2} \frac{\omega(D)}{D} + \sum_{\substack{d_\nu \mid P(z) \\ \nu = 1,2}} |\lambda_{d_1} \lambda_{d_2}| \, |R_{qD}|$$

$$= \frac{\omega(q)}{q} X \Sigma_1 + \Sigma_2, \tag{1.1}$$

say.

We shall require the function (cf. Section 5.3 with $k = p$)

$$G_k(x, z) = \sum_{\substack{d < x \\ d \mid P(z) \\ (d,k) = 1}} g(d); \tag{1.2}$$

and we recall that g is a non-negative, multiplicative function. It is clear that $G_k(x, z) \geqslant 1$ if $x > 1$.

In Chapter 3 we imposed at this stage the condition (see (3.1.2)) that the numbers λ_d with $d \geqslant z$ should be zero, with the object of keeping Σ_2 (or rather, the sum taking the place of our Σ_2 here in the simpler situation $q = 1$) small. We still need to keep Σ_2 under control, of course, but now give ourselves one more degree of freedom (which before we did not need) by introducing an arbitrary real number

$$\xi > 1$$

and requiring that

$$\lambda_d = 0 \quad \text{if} \quad d \geqslant \xi. \tag{1.3}$$

We use the function $G_k(x, z)$ defined in (1.2) to choose

$$\lambda_d = \frac{\mu(d)}{\prod_{p|d} (1 - \omega(p)/p)} \frac{G_d(\xi/d, z)}{G(\xi, z)} \tag{1.4}$$

where $G(\xi, z)$ $(= G_1(\xi, z))$ is the function that was introduced in (1.4.21); and we see immediately that both, the condition $\lambda_1 = 1$ and (1.3), are satisfied.

It is appropriate here to recall the discussion preceding the imposition of condition (3.1.2) in Chapter 3. Now as then, the problem of choosing the λ_d's in such a way as to minimize the right-hand side of (1.1), namely $(\omega(q)/q) X \Sigma_1 + \Sigma_2$, is unsolved. However, if we impose the additional condition (1.3) in order to keep Σ_2 sufficiently small (this time we can do this more effectively than in Chapter 3 because ξ can be chosen independently of z), the above choice of the λ_d's actually does minimize the sum Σ_1. As we do not need this fact, we refer the interested reader wishing to see a proof to the literature.

Our first step is to evaluate Σ_1 at (1.4). The procedure is in all ways similar to that at the corresponding stage in Chapter 3. We may exclude from Σ_1 all terms for which $\omega([d_1, d_2]) = 0$; then $\omega(p) \neq 0$ for every prime p dividing d_1 or d_2, and we may assume therefore that, in Σ_1,

$$\omega(d_1) \neq 0, \qquad \omega(d_2) \neq 0, \qquad \omega((d_1, d_2)) \neq 0.$$

Moreover,

$$\frac{\omega(D)}{D} = \frac{d_1 d_2}{\omega(d_1)\,\omega(d_2)} \frac{(d_1, d_2)}{\omega((d_1, d_2))}$$

since ω is a multiplicative function, and, by (1.4.17),

$$\frac{p}{\omega(p)} = 1 + \frac{p - \omega(p)}{\omega(p)} = 1 + \frac{1}{g(p)} \quad \text{if} \quad \omega(p) \neq 0.$$

It follows that

$$\frac{d}{\omega(d)} = \prod_{p|d} \left(1 + \frac{1}{g(p)}\right) = \sum_{l|d} \frac{1}{g(l)} \quad \text{if} \quad d = (d_1, d_2),$$

whence, using (1.3),

$$\Sigma_1 = \sum_{d_1|P(z)}' \sum_{d_2|P(z)}' \lambda_{d_1}\lambda_{d_2} \frac{\omega(d_1)\,\omega(d_2)}{d_1 d_2} \sum_{\substack{l|d_1\\ l|d_2}}' \frac{1}{g(l)}$$

$$= \sum_{\substack{l<\xi\\ l|P(z)}}' \frac{1}{g(l)} \Bigg(\sum_{\substack{d|P(z)\\ d\equiv 0 \bmod l}} \lambda_d \frac{\omega(d)}{d} \Bigg)^2;$$

here Σ' indicates that summation is confined to numbers at which ω does not vanish. But if

$$1 \leqslant l < \xi \quad \text{and} \quad l \mid P(z),$$

then

$$\sum_{\substack{d|P(z)\\ d\equiv 0 \bmod l}} \lambda_d \frac{\omega(d)}{d} = \sum_{\substack{m|P(z)\\ (m,l)=1}} \mu(lm)\, g(lm) \frac{G_{lm}(\xi/lm, z)}{G(\xi, z)}$$

$$= \frac{\mu(l)g(l)}{G(\xi,z)} \sum_{\substack{m|P(z)\\ (m,l)=1}} \mu(m)\, g(m) \sum_{\substack{d<\xi/lm\\ d|P(z)\\ (d,lm)=1}} g(d)$$

$$= \frac{\mu(l)g(l)}{G(\xi,z)} \sum_{\substack{n<\xi/l\\ n|P(z)\\ (n,l)=1}} g(n) \sum_{m|n} \mu(m) = \frac{\mu(l)g(l)}{G(\xi,z)}.$$

We substitute this back in the last expression for Σ_1 and obtain

$$\Sigma_1 = \frac{1}{G^2(\xi,z)} \sum_{\substack{l<\xi\\ l|P(z)}}' g(l) = \frac{1}{G(\xi,z)}. \tag{1.5}$$

Thus the choice (1.4) of the parameters λ_d leads to an (apparently) simple expression for Σ_1. It is no less important that (1.4) should be a good choice from the point of view of Σ_2; and here we shall prove that $|\lambda_d| \leqslant 1$. We recall that g is non-negative, and if d is any divisor of $P(z)$ we have that

$$G(\xi,z) = \sum_{l|d} \sum_{\substack{m<\xi\\ m|P(z)\\ (m,d)=l}} g(m) = \sum_{l|d} \sum_{\substack{h<\xi/l\\ h|P(z)\\ (h,d/l)=1\\ (h,l)=1}} g(lh)$$

$$= \sum_{l|d} g(l)\, G_d\left(\frac{\xi}{l}, z\right)$$

$$\geqslant \sum_{l|d} g(l) G_d\left(\frac{\xi}{d}, z\right);$$

but, by (2.3.10),

$$\sum_{l|d} g(l) = \prod_{p|d} (1 + g(p)) = 1\Big/\prod_{p|d}\left(1 - \frac{\omega(p)}{p}\right),$$

so that

$$\frac{1}{\prod_{p|d} (1 - \omega(p)/p)} \frac{G_d(\xi/d, z)}{G(\xi, z)} \leqslant 1,$$

and therefore

$$|\lambda_d| \leqslant 1 \quad \text{if} \quad d \mid P(z).$$

Hence, by (1.3),

$$\Sigma_2 \leqslant \sum_{\substack{d_\nu < \xi \\ d_\nu | P(z) \\ \nu = 1,2}} |R_{q[d_1, d_2]}|. \tag{1.6}$$

Now the numbers $d = [d_1, d_2]$ arising from the pairs d_1, d_2 in Σ_2 divide $P(z)$ and are $< \xi^2$, and, d being squarefree, the number of pairs with the same least common multiple d is (see (3.1.11)) at most $3^{\nu(d)}$. Hence we may deduce from (1.6) a second estimate of Σ_2, namely

$$\Sigma_2 \leqslant \sum_{\substack{d < \xi^2 \\ d | P(z)}} 3^{\nu(d)} |R_{qd}|. \tag{1.7}$$

We now combine (1.1), (1.5) and (1.7) to give the following result (cf. Theorem 3.2).

THEOREM 6.1. (Ω_1): *We have for any real number* $\xi > 1$, *that*

$$S(\mathscr{A}_q; \mathfrak{P}, z) \leqslant \frac{\omega(q)}{q} \frac{X}{G(\xi, z)} + \sum_{\substack{d < \xi^2 \\ d | P(z)}} 3^{\nu(d)} |R_{qd}|. \tag{1.8}$$

Comparison with Theorem 3.2 makes it clear that the new parameter ξ has taken over from z the role of regulating the size of the remainder term.

2. An Upper Estimate

With the extended upper Selberg sieve at our disposal, we shall now study the consequences of introducing our general ω-conditions $(\Omega_2(\kappa))$ and $(\Omega_2(\kappa, L))$. We shall begin with the weaker of the two, namely

$$(\Omega_2(\kappa)) \qquad \sum_{w \leqslant p < z} \frac{\omega(p)}{p} \log p \leqslant \kappa \log \frac{z}{w} + A_2 \quad \text{if} \quad 2 \leqslant w \leqslant z.$$

Subject to a suitable remainder condition, Theorem 6.2 will provide an *explicit* upper estimate when

$$\frac{\log \xi}{\log z}$$

is large, and the form of the leading term suggests a (one-sided) comparison with the Fundamental Lemma (see e.g. Theorem 2.5). In fact, we shall use it in the next chapter (by matching it with the corresponding lower estimate, cf. proof of Theorem 7.1) to derive a Selberg form of the Fundamental Lemma (Theorem 7.1).

To make Theorem 6.1 effective, we have to estimate the function $G(\xi, z)$ from below. We have already prepared the ground for this in Lemma 4.1 of Chapter 4. We shall see that the new parameter ξ, apart from its regulating effect on the magnitude of the remainder term, will also permit us to utilize the full power of Lemma 4.1.

THEOREM 6.2. (Ω_1), $(\Omega_2(\kappa))$: *If*

$$\xi \geqslant z,$$

we have

$$S(\mathscr{A}_q; \mathfrak{P}, z) \leqslant \frac{\omega(q)}{q} X W(z)\{1 + O(\exp\{-\tau(\log\tau + 2)\})\} \qquad (2.1)$$
$$+ \sum_{\substack{d < \xi^2 \\ d | P(z)}} 3^{\nu(d)} |R_{qd}|,$$

where

$$\tau = \frac{\log \xi}{\log z}.$$

Remark. Lemma 2.2 shows that this result holds even under the simpler condition (Ω_0).

Proof. Suppose first that

$$\tau > z.$$

Then, arguing quite crudely,

$$\log \xi > z \log z > \sum_{p<z} \log p \geqslant \log P(z),$$

so that $\xi > P(z)$ and, in view of (4.1.1),

$$G(\xi, z) = \sum_{d | P(z)} g(d) = 1/W(z) \quad \text{if} \quad \tau > z.$$

Hence Theorem 6.2 in this case follows immediately from Theorem 6.1.

We may suppose now that $\tau \leqslant z$. Also, because $z \leqslant \xi$ by hypothesis, we may apply Lemma 4.1. Hence, by Theorem 6.1 in conjunction with Lemma 4.1,

$$S(\mathscr{A}_q; \mathfrak{P}, z) \leqslant \frac{\omega(q)}{q} XW(z) \left\{1 + O\left(\exp\left\{-\lambda\tau + \left(\frac{2}{\lambda} + \frac{1}{\log z}\right) Ae^{\lambda}\right\}\right)\right\}$$
$$+ \sum_{\substack{d < \xi^2 \\ d | P(z)}} 3^{\nu(d)} |R_{qd}| \qquad (\lambda > 0), \tag{2.2}$$

where we have taken $A = \max(\kappa, A_2)$.

We may assume also that

$$e^{3A+2} \leqslant \log \tau; \tag{2.3}$$

for if τ is bounded, the result follows at once from (2.2) on taking $\lambda = 1$ (and is then a result of the quality of Theorem 4.1).

We now choose

$$\lambda = \log \tau + \log\log \tau.$$

Then, by (2.3), we have, in the first remainder term on the right,

$$-\lambda\tau = -\tau\log\tau - \tau\log\log\tau \leqslant -\tau\log\tau - (3A + 2)\tau$$

and, since $\log z \geqslant \log \tau$,

$$\left(\frac{2}{\lambda} + \frac{1}{\log z}\right) Ae^{\lambda} = \left(\frac{2}{\log\tau + \log\log\tau} + \frac{1}{\log z}\right) A\tau\log\tau \leqslant 3A\tau;$$

it will be clear from adding these that (2.1) follows at once.

The proof shows that we could have replaced $\tau\log\tau + 2\tau$ by a function growing more rapidly with τ – the proof actually gives

$$\tau\log\tau + \tau\log\log\tau - 3A\tau$$

—but we retain the simpler form because it will be easier to use later when we come to apply Theorem 6.2.

3. The Function σ_κ

We have already pointed out that Theorem 6.2 gives new information only if

$$\frac{\log \xi}{\log z}$$

is large. In this and the next section we shall study $G(\xi, z)$ without making this assumption, but we shall impose the stronger condition $(\Omega_2(\kappa, L))$; our ultimate objective will be to prove Theorem 6.3 below.

Functions of type $G(\xi, z)$, sometimes referred to in the literature as "incomplete" sums, are known to require for their precise description certain associated functions which arise as solutions of differential-difference equations (cf. (3.2) below). In this section we shall introduce the particular function of this kind, $\sigma_\kappa(u)$, that is associated with $G(\xi, z)$ when ω satisfies the condition $(\Omega_2(\kappa, L))$; the nature of the connection between G and σ_κ will be apparent from Lemma 6.1, stated at the beginning of the next section.

We define $\sigma_\kappa(u)$ as follows: Let

$$\sigma_\kappa(u) = 2^{-\kappa} \frac{e^{-\gamma\kappa}}{\Gamma(\kappa + 1)} u^\kappa \quad \text{if} \quad 0 \leqslant u \leqslant 2 \tag{3.1}$$

and

$$(u^{-\kappa}\sigma_\kappa(u))' = -\kappa u^{-\kappa-1}\sigma_\kappa(u - 2), \; u > 2, \tag{3.2}$$

with σ_κ required to be continuous at $u = 2$.

It can be proved that $\sigma_\kappa(u)$ is non-negative, increasing with u, and that

$$\lim_{u \to \infty} \sigma_\kappa(u) = 1, \tag{3.3}$$

so that we have that

$$0 \leqslant \sigma_\kappa(u) \leqslant 1, \qquad u \geqslant 0, \tag{3.4}$$

$$\sigma_\kappa'(u) > 0, \qquad u > 0, \tag{3.5}$$

and in particular that

$$\sigma_\kappa(u) \geqslant \sigma_\kappa(2) = \frac{e^{-\gamma\kappa}}{\Gamma(\kappa + 1)} \quad \text{if} \quad u \geqslant 2. \tag{3.6}$$

Rewriting (3.2), we have also that

$$u\sigma_\kappa'(u) = \kappa\{\sigma_\kappa(u) - \sigma_\kappa(u - 2)\}, \qquad u > 2. \tag{3.7}$$

Next, we need an upper bound for $\sigma_\kappa'(u)/\sigma_\kappa(u)$; using (3.1) for the range $0 < u \leqslant 2$, and (3.7) with (3.4) for $u > 2$, we obtain

$$\frac{\sigma_\kappa'(u)}{\sigma_\kappa(u)} \leqslant \frac{\kappa}{u}, \qquad u > 0. \tag{3.8}$$

Hence if u_1, u_2 are any two numbers satisfying $0 < \delta_1 \leqslant u_1 < u_2$, and u is a suitable number in the interval $u_1 < u < u_2$, we deduce from the mean value theorem that

$$\frac{1/\sigma_\kappa(u_1) - 1/\sigma_\kappa(u_2)}{u_2 - u_1} = \frac{\sigma_\kappa'(u)}{\sigma_\kappa^2(u)} \leqslant \frac{\kappa}{u\sigma_\kappa(u)},$$

and hence, using (3.5), that

$$(0 <)\frac{1}{\sigma_\kappa(u_1)} - \frac{1}{\sigma_\kappa(u_2)} \leqslant \frac{u_2 - u_1}{u_1}\frac{\kappa}{\sigma_\kappa(\delta_1)} \quad \text{for} \quad 0 < \delta_1 \leqslant u_1 < u_2. \tag{3.9}$$

With $\kappa > 0$, $\lambda > 0$ and $u > 0$, let us now consider the integral

$$I_{\kappa,\lambda}(u) := \int_0^u \sigma_\kappa'(t)\,\sigma_\lambda(u-t)\,dt.$$

We begin by remarking that $I_{\kappa,\lambda}(u)$ is symmetrical in κ and λ; for if we replace t by $u - t$ and integrate by parts we obtain, using that $\sigma_\kappa(0) = 0$, that

$$\begin{aligned} I_{\kappa,\lambda}(u) &= \int_0^u \sigma_\lambda(t)\sigma_\kappa'(u-t)\,dt = \int_0^u \sigma_\lambda'(t)\,\sigma_\kappa(u-t)\,dt \\ &= I_{\lambda,\kappa}(u). \end{aligned} \tag{3.10}$$

Our objective is to prove that $I_{\kappa,\lambda}(u) = \sigma_{\kappa+\lambda}(u)$ in the specified range (cf. (3.13) below). Let us suppose first of all that $u > 2$. Then we have

$$\begin{aligned} uI_{\kappa,\lambda}'(u) &= \int_0^u (t + u - t)\,\sigma_\kappa'(t)\,\sigma_\lambda'(u-t)\,dt \\ &= \int_0^u t\sigma_\kappa'(t)\,\sigma_\lambda'(u-t)\,dt + \int_0^u t\sigma_\lambda'(t)\,\sigma_\kappa'(u-t)\,dt; \end{aligned} \tag{3.11}$$

and we treat the two integrals on the right separately. For the first integral we obtain, by (3.7) and (3.1),

$$\begin{aligned} &\int_0^u t\sigma_\kappa'(t)\,\sigma_\lambda'(u-t)\,dt \\ &= \kappa\int_2^u \{\sigma_\kappa(t) - \sigma_\kappa(t-2)\}\,\sigma_\lambda'(u-t)\,dt + \kappa\int_0^2 \sigma_\kappa(t)\,\sigma_\lambda'(u-t)\,dt \\ &= \kappa\int_0^u \sigma_\kappa(t)\,\sigma_\lambda'(u-t)\,dt - \kappa\int_0^{u-2} \sigma_\kappa(t)\,\sigma_\lambda'(u-2-t)\,dt \\ &= \kappa\,\{I_{\kappa,\lambda}(u) - I_{\kappa,\lambda}(u-2)\} \end{aligned} \tag{3.12}$$

using (3.10) in the final step. The second integral on the right of (3.11) differs from the first only in having κ and λ interchanged, and therefore, by (3.12), (3.10) and (3.11) we conclude that

$$uI'_{\kappa,\lambda}(u) = (\kappa + \lambda)\{I_{\kappa,\lambda}(u) - I_{\kappa,\lambda}(u-2)\}.$$

If we compare this relation with (3.7) we observe that $I_{\kappa,\lambda}(u)$ satisfies, for $u > 2$, the same differential-difference equation as $\sigma_{\kappa+\lambda}(u)$. Suppose next that $0 < u \leqslant 2$. Here we apply (3.1) in the definition of $I_{\kappa,\lambda}(u)$, to obtain

$$\begin{aligned} I_{\kappa,\lambda}(u) &= \frac{(2e^{\gamma})^{-(\kappa+\lambda)}}{\Gamma(\kappa+1)\Gamma(\lambda+1)}\kappa \int_0^u t^{\kappa-1}(u-t)^{\lambda}\,dt \\ &= \frac{(2e^{\gamma})^{-(\kappa+\lambda)}}{\Gamma(\kappa)\Gamma(\lambda+1)} u^{\kappa+\lambda} \int_0^1 v^{\kappa-1}(1-v)^{\lambda}\,dv \\ &= \frac{(2e^{\gamma})^{-(\kappa+\lambda)}}{\Gamma(\kappa+\lambda+1)} u^{\kappa+\lambda} = \sigma_{\kappa+\lambda}(u). \end{aligned}$$

Hence we have proved what we set out to prove, namely that

$$\sigma_{\kappa+\lambda}(u) = \int_0^u \sigma_{\kappa}'(t)\,\sigma_{\lambda}(u-t)\,dt \quad \text{for} \quad \kappa > 0,\ \lambda > 0,\ u > 0. \tag{3.13}$$

From (3.13) we derive at once, in view of (3.4), the inequality

$$\sigma_{\kappa+\lambda}(u) \leqslant \sigma_{\kappa}(u) \quad \text{for} \quad \kappa > 0,\ \lambda > 0,\ u > 0, \tag{3.14}$$

from which we infer that for each fixed u, *$\sigma_{\kappa}(u)$ is a decreasing function of κ.*

We next introduce also the function

$$\bar{\sigma}_{\kappa}(u) = \int_0^u \sigma_{\kappa}(t)\,dt; \tag{3.15}$$

clearly $\bar{\sigma}_{\kappa}(u)$ too is non-negative, increasing and, from (3.1),

$$\bar{\sigma}_{\kappa}(u) = \frac{2^{-\kappa}e^{-\gamma\kappa}}{\Gamma(\kappa+2)} u^{\kappa+1} \quad \text{if} \quad 0 \leqslant u \leqslant 2. \tag{3.16}$$

Moreover, $\bar{\sigma}_{\kappa}(u)$ satisfies a differential-difference equation. To see this, we first deduce from (3.7), by integration, that

$$\sigma_{\kappa}(u) = (\kappa+1)\frac{\bar{\sigma}_{\kappa}(u)}{u} - \kappa\frac{\bar{\sigma}_{\kappa}(u-2)}{u}, \qquad u > 2; \tag{3.17}$$

if we multiply (3.17) by $u^{-\kappa-1}$ and rearrange the terms suitably, we find that $\bar{\sigma}_\kappa$ satisfies (cf. (3.2))

$$(u^{-\kappa-1}\,\bar{\sigma}_\kappa(u))' = -\kappa u^{-\kappa-2}\,\bar{\sigma}_\kappa(u-2), \qquad u > 2. \tag{3.18}$$

Integrating (3.18), we deduce that

$$\frac{\bar{\sigma}_\kappa(2\tau)}{\tau^{k+1}} = \frac{\bar{\sigma}_\kappa(2u)}{u^{\kappa+1}} - \kappa \int_u^\tau \frac{\bar{\sigma}_\kappa(2t-2)}{t^{\kappa+2}}\,dt, \qquad 1 \leqslant u \leqslant \tau. \tag{3.19}$$

In conclusion, we mention that (3.13) gives, on integration, the convolution formula

$$\bar{\sigma}_{\kappa+\lambda}(u) = \int_0^u \sigma_\kappa(t)\,\sigma_\lambda(u-t)\,dt \quad \text{for} \quad \kappa > 0, \ \ \lambda > 0, \ \ u > 0. \tag{3.20}$$

4. Asymptotic Formula for $G(\xi, z)$

We shall devote this section to the proof of the following result.

Lemma 6.1. (Ω_1), $(\Omega_2(\kappa, L))$: *We have*

$$\frac{1}{G(\xi, z)} = W(z)\left\{\frac{1}{\sigma_\kappa(2\tau)} + O\left(\frac{L\tau^{2\kappa+1}}{\log z}\right)\right\} \quad \text{if} \quad z \leqslant \xi, \tag{4.1}$$

where

$$\tau = \frac{\log \xi}{\log z} \tag{4.2}$$

and σ_κ is the function introduced in the preceding section.

Remark 1. Inspection of the error term in (4.1) shows that Lemma 6.1 becomes rather weak when τ assumes large values. However, for these we have Lemma 4.1 which, whilst it gives only an upper bound for $1/G(\xi, z)$, does so under the weaker condition $(\Omega_2(\kappa))$ (or one could use (Ω_0)). Moreover, the upper estimate given by Lemma 4.1 is very good indeed if τ is large, and we saw in Section 2 how Theorem 6.2—where we are concerned with large τ only—was derived from it.

If we take $\lambda = 1$ in Lemma 4.1 we see (cf. a similar procedure at the beginning of the proof of Lemma 5.4), in view of (3.6), that we may henceforward assume that

$$L\tau^{2\kappa+1} \leqslant \frac{1}{B_9}\log z, \tag{4.3}$$

say, where B_9 is a sufficiently large constant.

Remark 2. There are various ways of proving Lemma 6.1. Most methods start from some identity satisfied by G, and proceed by induction on the range of τ; that is, one assumes that the result is true if $\nu - 1 < \tau \leqslant \nu$ and proves it then also for $\nu < \tau \leqslant \nu + 1$. For example, one method starts from the formula

$$G(\xi, z) = G(\xi, x) - \sum_{z \leqslant p < x} g(p)\, G(\xi/p, p),$$

valid for $2 \leqslant z \leqslant x$; and we shall use some features of this method in later chapters. Here, in the interests of variety and because the method is rather illuminating, we set out from the quite different formula (5.3.5) of Chapter 5.

Proof. Since

$$G(x, z) \leqslant G(x) \ll \log^{\kappa} x \tag{4.4}$$

(see (5.3.11)), formula (5.3.5) may be written in the form

$$G(x, z) \log x = (\kappa + 1) T(x, z) - \kappa T(x/z, z) + O(L \log^{\kappa} x) \tag{4.5}$$

where

$$T(x, z) = \int_1^x G(t, z)\, dt/t. \tag{4.6}$$

If we divide (4.5) throughout by $x \log^{\kappa+2} x$ and integrate with respect to x from w to ξ, we obtain

$$\int_w^{\xi} \frac{G(x, z)}{x \log^{\kappa+1} x}\, dx = (\kappa + 1) \int_w^{\xi} \frac{T(x, z)}{x \log^{\kappa+2} x}\, dx$$
$$- \kappa \int_w^{\xi} \frac{T(x/z, z)}{x \log^{\kappa+2} x}\, dx + O\left(\frac{L}{\log w}\right), \qquad 2 \leqslant w \leqslant \xi; \tag{4.7}$$

but by (4.6)

$$\frac{\partial}{\partial x}\left\{\frac{T(x, z)}{\log^{\kappa+1} x}\right\} = \frac{G(x, z)}{x \log^{\kappa+1} x} - (\kappa + 1) \frac{T(x, z)}{x \log^{\kappa+2} x},$$

and integrating we see that (4.7) can be written as

$$\frac{T(\xi, z)}{\log^{\kappa+1} \xi} = \frac{T(w, z)}{\log^{\kappa+1} w} - \kappa \int_w^{\xi} \frac{T(x/z, z)}{x \log^{\kappa+2} x}\, dx + O\left(\frac{L}{\log w}\right) \tag{4.8}$$
$$\text{if} \quad 2 \leqslant w \leqslant \xi.$$

We now put

$$T(\xi, z) = \frac{1}{2} c_4 \bar{\sigma}_\kappa(2\tau) \log^{\kappa+1} z + R(\xi, z), \qquad \tau = \frac{\log \xi}{\log z}, \tag{4.9}$$

where

$$c_4 = e^{\gamma\kappa} \prod_p \left(1 - \frac{\omega(p)}{p}\right)^{-1} \left(1 - \frac{1}{p}\right)^\kappa \tag{4.10}$$

(for this product, see Lemma 5.3).

If we introduce (4.9) into (4.8) and utilize (3.19) we find that the leading terms disappear throughout and we are left with a relation just between the remainder terms, namely

$$\frac{R(\xi, z)}{\log^{\kappa+1} \xi} = \frac{R(w, z)}{\log^{\kappa+1} w} - \kappa \int_w^\xi \frac{R(x/z, z)}{x \log^{\kappa+2} x} dx + O\left(\frac{L}{\log w}\right), \qquad 2 \leqslant w \leqslant \xi. \tag{4.11}$$

We shall deduce from (4.11), for all positive integers $\nu \geqslant 2$, that

$$\frac{|R(\xi, z)|}{\log^{\kappa+1} \xi} \leqslant \frac{B_{12} L}{\log z} (\nu - 1)^{\kappa+1} \quad \text{if} \quad \nu - 1 < \tau \leqslant \nu, \tag{4.12}$$

and this, together with (4.5), will prove the Lemma.

Let us begin by assuming that

$$z < \xi \leqslant z^2, \tag{4.13}$$

so that we are with the case $\nu = 2$ and therefore $1 < \tau \leqslant 2$ (we observe that if $\tau = 1$, i.e. $\xi = z$, (4.1) is, in view of (3.6), the same as (5.3.1)). For $t \leqslant z$ we have, using the notation (5.3.6), that

$$T(t, z) = T(t) = \frac{e^{-\gamma\kappa}}{\Gamma(\kappa + 2)} c_4 \log^{\kappa+1} t \left\{1 + O\left(\frac{L}{\log t}\right)\right\} \quad (t > 1);$$

this relation follows from (5.3.9) subject to condition (5.3.3) or otherwise, trivially, from (4.4). By (4.9), and using (3.16), we may express this more simply as

$$|R(t, z)| \leqslant B_{10} L \log^\kappa t \quad \text{if} \quad 1 < t \leqslant z. \tag{4.14}$$

We now choose $w = z$ in (4.11) and note that then the conditions allow us to use (4.14) (with $t = w = z$ and $t = x/z$) on the right of (4.11). We obtain

$$\frac{|R(\xi, z)|}{\log^{\kappa+1} \xi} \leqslant B_{10}L \left\{\frac{1}{\log z} + \kappa \int_z^\xi \frac{\log^\kappa(x/z)}{x \log^{\kappa+2} x} dx + \frac{B_{11}}{B_{10}} \frac{1}{\log z}\right\}$$

where B_{11} is the O-constant in the error term on the right of (4.11). Putting $x = z^t$, the integral on the right is equal to

$$\frac{1}{\log z}\int_1^\tau \frac{(t-1)^\kappa}{t^{\kappa+2}} dt \leqslant \frac{1}{\log z},$$

and taking $B_{10} \geqslant B_{11}$, we see that (4.12) is satisfied if $v = 2$ with

$$B_{12} = B_{10}(\kappa + 2) \ \ (\geqslant B_{11}). \tag{4.15}$$

Now let r be a positive integer $\geqslant 2$. We assume that (4.12) has already been proved for $v \leqslant r$, and we consider the interval

$$z^r < \xi \leqslant z^{r+1},$$

so that $r < \tau \leqslant r + 1$. Then choosing $w = z^r$ in (4.11) we obtain by (4.12) (with $v = r$ – the induction hypothesis)

$$\frac{|R(\xi, z)|}{\log^{\kappa+1} \xi} \leqslant \frac{B_{12}\, L}{\log z} \left\{(r-1)^{\kappa+1} + \kappa(r-1)^{\kappa+1} \int_{z^r}^\xi \frac{\log^{\kappa+1}(x/z)}{x \log^{\kappa+2} x} dx + \frac{B_{11}}{B_{12}} \cdot \frac{1}{r}\right\}.$$

Putting $x = z^t$ in the integral on the right, we see that it is equal to

$$\int_r^\tau \frac{(t-1)^{\kappa+1}}{t^{\kappa+2}} dt \leqslant \frac{1}{r}$$

—note that $\tau \leqslant r + 1$—and therefore we obtain, in view of (4.15), that

$$\frac{|R(\xi, z)|}{\log^{\kappa+1} \xi} \leqslant \frac{B_{12}L}{\log z} \left\{(r-1)^{\kappa+1} + \frac{\kappa(r-1)^{\kappa+1}}{r} + \frac{1}{r}\right\}.$$

If now we use the easily verifiable inequality

$$(r-1)^{\kappa+1}\left(1 + \frac{\kappa+1}{r}\right) \leqslant r^{\kappa+1},$$

(4.12) follows for $v = r + 1$, and the inductive argument has been completed.

We now write $\log \xi = \tau \log z$ in (4.12) and note that $v - 1 < \tau$ there; we then derive at once, from (4.9), that

$$T(\xi, z) = \tfrac{1}{2}c_4\bar{\sigma}_\kappa(2\tau) \log^{\kappa+1} z + O(L\tau^{2\kappa+2} \log^\kappa z) \quad \text{if} \quad \xi > z. \tag{4.16}$$

With the behaviour of T thus determined, we can return to G and complete the proof of the Lemma. A combination of (4.5), with $x = \xi$, and (4.16) yields, with the aid of (3.17),

$$\begin{aligned} G(\xi, z) &= (\kappa + 1)\, c_4 \frac{\bar{\sigma}_\kappa(2\tau)}{2\tau} \log^\kappa z - \kappa c_4 \frac{\bar{\sigma}_\kappa(2\tau - 2)}{2\tau} \log^\kappa z \\ &\quad + O(L\tau^{2\kappa+1} \log^{\kappa-1} z) \\ &= c_4 \sigma_\kappa(2\tau) \log^\kappa z + O(L\tau^{2\kappa+1} \log^{\kappa-1} z); \end{aligned}$$

and Lemma 5.3 and (3.6) allow us to write this as

$$G(\xi, z) = c_4 \sigma_\kappa(2\tau) \log^\kappa z \left\{1 + O\left(\frac{L\tau^{2\kappa+1}}{\log z}\right)\right\}.$$

By (4.10), (4.3) and (3.6) we therefore arrive finally at

$$\frac{1}{G(\xi, z)} = \prod_p \left(1 - \frac{\omega(p)}{p}\right)\left(1 - \frac{1}{p}\right)^{-\kappa} \frac{e^{-\gamma\kappa}}{\log^\kappa z} \left\{\frac{1}{\sigma_\kappa(2\tau)} + O\left(\frac{L\tau^{2\kappa+1}}{\log z}\right)\right\}, \tag{4.17}$$

and (4.1) follows at once from (4.17) in view of (5.2.5).

We pointed out in the proof of Lemma 6.1 that if

$$\tau = \frac{\log \xi}{\log z} = 1,$$

the asymptotic formula (4.1) coincides with (5.3.1) of Lemma 5.4. Let us now consider, under the *same* conditions (Ω_1), $(\Omega_2(\kappa, L))$, the complementary case $0 < \tau \leqslant 1$. Here, by (1.4.23), (5.3.1) and (5.2.4), we have

$$\begin{aligned} \frac{1}{G(\xi, z)} = \frac{1}{G(\xi)} &= W(\xi) e^{\gamma\kappa}\, \Gamma(\kappa + 1) \left\{1 + O\left(\frac{L}{\log \xi}\right)\right\} \\ &= W(z) e^{\gamma\kappa} \Gamma(\kappa + 1) \left(\frac{\log z}{\log \xi}\right)^\kappa \left\{1 + O\left(\frac{L}{\log \xi}\right)\right\}, \end{aligned}$$

and we deduce, in view of (3.1), that

(Ω_1), $(\Omega_2(\kappa, L))$:

$$\frac{1}{G(\xi, z)} = W(z) \left\{\frac{1}{\sigma_\kappa(2\tau)} + O\left(\frac{L\tau^{-\kappa-1}}{\log z}\right)\right\} \quad \text{if} \quad 0 < \tau \leqslant 1. \tag{4.18}$$

5. The Main Result

Under the condition $(\Omega_2(\kappa, L))$, which is stronger than the condition $(\Omega_2(\kappa))$ used in Theorem 6.2, we apply (4.18) and Lemma 6.1 in Theorem 6.1 and obtain immediately the following result.

THEOREM 6.3. (Ω_1), $(\Omega_2(\kappa, L))$: *If*

$$\tau = \frac{\log \xi}{\log z},$$

we have

$$S(\mathscr{A}_q; \mathfrak{P}, z) \leqslant \frac{\omega(q)}{q} XW(z) \left\{ \frac{1}{\sigma_\kappa(2\tau)} + O\left(\frac{L}{\log z} (\tau^{-\kappa-1} + \tau^{2\kappa+1}) \right) \right\}$$
$$+ \sum_{\substack{d < \xi^2 \\ d | P(z)}} 3^{\nu(d)} |R_{qd}| \quad \textit{for} \quad \tau > 0. \tag{5.1}$$

Notes

6.1. In his lectures, Selberg [3, 4] indicated several ways of constructing lower bound sieves. A. I. Vinogradov [2] was the first to work one of these out in detail, and the method in question requires the introduction of the parameter ξ. Halberstam–Roth [1] follow him in their sketch of an approach to the lower bound sieve, as do all research papers that are based on the Selberg method and deal with estimates from below of $S(\mathscr{A}; \mathfrak{P}, z)$ or of similar sieve functions.

The optimal nature of the λ_d's is explained in the Notes for Chapter 3.

Theorem 6.1: If we impose the condition $|R_d| \leqslant 1$ in (1.6) (weaker conditions would also permit a similar argument) the remainder term in (1.8) is easily seen to be (cf. Jurkat–Richert [1]) at most

$$\omega(q) \Psi^2 (\xi, z)$$

where

$$\Psi(x, z) = \sum_{\substack{d < x \\ p | d \Rightarrow p < z}} 1;$$

this differs from the usual definition only in that we have $d < x$ as opposed to $d \leqslant x$ and $p < z$ as distinct from $p \leqslant z$. This function has been extensively studied and we refer the reader to de Bruijn [4, 5]; in the first of these it is proved, for example, that

$$\Psi(x, z) < x \log^2 z \exp(-u \log u - u \log\log u + O(u)), \qquad u = \log x / \log z,$$

cf. Theorem 6.2 below.

6.3. This section is based on Ankeny–Onishi [1]. In their notation

$$\sigma_\kappa(u) = e^{-\gamma\kappa} J_\kappa(\tfrac{1}{2}u)/\Gamma(\kappa)$$

(cf. our (3.1)). This paper contains a careful study of J_κ (and so of σ_κ) and we quote from it, both here and later on, several results concerning these and related functions.
(3.13): implicit in Ankeny–Onishi [1].

6.4. $G(\xi, z)$ is an instance of a so-called "incomplete" average of the arithmetic function g. Incomplete averages of a similar kind have been studied e.g. by van Lint–Richert [1] ($g(n) = 1/\phi(n)$); Levin–Faĭnleĭb [2].

Lemma 6.1: based on Halberstam–Richert [1], cf. Levin–Faĭnleĭb [2].

Remark 2: The argument indicated here is to be found in Ankeny–Onishi [1], although not quite with the degree of uniformity required. For a complex variable approach see Vinogradov [2]; Levin [1, 5, 6].

6.5. Theorem 6.3: cf. Ankeny–Onishi [1], Theorem 1.

Chapter 7

Selberg's Sieve Method (continued): A First Lower Bound

1. Combinatorial Identities

In our exploration of the Selberg sieve we have been, up to now, concerned solely with *upper* estimates. However, when introducing Chapter 6 we indicated that the results of that chapter were primarily intended for use in the construction of a *lower bound sieve*; and we shall now embark on this task.

Our method in this and the next Chapter will be based on the systematic use of a class combinatorial identities and in this respect goes back to a theme we touched upon in Section 2.1. The simplest of these identities is contained in Lemma 7.1 below. Formula (1.1) (which arose in another way in Section 2.1—see (2.1.10)) is really a result about arrangements from mathematical logic, and sometimes appears in the literature under Sylvesters' name. In the present context, however, this identity will always be linked with the name of Buchstab, who was the first to apply it, most fruitfully, in sieve theory.

Lemma 7.1. *If*

$$2 \leqslant z_1 \leqslant z,$$

we have

$$S(\mathscr{A}_q; \mathfrak{P}, z) = S(\mathscr{A}_q; \mathfrak{P}, z_1) - \sum_{\substack{z_1 \leqslant p < z \\ p \in \mathfrak{P}}} S(\mathscr{A}_{qp}; \mathfrak{P}, p) \tag{1.1}$$

and

$$W(z) = W(z_1) - \sum_{z_1 \leqslant p < z} \frac{\omega(p)}{p} W(p). \tag{1.2}$$

Proof. The basic idea in (1.1) is to arrange the elements of $\mathscr{A}_q$ that are counted in $S(\mathscr{A}_q; \mathfrak{P}, z_1)$ but not in $S(\mathscr{A}_q; \mathfrak{P}, z)$ into groups, one group to each prime p of $\mathfrak{P}$ between z_1 (inclusive) and z (exclusive); so that the elements of $\mathscr{A}_q$ which are in the group associated with p have p as their least prime factor. A moment's reflection shows that the number of elements in the group of p is precisely $S(\mathscr{A}_{qp}; \mathfrak{P}, p)$.

We can put this more formally as follows, proving both results simultaneously. Let $p_1, p_2, \ldots$ be all the primes of $\mathfrak{P}$, written in their natural order, that are greater than or equal to z_1. If $z \leqslant p_1$ we have $P(z) = P(z_1)$, so that

$$S(\mathscr{A}_q; \mathfrak{P}, z) = S(\mathscr{A}_q; \mathfrak{P}, z_1), \quad W(z) = W(z_1) \quad \text{if} \quad z \leqslant p_1; \tag{1.3}$$

and in this case $(z_1 \leqslant z \leqslant p_1)$, Lemma 7.1 is trivially true.

Now suppose that $p_1 < z$; to fix ideas, suppose that $p_N < z \leqslant p_{N+1}$ $(N \geqslant 1)$. Then for each integer $\nu = 1, \ldots, N$ we have

$$\begin{aligned} S(\mathscr{A}_q; \mathfrak{P}, p_{\nu+1}) &- S(\mathscr{A}_q; \mathfrak{P}, p_\nu) \\ &= -\,|\{a\colon a \in \mathscr{A}_q,\ a \equiv 0 \bmod p_\nu,\ (a, P(p_\nu)) = 1\}| \\ &= -\,S(\mathscr{A}_{qp_\nu}; \mathfrak{P}, p_\nu), \end{aligned}$$

and

$$W(p_{\nu+1}) - W(p_\nu) = -\frac{\omega(p_\nu)}{p_\nu} W(p_\nu).$$

If now we sum these identities from $\nu = 1$ to $\nu = N$, using (1.3) and noting that $S(\mathscr{A}_q; \mathfrak{P}, z)$ and $W(z)$ are constant on each interval, we obtain both results of Lemma 7.1.

We might note in passing that if q satisfies (1.4.5), then each qp on the right of (1.1) satisfies a corresponding set of conditions, namely that qp is squarefree, $(qp, P(p)) = 1$ and $(qp, \bar{\mathfrak{P}}) = 1$.

It is easily seen, in principle, how one might derive a *lower* bound for $S(\mathscr{A}_q; \mathfrak{P}, z)$ from (1.1): if we have a lower estimate for $S(\mathscr{A}_q; \mathfrak{P}, z_1)$—we know, for example, that $S(\mathscr{A}_q; \mathfrak{P}, 2) = |\mathscr{A}_q|$, and we have an asymptotic formula for $S(\mathscr{A}_q; \mathfrak{P}, z_1)$ for example, from Theorem 1.1 if z_1 is small enough compared with X—and estimate each term $S(\mathscr{A}_{qp}; \mathfrak{P}, p)$, $z_1 \leqslant p < z$ and $p \in \mathfrak{P}$, from above, we are led to a lower estimate for $S(\mathscr{A}_q; \mathfrak{P}, z)$ which is non-trivial provided it is positive. Here, of course, is the rub; in order to achieve a positive lower bound in a given problem, we shall find that we are forced, by the deficiencies inherent in our method, to take z appreciably smaller than the problem demands.

We shall discuss the various difficulties as they arise; in the important, and at this stage of development by far the most interesting, class of problems that correspond to $\kappa = 1$—the "linear" class—and we shall even be able to present (in Chapter 8) a theory that is remarkably powerful and, in a technical sense, indeed best possible.

2. An Asymptotic Formula for S

We shall begin by deriving a new form of the Fundamental Lemma (cf. Theorems 2.5 and 2.5′); for we shall be able to prove, by means of Lemma 7.1 (taking $z_1 = 2$) and Theorem 6.2, a *lower* bound for $S(\mathscr{A}_q; \mathfrak{P}, z)$ which is only slightly weaker (in the error term) than Theorem 6.2; it will lead, in conjunction with Theorem 6.2, to an asymptotic formula ((2.1) below) for $S(\mathscr{A}_p; \mathfrak{P}, z)$ which, though weaker in one respect than Theorems 2.5 and 2.5′, is valid under general conditions that are appropriate for subsequent applications in our unified account of the Selberg theory.

Theorem 7.1. (Ω_1), $(\Omega_2(\kappa))$: *If*

$$\xi \geqslant z,$$

we have that

$$S(\mathscr{A}_q; \mathfrak{P}, z) = \frac{\omega(q)}{q} XW(z)\{1 + O(\exp\{-\tau(\log\tau + 1)\})\} + \vartheta \sum_{\substack{d<\xi^2 \\ d|P(z)}} 3^{\nu(d)}|R_{qd}|, \tag{2.1}$$

where

$$\tau = \frac{\log\xi}{\log z}$$

and $|\vartheta| \leqslant 1$.

Remark. We note by reference to Lemma 2.2 that Theorem 7.1 remains all the more true if condition $(\Omega_2(\kappa))$ is replaced by (Ω_0).

Proof. The fact that the expression on the right of (2.1) is an upper bound for $S(\mathscr{A}_q; \mathfrak{P}, z)$, follows at once from Theorem 6.2. We have to show, therefore, that it is also an estimate from below.

We begin with the help of (1.1) and (1.2) (both with $z_1 = 2$) by forming the relation

$$S(\mathscr{A}_q; \mathfrak{P}, z) - \frac{\omega(q)}{q} XW(z) = S(\mathscr{A}_q; \mathfrak{P}, 2) - \frac{\omega(q)}{q} XW(2) - \sum_{\substack{p<z \\ p\in\mathfrak{P}}} \left\{S(\mathscr{A}_{qp}; \mathfrak{P}, p) - \frac{\omega(qp)}{qp} XW(p)\right\}.$$

We apply Theorem 6.2 to each term in the sum on the right, choosing $\xi p^{-1/2}$ instead of ξ (and p instead of z, of course) for the term corresponding to p, and writing

$$\tau_y = \frac{\log(\xi y^{-\frac{1}{2}})}{\log y} \quad (y > 1).$$

We obtain

$$S(\mathscr{A}_q; \mathfrak{P}, z) - \frac{\omega(q)}{q} X W(z)$$

$$\geqslant -|R_q| - \sum_{\substack{p<z \\ p \in \mathfrak{P}}} \left\{ \frac{\omega(q)}{q} X \frac{\omega(p)}{p} W(p)\, O(\exp\{-\tau_p(\log \tau_p + 2)\}) \right.$$

$$\left. + \sum_{\substack{d<\xi^2/p \\ d|P(p)}} 3^{\nu(d)} |R_{qpd}| \right\}; \tag{2.2}$$

the condition "$\xi \geqslant z$" of Theorem 6.2 is satisfied for each term if

$$\tau_z = \tau - \tfrac{1}{2} \geqslant 1. \tag{2.3}$$

Let us take first the terms on the right of (2.2) involving the remainders R_{qm}; these, when combined, satisfy

$$|R_q| + \sum_{\substack{p<z \\ p \in \mathfrak{P}}} \sum_{\substack{d<\xi^2/p \\ d|P(p)}} 3^{\nu(d)} |R_{qpd}| \leqslant \sum_{\substack{n<\xi^2 \\ n|P(z)}} 3^{\nu(n)} |R_{qn}|. \tag{2.4}$$

The remaining sum on the right of (2.2), call it Σ_0, satisfies

$$\Sigma_0 \ll \frac{\omega(q)}{q} X W(z) \sum_{p<z} \frac{\omega(p)}{p} \frac{W(p)}{W(z)} \exp\{-\tau_p(\log \tau_p + 2)\};$$

and since

$$\frac{W(p)}{W(z)} \ll \left(\frac{\log z}{\log p}\right)^{\kappa} = \frac{\log p}{\log z} \left(\frac{\log z}{\log p}\right)^{\kappa+1},$$

we obtain the inequality

$$\Sigma_0 \ll \frac{\omega(q)}{q} X W(z) \frac{1}{\log z} \sum_{p<z} \frac{\omega(p)}{p} \log p$$

$$\times \exp\left\{ -\tau_p(\log \tau_p + 2) + (\kappa + 1) \log \frac{\log z}{\log p} \right\}. \tag{2.5}$$

We pause to study the exponential function on the right. Since

$$\frac{\partial \tau_y}{\partial y} = -\frac{\tau_y + \frac{1}{2}}{y \log y},$$

we have

$$\frac{\partial}{\partial y}\left\{-\tau_y(\log \tau_y + 2) + (\kappa + 1)\log\frac{\log z}{\log y}\right\}$$

$$= \frac{1}{y\log y}\{(\tau_y + \tfrac{1}{2})(\log \tau_y + 3) - \kappa - 1\}, \tag{2.6}$$

and if we impose the condition

$$\tau \geqslant \kappa + 2, \tag{2.7}$$

which includes (2.3), we see that the derivative (2.6) becomes positive. Hence the exponential term in (2.5) may be estimated from above by replacing p by z; and this is equivalent to replacing τ_p by $\tau - \frac{1}{2}$. Using $(\Omega_2(\kappa))$, we now derive from (2.5)

$$\Sigma_0 \ll \frac{\omega(q)}{q} XW(z)\exp\{-(\tau - \tfrac{1}{2})(\log(\tau - \tfrac{1}{2}) + 2)\};$$

and since, by the mean value theorem,

$$\begin{aligned} -(\tau - \tfrac{1}{2})(\log(\tau - \tfrac{1}{2}) + 2) &\leqslant -\tau(\log\tau + 2) + \tfrac{1}{2}(\log\tau + 3) \\ &\leqslant -\tau(\log\tau + 1) \quad \text{if} \quad \tau > 2, \end{aligned}$$

we obtain finally

$$\Sigma_0 \ll \frac{\omega(q)}{q} XW(z)\exp\{-\tau(\log\tau + 1)\}.$$

If we substitute from here and (2.4) in (2.2), we obtain the required lower bound counterpart of Theorem 6.2, at any rate subject to (2.7). But if (2.7) is not true, that is, if $\tau < \kappa + 2$, the lower bound part of (2.1) is trivially true since $S(\mathscr{A}_q; \mathfrak{P}, z) \geqslant 0$. This remark completes the proof of Theorem 7.1.

3. Fundamental Lemma

If now we add condition (R) to the other conditions of Theorem 7.1, we arrive at Theorem 7.2 below. This is clearly a fundamental lemma in the sense of Section 2.8, this time obtained by Selberg's method. Unexpectedly it is weaker than Theorem 2.5, but we present it here for the sake of completeness.

THEOREM 7.2. (Ω_1), $(\Omega_2(\kappa))$, (R): *If*

$$X \geqslant z$$

we have

$$S(\mathscr{A};\mathfrak{P},z) = XW(z)\{1 + O(e^{-\frac{1}{2}u\log u}) + O(e^{-\sqrt{(\log X)}})\} \tag{3.1}$$

and

$$S(\mathscr{A};\mathfrak{P},z) = XW(z)\{1 + O(e^{-\frac{1}{2}u})\} \tag{3.2}$$

where

$$u = \frac{\log X}{\log z}.$$

Proof. Our condition $X \geqslant z$ is equivalent to $u \geqslant 1$. Moreover, we may assume that (cf. (2.8.1))

$$u \geqslant B_{13} \tag{3.3}$$

where B_{13} is sufficiently large.

We shall apply Theorem 7.1 with $q = 1$ and

$$\xi = \frac{X^{(1-\lambda)/2}}{\log^{2\kappa} z}, \tag{3.4}$$

where

$$0 \leqslant \lambda \leqslant \tfrac{3}{4}.$$

Then

$$\tau = \frac{\log \xi}{\log z} = \frac{\frac{1}{2}(1-\lambda)\log X - 2\kappa\log\log z}{\log z} = \frac{1-\lambda}{2}u - 2\kappa\frac{\log\log z}{\log z},$$

and since

$$\frac{\log\log z}{\log z} \leqslant \frac{1}{e},$$

it follows from (3.3) that

$$\tau \geqslant \frac{u}{8} - \frac{2\kappa}{e} \geqslant 1,$$

so that the condition $\xi \geqslant z$ of Theorem 7.1 is satisfied.

By the mean value theorem we have also, with the aid of this same calculation, that

$$-\tau(\log\tau + 1) \leqslant -(1-\lambda)\frac{u}{2}\left(\log\left(\frac{(1-\lambda)u}{2}\right) + 1\right)$$

$$+\frac{2\kappa}{e}\left(\log\left(\frac{(1-\lambda)u}{2}\right) + 2\right). \tag{3.5}$$

Subject to our condition (R), the remainder term in Theorem 7.1 was estimated in (3.1.13) for $\xi = z$. Here, in the general case, we obtain by the same method

$$\sum_{\substack{d<\xi^2\\ d|P(z)}} 3^{\nu(d)}|R_d| \leqslant \xi^2 \sum_{d|P(z)} \frac{3^{\nu(d)}\omega(d)}{d}$$

$$\leqslant \xi^2 \prod_{p<z}\left(1+\frac{\omega(p)}{p}\right)^3 \leqslant \frac{\xi^2}{W^3(z)},$$

so that, by (2.3.6) and (3.4),

$$\sum_{\substack{d<\xi^2\\ d|P(z)}} 3^{\nu(d)}|R_d| \ll \xi^2 W(z)\log^{4\kappa} z \ll XW(z)\exp\{-\lambda\log X\}. \tag{3.6}$$

We now choose

$$\lambda = \frac{1}{\sqrt{\log X}} = \frac{1}{\sqrt{u\log z}} \leqslant \frac{1}{\sqrt{\log 2}}\frac{1}{\sqrt{u}}.$$

Then, by (3.3),

$$\log\frac{1-\lambda}{2} + 1 > \tfrac{1}{4},$$

say, whence (3.5) yields

$$-\tau(\log\tau + 1) \leqslant -\left(\frac{u}{2} + O(u^{\frac{1}{2}})\right)(\log u + \tfrac{1}{4}) + O(\log u)$$

$$\leqslant -\frac{u}{2}\log u,$$

and this together with (3.6) (with $\lambda = (\log X)^{-\frac{1}{2}}$), when substituted in Theorem 7.1, proves (3.1).

Turning to (3.2), we observe that, because $\lambda \leqslant \frac{3}{4}$, the right-hand side of (3.5) is always less than $-\frac{1}{2}u$ for all sufficiently large values of u. In particular, if we now choose $\lambda = 1/\log 4$ (which is $< \frac{3}{4}$) we have, in (3.6),

$$\lambda \log X \geqslant \tfrac{1}{2}\frac{\log X}{\log z} = \tfrac{1}{2}u,$$

and therefore Theorem 7.1 also implies (3.2).

4. The Function η_κ

Theorems 7.1 and 7.2 are significant only if z is small compared with X, that is, only if

$$u = \frac{\log X}{\log z}$$

is large. We shall now replace $(\Omega_2(\kappa))$ by the stronger condition $(\Omega_2\,(\kappa, L))$ and derive significant lower estimates for finite u also; it will no longer be possible to obtain results with (asymptotic) equality, but Theorem 7.3 is a lower bound analogue of the upper estimate in Theorem 6.3.

In this section we prepare the ground by introducing an auxiliary function required in the succeeding sections. This function, $\eta_\kappa(u)$, will be defined in terms of the function $\sigma_\kappa(u)$ of Section 6.3.

Let

$$\eta_\kappa(u) = \kappa u^{-\kappa}\int_u^\infty t^{\kappa-1}\left(\frac{1}{\sigma_\kappa(t-1)} - 1\right)dt, \qquad u > 1. \tag{4.1}$$

It follows from (6.3.4) that

$$\eta_\kappa(u) \geqslant 0, \qquad u > 1, \tag{4.2}$$

and that

$$(u^\kappa\eta_\kappa(u))' = -\kappa u^{\kappa-1}\left(\frac{1}{\sigma_\kappa(u-1)} - 1\right) \leqslant 0, \qquad u > 1, \tag{4.3}$$

and we may write (4.3) in the alternative form

$$(u^\kappa(1 - \eta_\kappa(u)))' = \frac{\kappa u^{\kappa-1}}{\sigma_\kappa(u-1)}, \qquad u > 1. \tag{4.4}$$

It is known that if $\kappa \geqslant 1$, the equation

$$\eta_\kappa(u) = 1$$

possesses a zero, which we shall denote by ν_κ; since $\eta_\kappa(u)$ is strictly decreasing, ν_κ is uniquely determined by

$$\eta_\kappa(\nu_\kappa) = 1, \tag{4.5}$$

and we have

$$\eta_\kappa(u) < 1 \quad \text{for} \quad u > \nu_\kappa. \tag{4.6}$$

The number ν_κ plays a critical part in the lower bound sieve. Ankeny and Onishi proved that

$$\nu_\kappa \leqslant 2(e-1)\kappa + 1 + 2\log\frac{e-1}{e-2} \quad \text{if} \quad \kappa \geqslant 1, \tag{4.7}$$

(a better inequality is described in the Notes) and also that

$$\lim_{\kappa\to\infty} \frac{\nu_\kappa}{\kappa} = 2.44\ldots; \tag{4.8}$$

in a table they give the values

$$\begin{array}{lll} \nu_1 = 2{\cdot}06\ldots, & \nu_{3/2} = 3{\cdot}22\ldots, & \nu_2 = 4{\cdot}42\ldots, \\ \nu_{5/2} = 5{\cdot}63\ldots, & \nu_3 = 6{\cdot}85\ldots, & \nu_{7/2} = 8{\cdot}09\ldots, \end{array} \tag{4.9}$$

and Porter has since extended this list as follows:

TABLE 1.

κ	4	5	6	7	8	9	10
ν_κ	9·32..	11·80..	14·28..	16·77..	19·25..	21·74..	24·22..

κ	11	12	13	14	15	16
ν_κ	26·70..	29·20..	31·68..	34·15..	36·62..	39·09..

Even from this limited evidence—and numerical evidence in this context is very troublesome to compile—it appears that ν_κ always exceeds 2 (if $\kappa \geqslant 1$) and that ν_κ increases with κ. We shall prove both these assertions. By (6.3.14) $\sigma_\kappa(u)$ is a decreasing function of κ and so, by (4.1), $\eta_\kappa(u)$ increases with κ; this in turn implies, in view of (4.5), that

$$\nu_\kappa \textit{ is an increasing function of } \kappa, \tag{4.10}$$

and since $\nu_1 = 2{\cdot}06\ldots$, (4.10) implies that

$$\nu_\kappa > 2 \quad \text{for} \quad \kappa \geqslant 1. \tag{4.11}$$

We conjecture even that ν_κ/κ increases with κ; (4.8) would then imply that

$$\nu_\kappa \leqslant 2{\cdot}44\ldots\kappa.$$

The numerical evidence above certainly supports this conjecture as Table 2 below shows:

TABLE 2

κ	1	2	3	4	5	6	7	8
ν_κ/κ	2·06..	2·21..	2·28..	2·33..	2·36..	2·38..	2·39..	2·40..

κ	9	10	11	12	13	14	15	16
ν_κ/κ	2·41..	2·42..	2·42..	2·43..	2·43..	2·43..	2·44..	2·44..

Finally, we shall need to have available the estimate (4.12) below. To this end we rewrite (4.3) in the form

$$\eta_\kappa'(u) = -\frac{\kappa}{u}\eta_\kappa(u) - \frac{\kappa}{u}\left(\frac{1}{\sigma_\kappa(u-1)} - 1\right), \qquad u > 1;$$

now η_κ and $1/\sigma_\kappa$ are decreasing functions of u, whence, if $1 < u_1 < u_2$,

$$\frac{\eta_\kappa(u_1) - \eta_\kappa(u_2)}{u_2 - u_1} \leqslant -\eta_\kappa'(u_1),$$

or

$$\eta_\kappa(u_1) - \eta_\kappa(u_2) \leqslant (u_2 - u_1)\frac{\kappa}{u_1}\left\{\eta_\kappa(2) + \frac{1}{\sigma_\kappa(1)} - 1\right\}$$

$$\ll \frac{u_2 - u_1}{u_1} \quad \text{if} \quad 2 \leqslant u_1 \leqslant u_2. \tag{4.12}$$

5. A LOWER BOUND

For the most delicate part of the argument leading to our lower estimate we shall use the following general result.

LEMMA 7.2. (Ω_1), $(\Omega_2(\kappa, L))$: *Suppose that*

$$2 \leqslant z_1 \leqslant z_2 \leqslant \xi, \qquad z \geqslant z_2,$$

let $\psi(t)$ *be a non-negative, monotonic and continuous function for* $t \geqslant 1$, *and define*

$$M := \max_{z_1 \leqslant w \leqslant z_2} \psi\left(\frac{\log \xi^2/w}{\log w}\right). \tag{5.1}$$

Then

$$\sum_{z_1 \leqslant p < z_2} \frac{\omega(p)}{p} W(p)\, \psi\left(\frac{\log \xi^2/p}{\log p}\right)$$

$$= W(z)\kappa \left(\frac{\log z}{\log \xi^2}\right)^\kappa \int_{\log \xi^2/\log z_2}^{\log \xi^2/\log z_1} t^{\kappa-1}\psi(t-1)\,dt + O\left(\frac{LMW(z)\log^\kappa z}{\log^{\kappa+1} z_1}\right). \tag{5.2}$$

Proof. By (1.2) and (5.2.4) we have

$$D(w) := \frac{1}{W(z)} \sum_{z_1 \leqslant p < w} \frac{\omega(p)}{p} W(p) = \frac{W(z_1)}{W(z)} - \frac{W(w)}{W(z)}$$

$$= \log^\kappa z \left(\frac{1}{\log^\kappa z_1} - \frac{1}{\log^\kappa w}\right) + O\left(\frac{L \log^\kappa z}{\log^{\kappa+1} z_1}\right) \quad \text{for} \quad z_1 \leqslant w \leqslant z_2. \tag{5.3}$$

Let

$$E(w) = \psi\left(\frac{\log \xi^2/w}{\log w}\right) \quad \text{if} \quad z_1 \leqslant w \leqslant z_2.$$

Since $z_2 \leqslant \xi$ the argument of ψ here is at least 1. Hence $E(w)$ is non-negative, monotonic and continuous in $z_1 \leqslant w \leqslant z_2$. It follows from (5.3) that

$$\frac{1}{W(z)} \sum_{z_1 \leqslant p < z_2} \frac{\omega(p)}{p} W(p)\, \psi\left(\frac{\log \xi^2/p}{\log p}\right) = D(z_2)\,E(z_2) - \int_{z_1}^{z_2} D(w)\,dE(w)$$

$$= \log^\kappa z \left(\frac{1}{\log^\kappa z_1} - \frac{1}{\log^\kappa z_2}\right) E(z_2) + O\left(\frac{LM \log^\kappa z}{\log^{\kappa+1} z_1}\right)$$

$$- \log^\kappa z \int_{z_1}^{z_2} \left(\frac{1}{\log^\kappa z_1} - \frac{1}{\log^\kappa w}\right) dE(w)$$

$$= \kappa \log^\kappa z \int_{z_1}^{z_2} \frac{E(w)}{\log^{\kappa+1} w} \frac{dw}{w} + O\left(\frac{LM \log^\kappa z}{\log^{\kappa+1} z_1}\right),$$

and if we now change the variable of integration from w to t given by $w = (\xi^2)^{1/t}$, we find that the integral becomes

$$\frac{1}{(\log \xi^2)^\kappa} \int_{\log \xi^2/\log z_1}^{\log \xi^2/\log z_1} t^{\kappa-1} \psi(t-1)\, dt;$$

this completes the proof of (5.2).

We are now able to derive

THEOREM 7.3. (Ω_1), $(\Omega_2\ (\kappa, L))$: *Suppose that* $z \leqslant \xi$, *and put*

$$\tau = \frac{\log \xi}{\log z}.$$

Then we have

$$S(\mathscr{A}_q; \mathfrak{P}, z) \geqslant \frac{\omega(q)}{q} X W(z) \left\{1 - \eta_\kappa(2\tau) + O\left(\frac{L(\log\log 3\xi)^{3\kappa+2}}{\log \xi}\right)\right\}$$
$$- \sum_{\substack{d<\xi^2 \\ d|P(z)}} 3^{\nu(d)} |R_{qd}|. \tag{5.4}$$

Remark. Following our remarks in the previous section about the zero ν_κ of the equation $\eta_\kappa(u) = 1$, it becomes clear that our result is trivial unless

$$2\tau > \nu_\kappa, \tag{5.5}$$

since $S(\mathscr{A}_q; \mathfrak{P}, z)$ is always non-negative. The latter implies also that we lose nothing if we assume that

$$\xi \geqslant B_{14}, \tag{5.6}$$

where B_{14} is a sufficiently large constant.

Proof. We take

$$z_1 = \exp\left\{\frac{\log \xi}{\log\log \xi}\right\}, \tag{5.7}$$

and we may assume that

$$2 \leqslant z_1 \leqslant z \leqslant \xi; \tag{5.8}$$

for if z were less than z_1 it would follow that $\tau > \log\log \xi$, and Theorem 7.1 would then give a much stronger result than (5.4).

As in the proof of Theorem 7.1 we take as our starting point the identity

$$S(\mathscr{A}_q; \mathfrak{P}, z) = S(\mathscr{A}_q; \mathfrak{P}, z_1) - \sum_{\substack{z_1 \leqslant p < z \\ p \in \mathfrak{P}}} S(\mathscr{A}_{qp}; \mathfrak{P}, p) \tag{5.9}$$

given in (1.1). This time the choice $z_1 = 2$, used in the proof of Theorem 7.1, does not suffice, for it would spoil the leading term. For the term on the right of (5.9) we use Theorem 7.1 with z_1 in place of z, and we obtain, using only a weak form of the O-term: $\exp(-\tau)$ in place of $\exp\{-\tau(\log\tau + 1)\}$,

$$S(\mathscr{A}_q; \mathfrak{P}, z_1) \geqslant \frac{\omega(q)}{q} XW(z_1)\{1 + O(e^{-\tau'})\}$$

$$- \sum_{\substack{d < \xi^2 \\ d \mid P(z_1)}} 3^{\nu(d)} |R_{qd}|, \quad \tau' = \frac{\log\xi}{\log z_1},$$

where, by (5.7),

$$\tau' = \frac{\log\xi}{\log z_1} = \log\log\xi. \tag{5.10}$$

Using this, in conjunction with (5.2.4) and the fact that $z \leqslant \xi$, we see that

$$\frac{W(z_1)}{W(z)}\{1 + O(e^{-\tau'})\} = \left(\frac{\log z}{\log z_1}\right)^{\kappa} \left\{1 + O\left(\frac{L}{\log z_1}\right)\right\} \left\{1 + O\left(\frac{1}{\log\xi}\right)\right\}$$

$$= \left(\frac{\log z}{\log z_1}\right)^{\kappa} + O\left(\frac{L(\log\log\xi)^{\kappa+1}}{\log\xi}\right);$$

and it follows therefore that

$$S(\mathscr{A}_q; \mathfrak{P}, z_1) \geqslant \frac{\omega(q)}{q} XW(z) \left\{\left(\frac{\log z}{\log z_1}\right)^{\kappa}\right.$$

$$\left. + O\left(\frac{L(\log\log\xi)^{\kappa+1}}{\log\xi}\right)\right\} - \sum_{\substack{d < \xi^2 \\ d \mid P(z_1)}} 3^{\nu(d)} |R_{qd}|. \tag{5.11}$$

The sum on the right of (5.9) has to be estimated from above, term by term. To this end we apply Theorem 6.3 to each sequence $\mathscr{A}_{qp}$, with $\xi p^{-\frac{1}{2}}$ instead of ξ and p in place of z; so that the τ of Theorem 6.3 becomes $\log(\xi p^{-\frac{1}{2}})/\log p$ in the case of $\mathscr{A}_{qp}$. Since $z \leqslant \xi$ and (by (5.10)) $\tau' = \log\log\xi$, we have

$$\frac{1}{2} \leqslant \tau - \frac{1}{2} \leqslant \frac{\log(\xi p^{-\frac{1}{2}})}{\log p} \leqslant \tau' - \frac{1}{2} \leqslant \log\log\xi. \tag{5.12}$$

Hence this multiple application of Theorem 6.3 yields

$$\sum_{\substack{z_1 \leqslant p < z \\ p \in \mathfrak{P}}} S(\mathscr{A}_{qp}; \mathfrak{P}, p)$$

$$\leqslant \frac{\omega(q)}{q} X \sum_{z_1 \leqslant p < z} \frac{\omega(p)}{p} W(p) \left\{ \frac{1}{\sigma_\kappa(\log(\xi^2/p)/\log p)} + O\left(\frac{L(\log\log\xi)^{2\kappa+1}}{\log p}\right)\right\}$$

$$+ \sum_{\substack{z_1 \leqslant p < z \\ p \in \mathfrak{P}}} \sum_{\substack{d < \xi^2/p \\ d | P(p)}} 3^{\nu(d)} |R_{qpd}|. \tag{5.13}$$

For the leading sum on the right we apply Lemma 7.2 with $z_2 = z$ and $\psi(t) = 1/\sigma_\kappa(t)$; the conditions of the Lemma are satisfied by virtue of the information in Section 6.3, and we note also that, by (5.12) and (6.3.1),

$$M \leqslant 1/\sigma_\kappa(1) = 2^\kappa e^{\gamma\kappa} \Gamma(\kappa + 1) = O(1). \tag{5.14}$$

Thus we obtain

$$\sum_{z_1 \leqslant p < z} \frac{\omega(p)}{p} \frac{W(p)}{\sigma_\kappa(\log(\xi^2/p)/\log p)}$$

$$= W(z) \left\{ \kappa \left(\frac{\log \xi^2}{\log z}\right)^{-\kappa} \int_{\log \xi^2/\log z}^{\log \xi^2/\log z_1} \frac{t^{\kappa-1}}{\sigma_\kappa(t-1)}\, dt + O\left(\frac{L \log^\kappa z}{\log^{\kappa+1} z_1}\right)\right\}. \tag{5.15}$$

The integral on the right is equal to

$$\int_{\log \xi^2/\log z}^{\log \xi^2/\log z_1} t^{\kappa-1} \left(\frac{1}{\sigma_\kappa(t-1)} - 1\right) dt + \frac{1}{\kappa}\left(\frac{\log \xi^2}{\log z_1}\right)^\kappa - \frac{1}{\kappa}\left(\frac{\log \xi^2}{\log z}\right)^\kappa$$

$$\leqslant \int_{\log \xi^2/\log z}^{\infty} t^{\kappa-1} \left(\frac{1}{\sigma_\kappa(t-1)} - 1\right) dt + \frac{1}{\kappa}\left(\frac{\log \xi^2}{\log z_1}\right)^\kappa - \frac{1}{\kappa}\left(\frac{\log \xi^2}{\log z}\right)^\kappa,$$

since, by (6.3.4), $\sigma_\kappa(u) \leqslant 1$ always. We apply this inequality in (5.15) and obtain, by (4.1) and (5.7), that

$$\sum_{z_1 \leqslant p < z} \frac{\omega(p)}{p} \frac{W(p)}{\sigma_\kappa(\log(\xi^2/p)/\log p)}$$

$$\leqslant W(z) \left\{ \eta_\kappa\left(\frac{\log \xi^2}{\log z}\right) + \left(\frac{\log z}{\log z_1}\right)^\kappa - 1 + O\left(\frac{L(\log\log\xi)^{\kappa+1}}{\log \xi}\right)\right\}.$$

The sum on the right of (5.13) deriving from the O-remainder terms is estimated by means of (1.2) and (2.3.5) which lead, in view of (5.7), to

$$\sum_{z_1 \leqslant p < z} \frac{\omega(p)}{p} W(p)\, O\left(\frac{L(\log\log\xi)^{2\kappa+1}}{\log p}\right) \ll \frac{L(\log\log\xi)^{2\kappa+1}}{\log z_1} W(z_1)$$

$$\ll W(z)\frac{L(\log\log\xi)^{2\kappa+1}}{\log z_1}\left(\frac{\log z}{\log z_1}\right)^{\kappa} \ll W(z)\frac{L(\log\log\xi)^{3\kappa+2}}{\log\xi}.$$

Substitution of the last two inequalities in (5.13) results in

$$\sum_{\substack{z_1 \leqslant p < z \\ p \in \mathfrak{P}}} S(\mathscr{A}_{qp}; \mathfrak{P}, p)$$

$$\leqslant \frac{\omega(q)}{q} X W(z)\left\{\eta_\kappa(2\tau) + \left(\frac{\log z}{\log z_1}\right)^{\kappa} - 1 + O\left(\frac{L(\log\log\xi)^{3\kappa+2}}{\log\xi}\right)\right\}$$

$$+ \sum_{\substack{z_1 \leqslant p < z \\ p \in \mathfrak{P}}} \sum_{\substack{d < \xi^2/p \\ d \mid P(p)}} 3^{\nu(d)} |R_{qpd}|. \tag{5.16}$$

If we now combine (5.9) with (5.11) and (5.16) we see that Theorem 7.3 follows at once from the remark (cf. (2.4) where we had $z_1 = 2$) that

$$\sum_{\substack{d < \xi^2 \\ d \mid P(z_1)}} 3^{\nu(d)} |R_{qd}| + \sum_{\substack{z_1 \leqslant p < z \\ p \in \mathfrak{P}}} \sum_{\substack{d < \xi^2/p \\ d \mid P(p)}} 3^{\nu(d)} |R_{qpd}| \leqslant \sum_{\substack{n < \xi^2 \\ n \mid P(z)}} 3^{\nu(d)} |R_{qn}|.$$

It is natural at this point to compare the estimate from above in Theorem 6.3 and the estimate from below in Theorem 7.3. Both results are valid under the same pair of conditions, and the error terms are essentially of the same quality. Comparison focuses therefore on the leading terms, and in particular on the functions $1/\sigma_\kappa(2\tau)$ in Theorem 6.3 and $1 - \eta_\kappa(2\tau)$ in Theorem 7.3. Both functions tend to 1 as $\tau \to \infty$, $1/\sigma_\kappa(2\tau)$ from above and $1 - \eta_\kappa(2\tau)$ from below (see sections 6.3 and 7.4); but they are rather far apart for "intermediate" and "small" values of τ. In this latter respect they differ from the corresponding functions F and f in the case $\kappa = 1$ (see Theorem 8.3 and also the final result, Theorem 8.4), where, as we shall see, we shall be able to improve considerably on the results of Theorems 6.3 and 7.3.

6. The Main Result

In order to derive from Theorem 7.3 a result which is in a form convenient for direct application, we require some information about the remainders R_d. Here we impose a new condition (to be called $(R(\kappa, \alpha))$) on the sum of the remainder terms (which is less restrictive than (R)):

Suppose there exist constants α,

$$0 < \alpha < 1,$$

$A_4(\geqslant 1)$ and $A_5(\geqslant 1)$ such that

$$(R(\kappa, \alpha)) \qquad \sum_{\substack{d < X^\alpha/(\log X)^{A_4} \\ (d, \overline{\mathfrak{P}}) = 1}} \mu^2(d)\, 3^{\nu(d)}\, |R_d| \leqslant A_5 \frac{X}{\log^{\kappa+1} X} \quad (X \geqslant 2).$$

Then the case $q = 1$ of Theorem 7.3 implies the following result.

THEOREM 7.4. (Ω_1), $(\Omega_2(\kappa, L))$, $(R(\kappa, \alpha))$: *If*

$$z^2 \leqslant \frac{X^\alpha}{(\log X)^{A_4}}, \tag{6.1}$$

we have

$$S(\mathscr{A}; \mathfrak{P}, z) \leqslant XW(z) \left\{1 - \eta_\kappa\left(\alpha \frac{\log X}{\log z}\right) - B_{15} \frac{L(\log\log 3X)^{3\kappa+2}}{\log X}\right\}. \tag{6.2}$$

Proof. If X is bounded the result is trivially true because $S(\mathscr{A}; \mathfrak{P}, z) \geqslant 0$ so that we may assume that

$$X \geqslant B_{16},$$

where B_{16} is sufficiently large. We choose

$$\xi^2 = \frac{X^\alpha}{(\log X)^{A_4}} \tag{6.3}$$

in Theorem 7.3 (with $q = 1$) so that, in view of (6.1), condition $z \leqslant \xi$ of that theorem is satisfied. By (6.3) and (4.12)

$$\eta_\kappa(2\tau) - \eta_\kappa\left(\frac{\log X^\alpha}{\log z}\right) \ll \frac{\log\log X}{\log z}\,\frac{1}{\tau} \ll \frac{\log\log X}{\log X}, \tag{6.4}$$

so that we may replace $\eta_\kappa(2\tau)$ on the right of (5.4) by $\eta_\kappa(\log X^\alpha/\log z)$; and if we apply $(R(\kappa, \alpha))$ and (2.3.6) to the remainder term we obtain

$$\sum_{\substack{d < \xi^2 \\ d | P(z)}} 3^{\nu(d)} |R_d| \ll \frac{X}{\log^{\kappa+1} X} \ll XW(z) \frac{1}{\log X}. \tag{6.5}$$

This completes the proof of Theorem 7.4.

One might raise here the question whether the results we have obtained so far (relative to lower estimates) can be further improved by carrying out a second and third iteration of the Buchstab identity (1.1). For general κ ($\kappa \neq 1$) this has not been done; to mention only one of the obstacles, the technical difficulties arising from the non-elementary function η_κ are formidable. When we come to deal with the special case $\kappa = 1$, however (in Chapters 8 and 9), we shall find that the manifold iteration of (1.1) is, after some initial complications, completely successful in producing results which are not only sharp but also easy to apply to a large diversity of problems.

Remark on Theorem 7.4. If we consider the sequence

$$\mathscr{A} = \{p + 2; p \leqslant x\}$$

(cf. Example 5, $X = li\,x$), choose $\mathfrak{P} = \mathfrak{P}_2$ and take

$$z^2 = X^{1/u} \tag{6.6}$$

as well as

$$u = 2{\cdot}1,$$

say, then (cf. (4.9)) Theorem 7.4 applies with $\kappa = 1$ and $\alpha = \frac{1}{2}$, and we may readily deduce, for some $\delta > 0$, that for $x \geqslant x_0$

$$|\{p \leqslant x\colon (p + 2, P(z)) = 1\}| \geqslant \delta \frac{x}{\log^2 x}.$$

For each of these p's we obtain, in view of (6.6) and $(p + 2,\ P(z)) = 1$, that the total number $\Omega(p + 2)$ of prime factors of $p + 2$ satisfies the relation

$$x + 2 \geqslant z^{\Omega(p+2)} = X^{\Omega(p+2)/4\cdot 2}.$$

Therefore *there exist infinitely many primes p such that p + 2 has at most four prime factors.*

Although this is a better result than we obtained in Chapter 2, we shall derive a still sharper result in Chapter 9, and an even better result is implicit in Chapter 11.

Notes

7.1. Lemma 7.1: for (1.1) see Buchstab [1], p. 1241; also Notes for Chapter 2. Sylvester's combinatorial identity is quoted in G. Birkhoff–S. MacLane, "A survey of Modern Algebra," Macmillan N. Y. (1953) on pp. 347–348. The simple identity (1.2) occurs e.g. in de Bruijn [2] (for the special case $V(z)$), Barban [5], Vinogradov [2], Ankeny–Onishi [1].

7.2. Theorem 7.1: This result, which is plainly of the fundamental lemma type, represents an important component of the lower bound sieves of this and the next chapter. Taking $q = 1$ and imposing a suitable remainder condition, we could derive Theorem 2.6′, though with a slightly worse error term.

7.3. Theorem 7.2: cf. the sharper Theorem 2.5, proved under precisely the same conditions. Thus it is not true to say, as has been asserted from time to time in the literature, that Selberg's sieve is always better than Brun's. For a result similar to Theorem 7.2, but less general, see Jurkat–Richert [1], Theorem 3. (See also Iwaniec [1], Theorem 3, which derives from Rosser's sieve; this method, which is of the Brun genre, leads to a fundamental lemma even sharper than Theorem 2.5.) Theorem 7.2 clearly implies a slightly weaker form of Theorem 2.6.

7.4. (4.1): η_κ is introduced in Ankeny–Onishi [1]; in terms of their notation $\eta_\kappa(u) = G_\kappa(\frac{1}{2}u)$. We quote several basic properties of η_κ from there: for example, that the equation $\eta_\kappa(u) = 1$ possesses a (unique) zero ν_κ ($2\zeta_\kappa$ in their notation), and that the numbers ν_κ satisfy (4.7), (4.8). Recently Hagedorn [1] has improved substantially on (4.7) by showing that

$$\nu_\kappa \leqslant 2\max\left(\kappa + 1, \frac{\kappa}{\delta}\log M(\delta)\right) + 1 + \frac{2}{\delta}\log\frac{\log M(\delta)}{\log M(\delta) - 1}$$

for any δ, $\delta_0 < \delta < 2$, where

$$M(\delta) = \left(\frac{e^\delta - 1}{\delta}\right)^2 \exp\{2 + (\delta - 2)\,\delta^{-1}(e^\delta - 1)\},$$

and δ_0 is the unique root of $M(\delta) = e$. In fact, $\delta_0 = 0{\cdot}79089\ldots$. This implies that

$$\nu_\kappa \leqslant \min(2.54\,\kappa + 11{\cdot}58,\ 2{\cdot}63\,\kappa + 5{\cdot}15) \quad \text{if} \quad \kappa > 4.$$

For Tables 1, 2 see Porter [2].

7.5. Theorem 7.3: prepares the ground for the main result (Theorem 7.4), and could be used, with Theorem 6.3, as a basis for further Buchstab iterations.

7.6. The condition $(R(\kappa,\alpha))$ is modelled on Bombieri's theorem in the form of Lemma 3.5. A similar general condition is postulated in Ankeny–Onishi [1]. For earlier instances see Levin [9], Lemma 4, also Levin [5, 10]; Levin–Maksudov [1].

Note that (cf. Lemma 3.5) A_5 may not be an effective constant, and the same will then be true of all B-constants that depend on A_5.

Note also that $(R(\kappa, \alpha))$ is trivially true with $\alpha = 1$ whenever (R) and (Ω_0) hold, by virtue of Lemma 3.4 (cf. also p. 257). The new remainder condition is introduced with those problems in mind where an appeal to Bombieri's theorem or to some deep analogue is necessary.

Theorem 7.4: cf. Ankeny–Onishi [1], Theorem 1. For earlier sieve estimates based on Selberg's method see Vinogradov [2]; Levin [5]; Mientka [1]. Further iterations of Buchstab type, on the basis of Theorems 6.3, 7.3 may lead to further improvements. This is certainly true for $\kappa = 1$, as we show in Chapter 8. For $\kappa > 1$, Porter (Ph.D. thesis, Nottingham 1973) has confirmed this for a second iteration; the improvement is very small indeed. His calculations check with current research of Diamond and Jurkat (unpublished) which suggests that an appropriate scheme of repeated iterations leads to more substantial improvements.

Note that (6.2) is non-trivial only if

$$1 - \eta_\kappa\left(\alpha \frac{\log X}{\log z}\right) > 0,$$

i.e. (cf. (4.6)) if $X^\alpha > z^{\nu_\kappa}$. For this reason the number ν_κ is called the *sieving limit* of this method for κ-dimensional problems (see Table 1).

Remark on Theorem 7.4: this result appears to have been proved for the first time by Wang [2], but on the basis of GRH (see also Vinogradov [1]); Barban [6, 8] seems to have been the first to give an unconditional proof.

Chapter 8

The Linear Sieve

1. The Method

In this chapter we shall concern ourselves exclusively with *linear* sieve theory—that is, with the theory relating to the class of sieve problems having $\kappa = 1$. To this class belongs any problem in which $\omega(p)$ is about 1, at least on average (cf. condition $(\Omega_2(1, L))$); in particular, any problem in which $\mathscr{A}$ is a sequence of values assumed by an irreducible polynomial, with integer or prime arguments from some interval or arithmetic progression or both; or a "binary" problem which can be *linearized* (in the sense of Section 5.5) to the first kind (or a problem in which $\mathscr{A}$ indicates a dimension higher than 1 but $\mathfrak{P}$ is so "thin" that $\mathscr{A}$ and $\mathfrak{P}$ nevertheless satisfy $(\Omega_2(1, L))$). It is clear that this extensive class contains all those famous problems of classical prime number theory which first inspired the search for an effective sieve; and it is encouraging that for precisely this class we are able to go far beyond Theorems 6.3 and 7.3, and, indeed, to establish some results which have about them an air of finality (see, for example, Theorem 8.4).

We shall base our method on a manifold iteration of the identities of Lemma 7.1. The reader who has studied Chapter 7 will, inevitably, have asked himself whether further iterations of these identities might not lead to improvements (and not only in the cases $\kappa = 1$); and we have pointed out that there are practical difficulties. There is large scope of choice for deciding to which functions in (7.1.1) to apply this formula again, and with which parameters. One can find certain criteria which help to narrow the choice, and Theorem 8.1 below has proved to be the most suitable so far. This theorem is true for any κ but, contrary perhaps to first impressions, appears to be effective for $\kappa = 1$ only.

We therefore prove now the following generalization of Lemma 7.1.

Theorem 8.1. *Suppose that*

$$2 \leqslant z_1 \leqslant z \leqslant \xi,$$

and let

$$\xi_j^2 = \frac{\xi^2}{p_1 \cdots p_j} \quad (j = 1, \ldots).$$

Then, for any natural number b, we have

$$S(\mathscr{A}_q; \mathfrak{P}, z) = S(\mathscr{A}_q; \mathfrak{P}, z_1) + \sum_{i=1}^{b-1} (-1)^i \sum_{\substack{z_1 \leqslant p_i < \ldots < p_1 < z \\ p_j < \xi_j, p_j \in \mathfrak{P}(j=1,\ldots,i)}} S(\mathscr{A}_{qp_1 \ldots p_i}; \mathfrak{P}, z_1)$$

$$+ (-1)^b \sum_{\substack{z_1 \leqslant p_b < \ldots \leqslant p_1 < z \\ p_j < \xi_j, p_j \in \mathfrak{P}(j=1,\ldots,b)}} S(\mathscr{A}_{qp_1 \ldots p_b}; \mathfrak{P}, p_b)$$

$$+ \sum_{i=1}^{b} (-1)^i \sum_{\substack{z_1 \leqslant p_i < \ldots < p_1 < z \\ p_j < \xi_j, p_j \in \mathfrak{P}(j=1,\ldots,i-1) \\ \xi_i \leqslant p_i < \xi_i^2, p_i \in \mathfrak{P}}} S(\mathscr{A}_{qp_1 \ldots p_i}; \mathfrak{P}, p_i). \tag{1.1}$$

Remark. Before we give the proof, which is not difficult, a few comments on the structure of this rather complicated identity may help. As was the case with the application of Lemma 7.1 in the proof of Theorem 7.3, our ultimate intention will be to choose z_1 so small that all the functions S in the first expression on the right, and also all those in the second sum (with the p_b in the "z" positions), come within the orbit of the very accurate Theorem 7.1 (of the "fundamental lemma" type). This leaves only the functions S of the third expression; and here, depending on the sign, we shall either estimate from above by Theorem 6.3 (with τ small) or from below by zero. It appears as if here we shall lose accuracy, especially when using the trivial estimates from below; but the conditions $\xi_i \leqslant p_i < \xi_i^2$ actually take us into a range (of $\log \xi_i / \log p_i$) where these estimates cannot be improved upon.

Proof. We proceed by induction on b. By (7.1.1) we have

$$S(\mathscr{A}_q; \mathfrak{P}, z) = S(\mathscr{A}_q; \mathfrak{P}, z_1) - \sum_{\substack{z_1 \leqslant p < z \\ p < \sqrt{(\xi^2/p)}, p \in \mathfrak{P}}} S(\mathscr{A}_{qp}; \mathfrak{P}, p)$$

$$- \sum_{\substack{z_1 \leqslant p < z \\ \sqrt{(\xi^2/p)} \leqslant p < \xi^2/p, p \in \mathfrak{P}}} S(\mathscr{A}_{qp}; \mathfrak{P}, p) \tag{1.2}$$

which closer scrutiny reveals to be (1.1) for $b = 1$.

Suppose now that (1.1) has already been proved for b. Then we apply (1.1) in the first sum of (1.2), taking

$$\mathscr{A}_{qp},\ \xi^2/p,\ p \quad \text{for} \quad \mathscr{A}_q,\ \xi^2,\ z$$

respectively. Furthermore, we replace p_j by p_{j+1} and p by p_1, which changes

${\xi_j}^2/p$ into ξ_{j+1}^2 and ξ^2/p becomes ${\xi_1}^2$. Thus

$$\sum_{\substack{z_1 \leqslant p < z \\ p < \sqrt{(\xi^2/p)},\, p \in \mathfrak{P}}} S(\mathscr{A}_{qp}; \mathfrak{P}, p) = \sum_{\substack{z_1 \leqslant p_1 < z \\ p_1 < \xi_1,\, p_1 \in \mathfrak{P}}} S(\mathscr{A}_{qp_1}; \mathfrak{P}, z_1)$$

$$+ \sum_{i=1}^{b-1} (-1)^i \sum_{\substack{z_1 \leqslant p_{i+1} < \ldots < p_1 < z \\ p_j < \xi_j,\, p_j \in \mathfrak{P} (j = 1, \ldots, i+1)}} S(\mathscr{A}_{qp_1 \ldots p_{i+1}}; \mathfrak{P}, z_1)$$

$$+ (-1)^b \sum_{\substack{z_1 \leqslant p_{b+1} < \ldots < p_1 < z \\ p_j < \xi_j,\, p_j \in \mathfrak{P} (j = 1, \ldots, b+1)}} S(\mathscr{A}_{qp_1 \ldots p_{b+1}}; \mathfrak{P}, p_{b+1})$$

$$+ \sum_{i=1}^{b} (-1)^i \sum_{\substack{z_1 \leqslant p_{i+1} < \ldots < p_1 < z \\ p_j < \xi_j,\, p_j \in \mathfrak{P} (j = 1, \ldots, i) \\ \xi_{i+1} \leqslant p_{i+1} < \xi_{i+1}^2,\, p_{i+1} \in \mathfrak{P}}} S(\mathscr{A}_{qp_1 \ldots p_{i+1}}; \mathfrak{P}, p_{i+1}).$$

Using this in (1.2), we obtain (1.1) for $b + 1$.

2. The Functions F, f

It should be clear from the remarks following the statement of Theorem 8.1 how we plan to proceed; but our experience in Theorem 7.3 suggests that to find the appropriate estimating functions may be very complicated. Here we are in for a pleasant surprise: the two functions F and f, which we are about to describe, combine with the second identity of Lemma 7.1 to form, after iteration, an approximate identity matching (1.1) term by term (see Theorem 8.2 below); and after subtracting one identity from the other there remains only a computation with error terms (cf. (6.4.11), where a similar procedure was used, in a much simpler context).

We introduce the functions w and ρ, both solutions of differential-difference equations, by

$$w(u) = \frac{1}{u}, \quad \rho(u) = 1 \quad \text{if} \quad 0 < u \leqslant 2, \tag{2.1}$$

$$(uw(u))' = w(u-1), \quad (u-1)\rho'(u) = -\rho(u-1) \quad \text{if} \quad u \geqslant 2, \tag{2.2}$$

where in each case the right hand derivative has to be taken at $u = 2$.

Both functions have been investigated very thoroughly; indeed, they are associated with the Buchstab functions $\Phi(x, y)$ and $\Psi(x, y)$ respectively, so that it is perhaps not altogether surprising to encounter them in this context. De Bruijn has proved that

$$w(u) = e^{-\gamma} + O(e^{-u}), \quad \rho(u) = O(e^{-u}), \quad u \geqslant 1, \tag{2.3}$$

and that

$$\rho(u) > 0, \quad u > 0. \tag{2.4}$$

With these functions we now define

$$F(u) = e^{\gamma}\left\{w(u) + \frac{\rho(u)}{u}\right\}, \quad u > 0$$

and

$$f(u) = e^{\gamma}\left\{w(u) - \frac{\rho(u)}{u}\right\}, \quad u > 0;$$

and we proceed to list those properties of F, f that we shall need later.

By (2.1) and (2.2) we have

$$F(u) = \frac{2e^{\gamma}}{u}, \quad f(u) = 0 \quad \text{for} \quad 0 < u \leqslant 2, \tag{2.5}$$

and it is noteworthy, when comparing Theorem 7.3 ($\kappa = 1$) with Theorem 8.3 below, that $F(u)$ and $1/\sigma_{\kappa}(u)$ agree (the latter with $\kappa = 1$, of course) in this range of u. Also from (2.1) and (2.2) we see that F, f are solutions of the pair of simultaneous differential-difference equations

$$(uF(u))' = f(u-1), \quad (uf(u))' = F(u-1), \quad u \geqslant 2, \tag{2.6}$$

whence

$$\left.\begin{aligned} &\int_{u_1}^{u} f(t-1)\,dt = uF(u) - u_1F(u_1), \\ &\qquad\qquad\qquad\qquad\qquad\qquad 2 \leqslant u_1 \leqslant u, \\ &\int_{u_1}^{u} F(t-1)\,dt = uf(u) - u_1f(u_1). \end{aligned}\right\} \tag{2.7}$$

Taking $u_1 = 2$, we obtain, by (2.5),

$$uF(u) = 2F(2) + \int_2^u f(t-1)\,dt = 2e^{\gamma}, \quad 2 \leqslant u \leqslant 3,$$

and hence

$$F(u) = \frac{2e^{\gamma}}{u} \quad \text{if} \quad 0 < u \leqslant 3. \tag{2.8}$$

It follows from this that

$$uf(u) = 2f(2) + \int_2^u F(t-1)dt = 2e^\gamma \log(u-1), \quad 2 \leqslant u \leqslant 4. \tag{2.9}$$

For large u we have, by (2.3), that

$$F(u) = 1 + O(e^{-u}), \quad f(u) = 1 + O(e^{-u}), \quad u \geqslant 1, \tag{2.10}$$

and, in view of (2.4), F and f are also linked by

$$F(u) - f(u) = \frac{2e^\gamma}{u}\rho(u) > 0, \quad u > 0. \tag{2.11}$$

Suppose if possible that $F'(u) = 0$ for some u, and take u_0 to be the least root. Then, by (2.8),

$$F'(u_0) = 0 \quad \text{if} \quad u_0 > 3, \quad F'(u) < 0 \quad \text{for} \quad 0 < u < u_0. \tag{2.12}$$

However, if u' and u'' are suitable numbers satisfying

$$u_0 - 1 < u' < u_0, \quad u' - 1 < u'' < u',$$

we deduce from (2.6) and (2.11) that

$$0 = u_0 F'(u_0) = f(u_0 - 1) - F(u_0) < f(u_0 - 1) - f(u_0) = -f'(u')$$

$$= \frac{f(u') - F(u'-1)}{u'} < \frac{F(u') - F(u'-1)}{u'} = \frac{F'(u'')}{u'},$$

which contradicts (2.12). Hence $F'(u) < 0$ for $u > 0$, and therefore, using (2.6) and (2.11) again, we have

$$uf'(u) = F(u-1) - f(u) > F(u-1) - F(u) > 0 \quad \text{if} \quad u > 2.$$

Combining these results with (2.10) we derive that

$$\left.\begin{array}{l} F(u) \text{ is monotonically decreasing towards } 1, \\ f(u) \text{ is monotonically increasing towards } 1. \end{array}\right\} \tag{2.13}$$

One final set of facts about F and f: By the mean value theorem, and using (2.5), (2.6) and (2.13) we see readily that

$$\left.\begin{aligned} &0 < F(u_1) - F(u_2) \leqslant F(\delta_1)\frac{u_2 - u_1}{u_1}, \\ & \qquad\qquad\qquad\qquad\qquad\qquad \text{for } 0 < \delta_1 \leqslant u_1 < u_2. \\ &0 < f(u_2) - f(u_1) \leqslant 2e^{\gamma}\frac{u_2 - u_1}{u_1}, \end{aligned}\right\} \tag{2.14}$$

When we come to relate these functions to the sieve theory, as we shall do in the next section, we shall find it convenient to write

$$\phi_\nu(u) := \begin{cases} F(u) & \text{if} \quad \nu \equiv 1 \bmod 2, \\ f(u) & \text{if} \quad \nu \equiv 0 \bmod 2, \end{cases} \qquad u > 0. \tag{2.15}$$

3. An Approximate Identity for the Leading Terms

From now on we need some condition on our function $\omega(p)$ of the type $(\Omega_2(\kappa, L))$. However, since we can obtain the main result of this chapter (Theorem 8.4) only for $\kappa = 1$, there is no need to state the preparatory results for a general κ. Thus, we shall use the condition

$$(\Omega_2(1, L)) \qquad -L \leqslant \sum_{v \leqslant p < w} \frac{\omega(p)}{p} \log p - \log \frac{w}{v} \leqslant A_2, \quad 2 \leqslant v \leqslant w.$$

Lemma 8.1. (Ω_1), $(\Omega_2(1, L))$: *Let*

$$2 \leqslant z_1 \leqslant z \leqslant \xi.$$

Then, for any value of ν, *we have*

$$W(z)\phi_\nu\left(\frac{\log \xi^2}{\log z}\right)$$

$$= W(z_1)\phi_\nu\left(\frac{\log \xi^2}{\log z_1}\right) - \sum_{z_1 \leqslant p < z} \frac{\omega(p)}{p} W(p)\phi_{\nu+1}\left(\frac{\log \xi^2/p}{\log p}\right)$$

$$+ O\left(\frac{W(z)L \log z}{\log^2 z_1}\right). \tag{3.1}$$

Proof. We apply Lemma 7.2 with $\kappa = 1$, $z_2 = z$ and $\psi(t) = \phi_{\nu+1}(t)$. Using (2.7) and $M = O(1)$ we obtain for the sum on the right of (3.1)

$$W(z)\frac{\log z}{\log \xi^2}\left\{\frac{\log \xi^2}{\log z_1}\phi_\nu\left(\frac{\log \xi^2}{\log z_1}\right) - \frac{\log \xi^2}{\log z}\phi_\nu\left(\frac{\log \xi^2}{\log z}\right)\right\} + O\left(\frac{W(z)L\log z}{\log^2 z_1}\right). \tag{3.2}$$

By (5.2.4) we have

$$\frac{W(z_1)}{W(z)} = \frac{\log z}{\log z_1} + O\left(\frac{L\log z}{\log^2 z_1}\right),$$

and employing this in (3.2) completes the proof of Lemma 8.1.

With a similar proof we obtain

LEMMA 8.2. (Ω_1), $(\Omega_2(1, L))$: *Let*

$$z \leqslant \xi^2, \quad 2 \leqslant z_1 \leqslant \mu,$$

where

$$\mu = \min(z, \xi^{2/3}).$$

Then

$$\sum_{z_1 \leqslant p < \mu} \frac{\omega(p)}{p} W(p) e^{-\log(\xi^2/p)/\log p}$$

$$\leqslant W(z)e^{-\log \xi^2/\log z}\frac{e}{3}\left\{1 + B_{17}\frac{L\log \xi}{\log^2 z_1}\right\}. \tag{3.3}$$

Proof. Here we apply Lemma 7.2 with $\kappa = 1$, $z_2 = \mu$ and $\psi(t) = e^{-t}$. Then $M = e^{-\log(\xi^2/\mu)/\log \mu}$, and we obtain for the left of (3.3) the expression

$$W(z)\frac{\log z}{\log \xi^2}\int_{\log \xi^2/\log \mu}^{\log \xi^2/\log z_1} e^{1-t}\,dt + O\left(\frac{Le^{-\log(\xi^2/\mu)/\log \mu}W(z)\log z}{\log^2 z_1}\right)$$

$$\leqslant W(z)\frac{\log z}{\log \xi^2}e^{1-\log \xi^2/\log \mu}\left\{1 + B_{18}\frac{L\log \xi}{\log^2 z_1}\right\};$$

noting that, under our conditions,

$$\frac{\log z}{\log \xi^2}e^{1-\log \xi^2/\log \mu} \leqslant e^{-\log \xi^2/\log z}\frac{e}{3}$$

always, (3.3) follows.

THEOREM 8.2. (Ω_1), $(\Omega_2(1, L))$: *Let*

$$2 \leqslant z_1 \leqslant z \leqslant \xi$$

and set

$$\xi_j{}^2 = \frac{\xi^2}{p_1 \cdots p_j}, \quad j = 1, 2, \ldots.$$

Then, for any natural number b and any value of v, we have

$$W(z)\phi_v\left(\frac{\log \xi^2}{\log z}\right) = W(z_1)\phi_v\left(\frac{\log \xi^2}{\log z_1}\right)$$

$$+ \sum_{i=1}^{b-1} (-1)^i \sum_{\substack{z_1 \leqslant p_i < \dots < p_1 < z \\ p_j < \xi_j (j=1,\dots,i)}} \frac{\omega(p_1 \dots p_i)}{p_1 \dots p_i} W(z_1)\phi_{v+i}\left(\frac{\log \xi_i^2}{\log z_1}\right)$$

$$+ (-1)^b \sum_{\substack{z_1 \leqslant p_b < \dots < p_1 < z \\ p_j < \xi_j (j=1,\dots,b)}} \frac{\omega(p_1 \dots p_b)}{p_1 \dots p_b} W(p_b)\phi_{v+b}\left(\frac{\log \xi_b^2}{\log p_b}\right)$$

$$+ \sum_{i=1}^{b} (-1)^i \sum_{\substack{z_1 \leqslant p_i < \dots < p_1 < z \\ p_j < \xi_j (j=1,\dots,i-1) \\ \xi_i \leqslant p_i < \xi_i^2}} \frac{\omega(p_1 \dots p_i)}{p_1 \dots p_i} W(p_i)\phi_{v+i}\left(\frac{\log \xi_i^2}{\log p_i}\right)$$

$$+ O\left(\frac{W(z)L\log^2 z}{\log^3 z_1}\right).$$

The O-constant is independent of b.

Proof. Apart from the remainder term, the proof follows along the lines of the proof of Theorem 8.1. Here we start from Lemma 8.1, taking $v+1$, ξ^2/p, p for v, ξ^2, z respectively. The O-term, as can be seen by induction, becomes (with the same O-constant as in Lemma 8.1)

$$O\left(\frac{L}{\log^2 z_1}\right)\left\{W(z)\log z + \sum_{i=1}^{b-1} \sum_{z_1 \leqslant p_i < \dots < p_1 < z} \frac{\omega(p_1 \dots p_i)}{p_1 \dots p_i} W(p_i)\log p_i\right\};$$

using (2.3.5) (with $\kappa = 1$) we obtain

$$O\left(\frac{W(z)L\log z}{\log^2 z_1}\sum_{i=0}^{b-1}\frac{1}{i!}\left(\sum_{z_1 \leqslant p < z}\frac{\omega(p)}{p}\right)^i\right). \tag{3.4}$$

Finally, by (2.3.3) we see that our condition $(\Omega_2(1, L))$ implies that

$$\sum_{z_1 \leqslant p < z}\frac{\omega(p)}{p} \leqslant \log\frac{\log z}{\log z_1} + O\left(\frac{1}{\log z_1}\right). \tag{3.5}$$

Therefore, the error term in (3.4) is

$$O\left(\frac{W(z)L\log^2 z}{\log^3 z_1}\right),$$

and this completes the proof of the lemma.

4. Upper and Lower Bounds for S

We now prove our main result regarding the influence of condition $(\Omega_2(1, L))$ on the leading terms of our estimates—both from above and below.

The question of the remainder term is left open, i.e. we do not impose at this stage any condition on the remainders R_d.

THEOREM 8.3. (Ω_1), $(\Omega_2(1, L))$: *For*

$$\xi \geqslant z$$

we have

$$S(\mathscr{A}_q; \mathfrak{P}, z) \leqslant \frac{\omega(q)}{q} XW(z)\left\{F\left(\frac{\log \xi^2}{\log z}\right) + B_{19}\frac{L}{(\log \xi)^{1/14}}\right\} + \sum_{\substack{n<\xi^2 \\ n|P(z)}} 3^{\nu(n)} |R_{qn}| \tag{4.1}$$

and

$$S(\mathscr{A}_q; \mathfrak{P}, z) \geqslant \frac{\omega(q)}{q} XW(z)\left\{f\left(\frac{\log \xi^2}{\log z}\right) - B_{19}\frac{L}{(\log \xi)^{1/14}}\right\} - \sum_{\substack{n<\xi^2 \\ n|P(z)}} 3^{\nu(n)} |R_{qn}|; \tag{4.2}$$

this is also true if $1 < \xi < z$ *but* $z \ll \xi^\lambda$ *with a positive constant* λ, *and (only in* (4.1)) B_{19} *is replaced by some suitable* $B_{19}(\lambda)$.

Proof. If $L^{-1}(\log \xi)^{1/14}$ is bounded, our theorem is true because of Theorem 7.1 and (2.13). Therefore we may assume

$$\log \xi \geqslant B_{20} L^{14} \tag{4.3}$$

where B_{20} is sufficiently large.

We shall now apply Theorem 8.1 and Theorem 8.2 with

$$z_1 = \exp\{(\log \xi)^{7/10}\} \tag{4.4}$$

and b satisfying

$$2\frac{(\log \xi)^{3/10}}{\log\log \xi} \leqslant \left(\frac{3}{2}\right)^b \leqslant 3\frac{(\log \xi)^{3/10}}{\log\log \xi}. \tag{4.5}$$

If $z < z_1$, in view of (2.13) our theorem has already been proved by Theorem 7.1, so we can assume that the condition

$$2 \leqslant z_1 \leqslant z \leqslant \xi$$

is satisfied.

In Theorem 8.2 we multiply by $(\omega(q)/q)\ X$ and subtract this from the expression in Theorem 8.1, thus obtaining for any value of ν

$$(-1)^{\nu+1}\left\{S(\mathscr{A}_q;\mathfrak{P},z)-\frac{\omega(q)}{q}XW(z)\phi_\nu\left(\frac{\log\xi^2}{\log z}\right)\right\}$$

$$=(-1)^{\nu+1}\left\{S(\mathscr{A}_q;\mathfrak{P},z_1)-\frac{\omega(q)}{q}XW(z_1)\phi_\nu\left(\frac{\log z^2}{\log z_1}\right)\right\}$$

$$-\sum_{i=1}^{b-1}(-1)^{\nu+i}\sum_{\substack{z_1\leqslant p_i<\ldots<p_1<z\\ p_j<\xi_j,\,p_j\in\mathfrak{P}(j=1,\ldots,i)}}\left\{S(\mathscr{A}_{qp_1\ldots p_i};\mathfrak{P},z_1)\right.$$

$$\left.-\frac{\omega(q)}{q}\frac{\omega(p_1\ldots p_i)}{p_1\ldots p_i}XW(z_1)\phi_{\nu+i}\left(\frac{\log\xi_i^2}{\log z_1}\right)\right\}$$

$$-(-1)^{\nu+b}\sum_{\substack{z_1\leqslant p_b<\ldots<p_1<z\\ p_j<\xi_j,\,p_j\in\mathfrak{P}(j=1,\ldots,b)}}\left\{S(\mathscr{A}_{qp_1\ldots p_b};\mathfrak{P},p_b)\right.$$

$$\left.-\frac{\omega(q)}{q}\frac{\omega(p_1\ldots p_b)}{p_1\ldots p_b}XW(p_b)\phi_{\nu+b}\left(\frac{\log\xi_b^2}{\log p_i}\right)\right\}$$

$$-\sum_{i=1}^{b}(-1)^{\nu+i}\sum_{\substack{z_1\leqslant p_i<\ldots<p_1<z\\ p_j<\xi_j,\,p_j\in\mathfrak{P}(j=1,\ldots,i-1)\\ \xi_i\leqslant p_i<\xi_i^2,\,p_i\in\mathfrak{P}}}\left\{S(\mathscr{A}_{qp_1\ldots p_i};\mathfrak{P},p_i)\right.$$

$$\left.-\frac{\omega(q)}{q}\frac{\omega(p_1\ldots p_i)}{p_1\ldots p_i}XW(p_i)\phi_{\nu+i}\left(\frac{\log\xi_i^2}{\log p_i}\right)\right\}$$

$$+O\left(\frac{\omega(q)}{q}XW(z)\frac{L\log^2 z}{\log^3 z_1}\right). \tag{4.6}$$

According to (2.15), for a proof of both, (4.1) and (4.2), the right-hand side of (4.6) has to be estimated from above by

$$O\left(\frac{\omega(q)}{q}XW(z)\frac{L}{(\log\xi)^{1/14}}\right)+\sum_{\substack{n<\xi^2\\ n\mid P(z)}}3^{\nu(n)}\,|R_{qn}|\,. \tag{4.7}$$

For the first term and the first two sums on the right of (4.6) we apply Theorem 7.1, taking for ξ and z the pairs (note that always $\xi\geqslant z$)

$$\xi,z_1;\quad \xi_i,z_1;\quad \xi_b,p_b,$$

respectively, and it is both more convenient and sufficient to use for the O-term the estimate $O(e^{-2\tau})$. Then, making use of (2.10), we obtain the following upper estimates

$$O\left(\frac{\omega(q)}{q}XW(z_1)e^{-2\log\xi/\log z_1}\right)+\sum_{\substack{d<\xi^2\\ d|P(z_1)}}3^{\nu(d)}\,|R_{qd}|, \tag{4.8}$$

$$\sum_{i=1}^{b-1}\sum_{\substack{z_1\leqslant p_i<\dots<p_1<z\\ p_j<\xi_j,\,p_j\in\mathfrak{P}(j=1,\dots,i)}}\left\{O\left(\frac{\omega(q)}{q}\,\frac{\omega(p_1\dots p_i)}{p_1\dots p_i}XW(z_1)e^{-2\log\xi_i/\log z_1}\right)\right.$$
$$\left.+\sum_{\substack{d<\xi^2/p_1\dots p_i\\ d|P(z_1)}}3^{\nu(d)}\,|R_{qp_1\dots p_id}|\right\} \tag{4.9}$$

and

$$\sum_{\substack{z_1\leqslant p_b<\dots<p_1<z\\ p_j<\xi_j,\,p_j\in\mathfrak{P}(j=1,\dots,b)}}\left\{O\left(\frac{\omega(q)}{q}\,\frac{\omega(p_1\dots p_b)}{p_1\dots p_b}XW(p_b)e^{-2\log\xi_b/\log p_b}\right)\right.$$
$$\left.+\sum_{\substack{d<\xi^2/p_1\dots p_b\\ d|P(p_b)}}3^{\nu(d)}\,|R_{qp_1\dots p_bd}|\right\}. \tag{4.10}$$

The conditions $p_j<\xi_j,\quad j=1,\dots,i$ imply that

$$\xi_1{}^2=\xi^2/p_1>\xi^2/\xi_1$$

which gives

$$\log\xi_1>\tfrac{2}{3}\log\xi;$$

and for $i>1$ we have

$$\log\xi_i{}^2=\log\frac{\xi_{i-1}^2}{p_i}>\log\xi_{i-1}^2-\log\xi_i,$$

i.e.

$$\log\xi_i>\tfrac{2}{3}\log\xi_{i-1}.$$

Therefore we obtain by induction

$$\log\xi_i>(\tfrac{2}{3})^i\log\xi\quad\text{if}\quad p_j<\xi_j\quad\text{for}\quad j=1,\dots,i, \tag{4.11}$$

and using (4.5) we find that

$$\frac{\log\xi_i}{\log z_1}>(\tfrac{2}{3})^{b-1}(\log\xi)^{3/10}\geqslant\tfrac{1}{2}\log\log\xi\quad\text{for}\quad i=1,\dots,b-1. \tag{4.12}$$

For the last sum we have to deal with terms of the form

$$-(-1)^{\nu+i}\left\{S(\mathscr{A}_{qp_1\ldots p_i};\mathfrak{P},p_i)\right.$$

$$\left.-\frac{\omega(q)}{q}\frac{\omega(p_1\cdots p_i)}{p_1\cdots p_i}XW(p_i)\phi_{\nu+i}\left(\frac{\log\xi_i^2}{\log p_i}\right)\right\} \tag{4.13}$$

where $\xi_i \leqslant p_i < \xi_i^2$. If $\nu + i$ is even then, by (2.5), $\phi_{\nu+i}(\log\xi_i^2/\log p_i) = 0$, and hence (4.13) can be estimated from above by zero. If $\nu + i$ is odd, then, taking ξ_i and p_i for ξ and z respectively, Theorem 6.3 and (2.5) give the upper estimate

$$\sum_{i=1}^{b}\sum_{\substack{z_1\leqslant p_i<\ldots<p_1<z\\ p_j<\xi_j,p_j\in\mathfrak{P}(j=1,\ldots,i-1)\\ \xi_i\leqslant p_i<\xi_i^2,p_i\in\mathfrak{P}}}\left\{O\left(\frac{\omega(q)}{q}\frac{\omega(p_1\cdots p_i)}{p_1\cdots p_i}XW(p_i)\frac{L\log p_i}{\log^2 z_1}\right)\right.$$

$$\left.+\sum_{\substack{d<\xi^2/p_1\ldots p_i\\ d|P(p_i)}}3^{\nu(d)}|R_{qp_1\ldots p_id}|\right\}. \tag{4.14}$$

Now our problem has been reduced to estimating (4.8), (4.9), (4.10), (4.14) and the error term of (4.6) by the expression given in (4.7).

Let us first deal with sums containing the remainders R_{qn}. Their total contribution is

$$\sum_{\substack{d<\xi^2\\ d|P(z_1)}}3^{\nu(d)}|R_{qd}|+\sum_{i=1}^{b-1}\sum_{\substack{z_1\leqslant p_i<\ldots<p_1<z\\ p_j<\xi_j,p_j\in\mathfrak{P}(j=1,\ldots,i)}}\sum_{\substack{d<\xi^2/p_1\ldots p_i\\ d|P(z_1)}}3^{\nu(d)}|R_{qp_1\ldots p_id}|$$

$$+\sum_{\substack{z_1\leqslant p_b<\ldots<p_1<z\\ p_j<\xi_j,p_j\in\mathfrak{P}(j=1,\ldots,b)}}\sum_{\substack{d<\xi^2/p_1\ldots p_b\\ d|P(p_b)}}3^{\nu(d)}|R_{qp_1\ldots p_bd}|$$

$$+\sum_{i=1}^{b}\sum_{\substack{z_1\leqslant p_i<\ldots<p_1<z\\ p_j<\xi_j,p_j\in\mathfrak{P}(j=1,\ldots,i-1)\\ \xi_i\leqslant p_i<\xi_i^2,p_i\in\mathfrak{P}}}\sum_{\substack{d<\xi^2/p_1\ldots p_i\\ d|P(p_i)}}3^{\nu(d)}|R_{qp_1\ldots p_id}|. \tag{4.15}$$

In the arguments qn of the R's, each n is $< \xi^2$, is a squarefree number relatively prime to $\bar{\mathfrak{P}}$, and has no prime factors $\geqslant z$; i.e. $n \mid P(z)$. We shall prove that all occurring n's are distinct. This is obvious within each of the four sums. The first sum contains only n's that have no prime factors $\geqslant z_1$ whereas the n's in the other sums have at least one prime divisor $\geqslant z_1$. In the second sum all the n's have at most $b-1$ prime factors $\geqslant z_1$, while each n in the third sum has at least b prime factors. Also, if an n in the second sum were equal to an n in the fourth sum there would be an h, such that that the h-th largest prime factor $\geqslant z_1$ in the second sum satisfies $p_h < \xi_h$ whereas that prime factor in the fourth sum has, on the contrary, to satisfy $\xi_h \leqslant p_h < \xi_h^2$. A similar reasoning shows that the terms in the third and fourth sums contain distinct n's.

Hence, the whole expression in (4.15) is at most

$$\sum_{\substack{n<\xi^2\\ n|P(z)}} 3^{\nu(n)}|R_{qn}|, \tag{4.16}$$

where we have made use of the fact that, for any positive integer m,

$$3^{\nu(d)} \leqslant 3^{\nu(md)}.$$

We collect now the remaining error terms. Using (2.3.5) and (4.12) we obtain

$$O\left(\frac{\omega(q)}{q}XW(z)\frac{\log z}{\log z_1}\left\{e^{-2\log\xi/\log z_1} + \sum_{i=1}^{b-1}\frac{1}{i!}\left(\sum_{z_1\leqslant p<z}\frac{\omega(p)}{p}\right)^i\frac{1}{\log\xi}\right.\right.$$

$$\left.\left.+\sum_{i=1}^{b}\frac{1}{i!}\left(\sum_{z_1\leqslant p<z}\frac{\omega(p)}{p}\right)^i\frac{L}{\log z_1}+\frac{L\log z}{\log^2 z_1}\right\}\right)$$

$$+\,O\left(\frac{\omega(q)}{q}X\sum_{\substack{z_1\leqslant p_b<\ldots<p_1<z\\ p_j<\xi_j(j=1,\ldots,b)}}\frac{\omega(p_1\ldots p_b)}{p_1\ldots p_b}W(p_b)e^{-\log\xi_b^2/\log p_b}\right).$$

Since $z \leqslant \xi$ the first O-term is, by (3.5) and (4.4),

$$O\left(\frac{\omega(q)}{q}XW(z)\frac{L\log^2 z}{\log^3 z_1}\right) = O\left(\frac{\omega(q)}{q}XW(z)\frac{L}{(\log\xi)^{1/10}}\right). \tag{4.17}$$

Inside the second O-term we have to estimate

$$\sum_{\substack{z_1\leqslant p_b<\ldots<p_1<z\\ p_j<\xi_j(j=1,\ldots,b)}}\frac{\omega(p_1\ldots p_b)}{p_1\ldots p_b}W(p_b)e^{-\log\xi_b^2/\log p_b}. \tag{4.18}$$

The sum over p_b is

$$D_b := \sum_{z_1\leqslant p_b<\mu_b}\frac{\omega(p_b)}{p_b}W(p_b)e^{-\log\xi_b^2/\log p_b}, \qquad \mu_b = \min(p_{b-1}, \xi_{b-1}^{2/3}).$$

Using Lemma 8.2 with $z = p_{b-1}$ and $\xi = \xi_{b-1}$ we find that

$$D_b \leqslant W(p_{b-1})\,e^{-\log\xi_{b-1}^2/\log p_{b-1}}\,\varepsilon, \qquad \varepsilon = \frac{e}{3}\left\{1 + B_{17}\frac{L}{(\log\xi)^{2/5}}\right\}.$$

Introducing this in (4.18) we see that, apart from the factor ε, we obtain the same sum but with $b-1$ instead of b as an upper estimate. Thus, repeating this procedure and using $p_0 = z$, $\xi_0 = \xi$ in the last step we obtain

$$\sum_{\substack{z_1 \leqslant p_b < \ldots < p_1 < z \\ p_j < \xi_j (j=1,\ldots,b)}} \frac{\omega(p_1 \ldots p_b)}{p_1 \ldots p_b} W(p_b) e^{-\log \xi_b^2/\log p_b} \leqslant W(z) e^{-\log \xi^2/\log z}$$

Since, by (4.3), ε can be brought arbitrarily close to $e/3$, we find by (4.5) that

$$\varepsilon^b = O((\log \xi)^{-1/14}), \tag{4.19}$$

and this completes the proof of the main part of Theorem 8.3.

Finally, if $1 < \xi < z$ the arguments of F and f are less than 2. Then, by (2.5), the lower estimate (4.2) holds trivially. On the other hand, if also $\log z = O_\lambda(\log \xi)$, then by Theorem 6.3 ($\kappa = 1$) and (2.5), we obtain (4.1) immediately with some $B_{19}(\lambda)$.

5. The Main Result

We are now in a position to prove the main theorem of the present chapter where, beside our ω-conditions (Ω_1) and $(\Omega_2(1, L))$ we impose also $(R(1, \alpha))$:

There exist constants α with

$$0 < \alpha \leqslant 1,$$

A_4 ($\geqslant 1$) and A_5 ($\geqslant 1$) such that†

$$(R(1,\alpha)): \qquad \sum_{\substack{d < X^\alpha/(\log X)^{A_4} \\ (d,\overline{\mathfrak{P}})=1}} \mu^2(d)\, 3^{\nu(d)} |R_d| \leqslant A_5 \frac{X}{\log^2 X} \quad (X \geqslant 2).$$

Theorem 8.4. (Ω_1), $(\Omega_2(1, L))$, $(R(1, \alpha))$: *For $z \leqslant X$ we have*

$$S(\mathscr{A}; \mathfrak{P}, z) \leqslant XW(z)\left\{F\left(\alpha \frac{\log X}{\log z}\right) + B_{21} \frac{L}{(\log X)^{1/14}}\right\} \tag{5.1}$$

and

$$S(\mathscr{A}; \mathfrak{P}, z) \geqslant XW(z)\left\{f\left(\alpha \frac{\log X}{\log z}\right) - B_{21} \frac{L}{(\log X)^{1/14}}\right\}, \tag{5.2}$$

where F and f are the functions defined in Section 2 of this chapter.

† As the proof will show, it would suffice to have $\log^{-15/14} X$ on the right in place of $\log^{-2} X$.

Proof. If X is bounded the estimates are trivially true, so that we may assume $X \geqslant B_{22}$ where B_{22} is sufficiently large. Let

$$\xi^2 = \frac{X^\alpha}{(\log X)^{A_4}}. \tag{5.3}$$

Then, either $z \leqslant \xi$ or $1 < \xi < z \leqslant X = O(\xi^{3/\alpha})$, so that (4.1) and (4.2) hold; and we note that $q = 1$.

By (2.14) it follows that

$$F\left(\frac{\log \xi^2}{\log z}\right) - F\left(\frac{\log X^\alpha}{\log z}\right) = O\left(\frac{\log\log X}{\log z}\,\frac{\log z}{\log \xi^2}\right) = O\left(\frac{\log\log X}{\log X}\right) \tag{5.4}$$

and correspondingly that

$$f\left(\frac{\log \xi^2}{\log z}\right) - f\left(\frac{\log X^\alpha}{\log z}\right) = O\left(\frac{\log\log X}{\log X}\right).$$

Finally, by $(R(1, \alpha))$ and (2.3.6) we have

$$\sum_{\substack{n < \xi^2 \\ n \mid P(z)}} 3^{\nu(n)} |R_n| = O\left(XW(z)\frac{1}{(\log X)^{1/14}}\right). \tag{5.5}$$

This completes the proof of Theorem 8.4.

Notes

8.1. This chapter contains a generalization to a more extensive class of linear sieve problems of the method first developed in Jurkat–Richert [1] and later in Halberstam–Jurkat–Richert [1] (see also Uchiyama [7]). All the basic ideas derive from the first of these papers.

Theorem 8.1: The fundamental identity (1.1) (with $q = 1$) appears for the first time in Jurkat–Richert [1].

Theorem 8.1, Remark: The concluding sentence refers to Selberg [4, 5].

8.2. (2.1), (2.2): The function $w(u)$ was first studied by Buchstab [1, 7] (in the form $\psi(u) = uw(u)$). In particular, he derived an asymptotic formula for $w(u)$ as $u \to \infty$ which Hua [1] subsequently improved to (cf. (2.3))

$$|w(u) - e^{-\gamma}| < \exp\left(-u\left(\log u + \log\log u - \frac{\log\log u}{\log u} + O\left(\frac{1}{\log u}\right)\right)\right).$$

For further investigations of w see de Bruijn [2] (we have shifted the argument by 1 relative to de Bruijn's notation); Levin–Faĭnleĭb [2].

The function $\rho(u)$, sometimes referred to as Dickman's function, is discussed in Hua [1], and, at greater length, in Norton [1]. It is known (cf.

Norton [1], p.17) that

$$\rho(u) = \exp\left\{-u\left(\log u + \log\log u - 1 + \frac{\log\log u}{\log u} + O\left(\frac{1}{\log u}\right)\right)\right\}, \quad u \to \infty.$$

The weaker estimates (2.3) as stated by us can be found e.g. in de Bruijn [3].

The functions F, f occur for the first time in Jurkat–Richert [1]. Where the functions $1/\sigma_1$ and $1 - \eta_1$ arise from a single application of a Buchstab identity (cf. Chapters 6, 7—actually, such an identity is involved only in the case of the latter), the functions F, f are the result of an "infinite" iteration of Buchstab identities on the basis of Theorem 8.1. Numerical tables for the pairs of functions corresponding to small numbers of iterations can be found in Buchstab [10]. The limiting pair F, f not only correspond to a better sieve but are also much simpler in form than all intermediate pairs.

8.3. Lemma 8.1: a "weighted" approximation of Lemma 7.1, (1.2); the iteration procedure in the proof of Theorem 8.2 will be based on this result.

Lemma 8.2: this estimate is very important for the success of the iterative process in that it deals with a potentially dangerous accumulation of error terms (see the estimation of (4.18) in the next section). However, we shall see that it also limits the quality of the final error term in Theorem 8.3.

Theorem 8.2: an approximate "matching" of Theorem 8.1. These two theorems provide the combinatorial frame-work for the proof of Theorem 8.3.

8.4. Theorem 8.3: cf. Jurkat–Richert [1], Theorem 4. From the point of view of those applications where the weighted sieve procedure of the next chapter is appropriate, this is the main result of Chapter 8. The error terms $L(\log \xi)^{-1/14}$ derive from the use of Lemma 8.2. It would be interesting to know whether the iterative process constructed in Sections 8.1, 8.3 can be improved to lead to better error terms. It is clear from Iwaniec [1] that Rosser's sieve method improves on Jurkat–Richert [1] in just this respect; presumably it could be made to improve Theorem 8.3 also.

8.5. Theorem 8.4: cf. Jurkat–Richert [1], Theorem 5; Halberstam–Jurkat–Richert [1] (where a somewhat less general result is stated in Theorem 1); Iwaniec [1], Theorem 1, Corollary. Note that (5.2) gives a non-trivial result only if $f(\alpha \log X/\log z) > 0$, i.e. if $X^\alpha > z^2$. This shows (cf. Notes for Section 7.6) 2 to be the sieving limit of this particular method, as compared with 2·06 . . . , the limit of the method of Chapter 7 (so far as $\kappa = 1$ is concerned)—rather a modest improvement. In fact, for the quasi-prime-twin problem considered briefly at the end of Chapter 7 we cannot do better with Theorem 8.4 alone, given the present state of knowledge about admissible values of α (see discussion of Bombieri's theorem). Nevertheless, there is a sense in which Theorem 8.4 is actually best possible; taking

$$\mathscr{A} = \mathscr{B}_\nu = \{n: 1 \leqslant n \leqslant x,\ \Omega(n) \equiv \nu \bmod 2\} \qquad (\nu = 1, 2)$$

in turn, also $\mathfrak{P} = \mathfrak{P}_1$ and† $\alpha = 1$, it can be proved (Jurkat–Richert, unpublished manuscript) that (5.1) holds with equality for $\nu = 1$ and (5.2) holds with equality for $\nu = 2$, both for all values of $u > 0$. For $0 < u \leqslant 2$, this fact was established by Selberg [3]; he remarks that one may regard this as indicating that the sieve method (described here) "cannot distinguish between numbers with an odd or an even number of prime factors". Alternatively, one may point to these examples as illustrating the fact that the very generality of the sieve is a weakness rather than advantage when applied to any one specific sieve problem.

It is possible to improve Theorem 8.4 so far as the error terms in (5.1), (5.2) are concerned by means of the (Rosser) method of Iwaniec [1]; the exponent 1/14 can be replaced by 1. In general these error terms are not of major importance in the sense that they do not affect the quality of the results that can be derived from Theorem 8.4; but we refer below to one interesting application made by Iwaniec where the size of the error term is significant.

We have given no applications of Theorem 8.4 because, in the context of the problem area defined by Hypothesis H, the weighting procedure introduced in the next chapter will enhance, as we shall see, the efficiency of Theorem 8.4 to a surprising extent. Nevertheless, there are in the literature some interesting applications and we mention three of these now.

The first concerns a problem of Legendre taken up more recently by Jacobsthal (see Erdös [14]): For each natural number r, let $C_0(r)$ denote the maximal length of a sequence of consecutive integers each divisible by one of the first r primes. It is known (Rankin, *Proc. Edinburgh Math. Soc.* **13** (1962/3), 331–332) that

$$C_0(r) > e^{\gamma-\varepsilon} \frac{r \log^2 r \log\log\log r}{(\log\log r)^2},$$

and it follows from Theorem 8.4 that

$$C_0(r) < r^2 \exp\left((\log r)^{13/14}\right).$$

By Rosser's sieve (Iwaniec [1], p.2), the improved quality of the error term leads to the superior

$$C_0(r) \ll r^2 \log^2 r.$$

† Actually, the condition $(R(1, \alpha))$ of Theorem 8.4, with $\alpha = 1$, would have to be modified slightly for Theorem 8.4 to apply to $\mathscr{B}_1$ and $\mathscr{B}_2$.

Next, we mention an extraordinary application to latin squares. Let $N(s)$ denote the maximal number of pairwise orthogonal latin squares of order s. Euler conjectured that $N(s) = 1$ for $s > 10$ and $s \equiv 2 \bmod 4$, but his conjecture was disproved by Bose, Parker and Shrikhande (*Canadian J.* 7 (1960), 189–203) who showed that $N(s) \geqslant 2$ for $s > 6$. Using a lower bound sieve, several authors (Chowla–Erdös–Straus [1] were the first) have since proved inequalities of the type $N(s) \gg s^{\lambda}$ $(s > s_0)$; at present we know that $\lambda \geqslant 1/26$ (Wang [13]; this paper gives a bibliography of earlier results) but something better could well be accessible now. (According to P. Erdös, "The Art of Counting," MIT Press (1973), p. 639, Richard Wilson has proved $s \geqslant 1/17$.)

Finally, Halberstam–Rotkiewicz [1] apply Theorem 8.4 to show that, given $\varepsilon > 0$, there exist infinitely many primes p such that

$$\tfrac{1}{2} - \varepsilon \leqslant \phi(p-1)/(p-1) \leqslant 1/2.$$

This is the special case $k = 1$ of the following result (their Lemma 5): Suppose that $(l, k) = 1$, and that D is a given positive number. Then there exists a prime $p \equiv l \bmod k$ such that for $x \geqslant x_1(k, D)$ we have $x \leqslant p < x\,(1 + \log^{-D} x)$ and

$$c\left(1 - \frac{4}{x^{1/5}}\right) \leqslant \frac{\phi(p-1)}{p-1} \leqslant c, \qquad c = \frac{\phi(2(k, l-1))}{2(k, l-1)}.$$

For another interesting application concerning the least prime primitive root mod p see Elliott [2].

Chapter 9

A Weighted Sieve: The Linear Case

1. The Method

The most important way of distinguishing between different sieve problems is by means of the parameter κ appearing in condition $(\Omega_2(\kappa, L))$. Any problem associated with the number κ was described by us as being of dimension κ, and in the case of $\kappa = 1$ we speak of a "linear" problem. Broadly speaking, the higher its dimension the more difficult a sieve problem is. For this reason, the class of linear problem is, at the present time, of greatest interest; until we have succeeded with these, there is little hope that we shall do better, or even as well, in higher dimensions. Comparison between the principal results of Chapters 7 and 8 shows that we do, indeed, know much more about the linear case, and the object of this chapter is to apply the sharp results of Chapter 8 to a large variety of linear sieve problems. It should be clear from all that has gone before that we count among linear problems not only those that are linear by nature but also those that can, with advantage, be "linearized" (e.g. by working with a polynomial of lower degree, and prime instead of integer arguments—cf. Section 5.5).

Theorem 8.4, despite its quality, gives relatively poor results when applied directly. It was Kuhn who first noticed—actually in connection with Brun's method—that if weights of the form

$$1 - \sum_{p|a} w_p \tag{1.1}$$

are attached to the members of $\mathcal{A}$, and lower estimates are then derived for the weighted sum over the members a of $\mathcal{A}$, the power of the sieve method becomes appreciably increased. Indeed, most of the best results in the literature which have been proved by a sieve method derive from a combination of Selberg's sieve and Kuhn's weights.

Since the work of Kuhn, other mathematicians have experimented with various weight functions. Among these we may distinguish two broad categories. In the first category are those weights which are introduced at the very outset, and the associated extremal problem, implicit in any sieve method, is then somewhat different, and usually much more complicated. Nevertheless, this kind of approach has met with considerable success. Thus Rényi applied the Brun technique to the (weighted) sum

$$\sum_{\substack{p<2N \\ (2N-p,P(z))=1}} \log p \exp\left(-p\frac{\log 2N}{2N}\right)$$

in his pioneering work on the quasi-Goldbach problem. If P_r denotes a positive or negative number having at most r prime factors, multiple factors being counted multiply (one can express this by saying that $\Omega(P_r) \leqslant r$, and it is usual to speak of P_r as an *almost-prime* of order r), Rényi was the first to prove that there exists a constant c such that $2N = p + P_c$ for all sufficiently large N.

Another such method was devised by A. Selberg for the quasi-twin-prime problem, when he succeeded in showing that there exist infinitely many almost-primes P_2 such that $P_2 + 2 = P_3$. We shall touch on Selberg's method in Section 6 of Chapter 10. A variant of this method, capable of many applications, was given by Miech, and there is a further generalization due to Porter. We shall present an extended form of this method in Section 4 of Chapter 10.

We distinguish a second category of weight functions, with Kuhn's own weights as prototypes, which have the property that the weighted sum can be expressed in terms of the sums $S(\mathscr{A}; \mathfrak{P}, z)$. Ankeny and Onishi introduced a certain logarithmic weight function of this kind, and both this chapter and the next are based on an extended form (due to Richert) of their weight. Also to this category belongs the method of Buchstab who, in his proof of the solvability of $2N = p + P_3$ (cf. Theorem 9.2 below), formulated the procedure of choosing weights as a problem in linear programming, and so was lead to a very complicated set of constant weights. We shall not describe Buchstab's method in this book, largely because we are able to obtain his results without loss of quality and yet without the use of computers or numerical integration. Nevertheless, his work represents an important contribution, for in the long run it may well turn out that *each* sieve problem requires an optimization procedure *of its own* in order to reach the most precise results.

We shall now introduce our weight function in connection with the problem of sifting a sequence $\mathscr{A}$ by the primes of $\mathfrak{P}$ truncated at $X^{1/v}$. To this end we introduce a number $v > u$ and put

$$w_p = \begin{cases} \lambda\left(1 - \dfrac{u \log p}{\log X}\right) & \text{if} \quad X^{1/v} \leqslant p < X^{1/u}, \ p \in \mathfrak{P}, \\ 0 & \text{otherwise;} \end{cases}$$

here λ is a number that remains to be chosen, depending on the problem. With these weights we form the weighted sum

$$W(\mathscr{A};\mathfrak{P},v,u,\lambda) = \sum_{\substack{a\in\mathscr{A}\\(a,P(X^{1/v}))=1}} \left\{1-\lambda\sum_{\substack{X^{1/v}\leqslant p<X^{1/u}\\p|a,\,p\in\mathfrak{P}}}\left(1-u\frac{\log p}{\log X}\right)\right\}, \quad (1.2)$$

and prove for it the following basic result.

THEOREM 9.1. (Ω_1), $(\Omega_2(1,L))$, $(R(1,\alpha))$: *If u and v are two numbers (independent of X) satisfying*

$$1/\alpha < u < v, \quad (1.3)$$

and if

$$0 < \lambda \leqslant A_7, \quad (1.4)$$

we have

$$W(\mathscr{A};\mathfrak{P},v,u,\lambda) \geqslant XW(X^{1/v})$$
$$\times\left\{f(\alpha v)-\lambda\int_u^v F\left(v\left(\alpha-\frac{1}{t}\right)\right)\left(1-\frac{u}{t}\right)\frac{dt}{t}-\frac{C_2 L}{(\log X)^{1/14}}\right\}, \quad (1.5)$$

where C_2 depends at most on u and v (as well as on the A_i's and α).

Remark. We shall show later on in detail how to apply this result. For the present, let it suffice to say that the argument will run as follows: For a suitable (optimal) choice of u, v and λ (1.5) will imply that $W(\mathscr{A},\mathfrak{P},v,u,\lambda)$ is *positive*. This implies that $\mathscr{A}$ contains elements a for which

$$\lambda^{-1} > \sum_{\substack{X^{1/v}\leqslant p<X^{1/u}\\p|a,\,p\in\mathfrak{P}}}\left(1-u\frac{\log p}{\log X}\right); \quad (1.6)$$

and from this inequality it follows in turn that each such element a has relatively few prime factors. To be a little more precise, such an a has no prime factors (of $\mathfrak{P}$) less than $X^{1/v}$ because it is counted in the sum (1.2), it has only a few prime factors between $X^{1/v}$ and $X^{1/u}$ because of (1.6), and it can have only a few large prime factors, i.e. $\geqslant X^{1/u}$, for trivial reasons.

Proof of Theorem 9.1. We may assume immediately that

$$X \geqslant C_3$$

where C_3 is sufficiently large and may depend on u and v (as well as on the A_i's and α), since otherwise the theorem holds trivially in view of (8.2.13). (We shall require the constants $C_4, \ldots, C_{11}$ occurring below in this chapter to be similarly qualified, without saying so again explicitly.)

Using the abbreviations

$$z = X^{1/v}, \quad y = X^{1/u},$$

we may then write

$$W(\mathscr{A}; \mathfrak{P}, v, u, \lambda) = S(\mathscr{A}; \mathfrak{P}, z) - \lambda \sum_{\substack{z \leqslant p < y \\ p \in \mathfrak{P}}} \left(1 - \frac{\log p}{\log y}\right) S(\mathscr{A}_p, \mathfrak{P}, z). \tag{1.7}$$

Because $v > 1$, Theorem 8.4 yields the lower estimate

$$S(\mathscr{A}; \mathfrak{P}, z) \geqslant XW(z)\left\{f(\alpha v) - B_{21}\frac{L}{(\log X)^{1/14}}\right\}, \tag{1.8}$$

and for the remaining S-sums on the right of (1.7) we shall derive upper estimates from Theorem 8.3; note that upper estimates are in order since

$$1 - \frac{\log p}{\log y} > 0 \quad \text{if} \quad p < y.$$

First we set

$$\xi^2 = \frac{X^\alpha}{(\log X)^{A_4}},$$

and then we apply Theorem 8.3 to each sum $S(\mathscr{A}_p; \mathfrak{P}, z)$ but with ξ^2/p in place of ξ^2. We may do so because $u > 1/\alpha$ and

$$\frac{\xi^2}{p} \geqslant \frac{X^{\alpha - 1/u}}{(\log X)^{A_4}}. \tag{1.9}$$

We obtain

$$\sum_{\substack{z \leqslant p < y \\ p \in \mathfrak{P}}} \left(1 - \frac{\log p}{\log y}\right) S(\mathscr{A}_p; \mathfrak{P}, z) \leqslant \sum_{\substack{z \leqslant p < y \\ p \in \mathfrak{P}}} \left(1 - \frac{\log p}{\log y}\right)$$

$$\times \left(\frac{\omega(p)}{p} XW(z)\left\{F\left(\frac{\log \xi^2/p}{\log z}\right) + C_4 \frac{L}{(\log \xi/\sqrt{p})^{1/14}}\right\} + \sum_{\substack{n < \xi^2/p \\ n | P(z)}} 3^{\nu(n)} |R_{pn}|\right). \tag{1.10}$$

It is now necessary to develop the right-hand side of (1.10) into a more convenient (as well as agreeable) form. First of all, by† (8.2.14) and using (1.9) (cf. (8.5.4)) we have

$$F\left(\frac{\log \xi^2/p}{\log z}\right) - F\left(\frac{\log X^\alpha/p}{\log z}\right) \leqslant A_4 F\left(\frac{1}{2}v\left(\alpha - \frac{1}{u}\right)\right)\frac{\log\log X}{\log z}\,\frac{\log z}{\log(\xi^2/p)}$$

$$\leqslant \frac{B_{23}F(\tfrac{1}{2}v(\alpha - 1/u))}{\alpha - 1/u}\,\frac{\log\log X}{\log X}.$$

Hence, using (1.9) again, the right-hand side of (1.10) is at most

$$XW(z)\sum_{z\leqslant p<y}\left\{\left(1 - \frac{\log p}{\log y}\right)\frac{\omega(p)}{p}F\left(\frac{\log X^\alpha/p}{\log z}\right) + C_5\frac{\omega(p)}{p}\frac{L}{(\log X)^{1/14}}\right\}$$

$$+ \sum_{\substack{z\leqslant p<y\\ p\in\mathfrak{P}}}\ \sum_{\substack{n<\xi^2/p\\ n|P(z)}} 3^{\nu(n)}|R_{pn}|$$

$$\leqslant XW(z)\left\{\sum_{z\leqslant p<y}\left(1 - \frac{\log p}{\log y}\right)\frac{\omega(p)}{p}F\left(\frac{\log X^\alpha/p}{\log z}\right) + C_6\frac{L}{(\log X)^{1/14}}\right\}$$

$$+ \sum_{\substack{d<\xi^2\\ (d,\overline{\mathfrak{P}})=1}} \mu^2(d)3^{\nu(d)}|R_d|$$

since, by (8.3.5),

$$\sum_{z\leqslant p<y}\frac{\omega(p)}{p} \leqslant C_7.$$

If now we apply condition ($R(1, \alpha)$) and use (2.3.6) we may conclude that

$$\sum_{\substack{z\leqslant p<y\\ p\in\mathfrak{P}}}\left(1 - \frac{\log p}{\log y}\right)S(\mathscr{A}_p; \mathfrak{P}, z)$$

$$\leqslant XW(z)\left\{\sum_{z\leqslant p<y}\left(1 - \frac{\log p}{\log y}\right)\frac{\omega(p)}{p}F\left(\frac{\log X^\alpha/p}{\log z}\right) + C_8\frac{L}{(\log X)^{1/14}}\right\}. \quad (1.11)$$

Next we convert the sum on the right of (1.11) to integral form. By (5.2.1) and ($\Omega_2(1, L)$),

$$\sum_{z\leqslant p<w}\left(1 - \frac{\log p}{\log y}\right)\frac{\omega(p)}{p} = \log\frac{\log w}{\log z} - \frac{\log w}{\log y} + \frac{u}{v} + O\left(\frac{C_9 L}{\log X}\right) \quad (1.12)$$

for $z \leqslant w \leqslant y$.

† with $u_1 = \dfrac{\log(\xi^2/p)}{\log z} = v\left(\alpha - \dfrac{1}{u}\right) - vA_4\dfrac{\log\log X}{\log X} \geqslant \tfrac{1}{2}v\left(\alpha - \dfrac{1}{u}\right) = \delta_1$, using $X \geqslant C_3$.

If now we put (cf. Lemma 7.2)

$$D(w) = \sum_{z \leqslant p < w} \left(1 - \frac{\log p}{\log y}\right) \frac{\omega(p)}{p} \quad \text{for} \quad z \leqslant w \leqslant y$$

and

$$E(w) = F\left(\frac{\log X^{\alpha}/w}{\log z}\right) \quad \text{for} \quad z \leqslant w \leqslant y,$$

we obtain by (1.12), in view of (8.2.13),

$$\sum_{z \leqslant p < y} \left(1 - \frac{\log p}{\log y}\right) \frac{\omega(p)}{p} F\left(\frac{\log X^{\alpha}/p}{\log z}\right) = \sum_{z \leqslant p < y} \left(1 - \frac{\log p}{\log y}\right) \frac{\omega(p)}{p} E(p)$$

$$= \left\{\log \frac{v}{u} - 1 + \frac{u}{v} + O\left(\frac{C_9 L}{\log X}\right)\right\} E(y) - \int_z^y D(w)\, dE(w)$$

$$= \log\left\{\frac{v}{u} - 1 + \frac{u}{v}\right\} E(y) + O\left(\frac{C_{10} L}{\log X}\right)$$

$$- \int_z^y \left\{\log \frac{\log w}{\log z} - \frac{\log w}{\log y} + \frac{u}{v}\right\} dE(w)$$

$$= \int_{X^{1/v}}^{X^{1/u}} E(w)\left(1 - u\frac{\log w}{\log X}\right) \frac{dw}{w \log w} + O\left(\frac{C_{10} L}{\log X}\right)$$

$$= \int_u^v F\left(v\left(\alpha - \frac{1}{t}\right)\right)\left(1 - \frac{u}{t}\right)\frac{dt}{t} + O\left(\frac{C_{10} L}{\log X}\right); \tag{1.13}$$

the last step involved the change of variable $w = X^{1/t}$.

Our theorem now follows from (1.7), (1.8), (1.11) and (1.13).

The expression in curly brackets on the right of (1.5) appears to present considerable difficulties because f and F are not known explicitly except for small values of the argument. Fortunately it turns out that we can get very good results even by restricting ourselves to these small values. The following lemma then enables us to express (1.5) in an explicit form having only elementary functions on the right.

LEMMA 9.1. *If*

$$\frac{1}{\alpha} < u < v, \qquad \frac{2}{\alpha} \leqslant v \leqslant \frac{4}{\alpha}, \qquad \lambda \geqslant 0,$$

we have

$$f(\alpha v) - \lambda \int_u^v F\left(v\left(\alpha - \frac{1}{t}\right)\right)\left(1 - \frac{u}{t}\right)\frac{dt}{t}$$

$$= \frac{2\,e^\gamma}{\alpha v}\left\{\log(\alpha v - 1) - \lambda\,\alpha u \log\frac{v}{u} + \lambda(\alpha u - 1)\log\frac{\alpha v - 1}{\alpha u - 1}\right\}. \qquad (1.14)$$

Proof. To take the argument of F first, we note that

$$0 < v\left(\alpha - \frac{1}{u}\right) \leqslant v\left(\alpha - \frac{1}{t}\right) \leqslant v\left(\alpha - \frac{1}{v}\right) = \alpha v - 1.$$

Hence, by (8.2.9) and (8.2.8), the left-hand side of (1.14) equals

$$\frac{2\,e^\gamma}{\alpha v}\left\{\log(\alpha v - 1) - \lambda\alpha\int_u^v \frac{1 - u/t}{\alpha t - 1}\,dt\right\}.$$

If now we use the relation

$$\frac{1 - u/t}{\alpha t - 1} = \frac{u}{t} - \frac{\alpha u - 1}{\alpha t - 1},$$

and integrate, the result follows at once.

2. Application to the Prime Twins and Goldbach Problems

Our first application of Theorem 9.1 is to the problem of prime twins and to Goldbach's conjecture. Until recently Theorem 9.2 below was perhaps the most striking achievement of the modern sieve method and, as a quantitative statement about the representation of $p + h$ (or $N - p$) as a P_3, is still unsurpassed. However, in a qualitative sense, Chapter 11 contains a better result.

Theorem 9.2. *There is an absolute constant N_0 such that for $N \geqslant N_0$ and each even number h satisfying $0 < |h| \leqslant N$,*

$$|\{p : p \leqslant N,\, p + h = P_3\}| \geqslant \frac{13}{3}\prod_{p>2}\left(1 - \frac{1}{(p-1)^2}\right)\prod_{2<p|h}\frac{p-1}{p-2}\cdot\frac{N}{\log^2 N}; \qquad (2.1)$$

in particular (letting $N \to \infty$) it follows that, given any even number $h \neq 0$, the equation

$$p + h = P_3$$

has infinitely many solutions.

Thus, if $N \geqslant N_0$ and N is even,

$$|\{p : p \leqslant N,\ N - p = P_3\}| \geqslant \frac{13}{3} \prod_{p>2} \left(1 - \frac{1}{(p-1)^2}\right) \prod_{2<p|N} \frac{p-1}{p-2} \cdot \frac{N}{\log^2 N} \ (2 \mid N); \qquad (2.2)$$

in other words, every sufficiently large even number N can be represented in the form

$$N = p + P_3.$$

Remark. We have deliberately stated the result in a simple form, but it will be clear from the proof that we can be more precise in a number of respects. First of all, we can say more about the prime decomposition of the almost-primes P_3 occurring in the statement: namely, we know that each prime factor of P_3 is $\geqslant (li\,N)^{1/8}$, and that either P_3 has at most *two* prime factors (so that it is, in fact, a P_2), or P_3 has a prime factor between $(li\,N)^{1/8}$ and $(li\,N)^{3/8}$. One consequence of this closer specification of the P_3's is that the expressions on the right of (2.1) and (2.2) have the "right" order of magnitude. Indeed, the expressions on the right of (2.1) and (2.2) have precisely the form conjectured by Hardy and Littlewood on the basis of their circle method, for the *actual* twin prime and Goldbach problems, and differ only in having 13/3 where Hardy and Littlewood had the factor 2.

Actually, 13/3 enters the proof as $4 \log 3 - \varepsilon$ for any $\varepsilon > 0$, and there is little doubt that we could arrive at an even bigger constant by resorting to numerical integration instead of "making do" with the elementary Lemma 9.1. Such an improvement would be by no means without interest; for one possible attack on the problem of

$$N = p + P_2$$

is to attempt to show that the number of numbers $N - p$ having *exactly* three large (in some precise sense) prime factors, and no small ones, is smaller than the expression on the right of (2.2)—and it would therefore be important to have in (2.2) as good a lower estimate as possible. The problem we have just mentioned appears to us to be, in any case, of great interest (see Chapter 11).

Proof of Theorem 9.2. We shall deal only with (2.1), since (2.2) follows from (2.1) on taking $h = -N$ (remember that an almost prime P_r is not necessarily positive). Let $\mathscr{A}$ be the sequence $\mathscr{L} = \{p + h : p \leqslant N\}$ $(2 \mid h)$; this was studied in Theorem 3.11, for the purpose of obtaining an upper bound, and we recall that we then took K to be h. We see from Example 5 of

Chapter 1 (as we did also in the proof of Theorem 3.11) that we should take

$$X = li\,N, \qquad \omega(p) = \frac{p}{p-1} \quad \text{if} \quad p \nmid h,$$

and we then obtain (cf. (3.6.8))

$$|R_d| \leqslant E(N, d).$$

Since h is even, (Ω_1) and $(\Omega_2(1, L))$ are satisfied, by virtue of the elementary estimates (5.1.1) and (5.1.5), with some absolute constants A_1, A_2 and with

$$L \leqslant O(1) + \sum_{p|h} \frac{\log p}{p-1} \ll \log\log 3|h| \ll \log\log 3N$$

using the fact that $|h| \leqslant N$.

We have also to check on condition $(R(1, \alpha))$. By (3.7.4) we have, with a suitable constant C_1, that

$$\sum_{d < N^{1/2}/\log^{C_1} N} \mu^2(d)\, 3^{\nu(d)}\, E(N, d) = O\left(\frac{N}{\log^3 N}\right) = O\left(\frac{X}{\log^2 X}\right),$$

so that $(R(1, \alpha))$ is satisfied with

$$\alpha = \tfrac{1}{2}.$$

We are now ready to apply Theorem 9.1 with Lemma 9.1; the Lemma first. Taking

$$u = \tfrac{8}{3}, \qquad v = 8,$$

Lemma 9.1 with $\alpha = \frac{1}{2}$ yields

$$f(\alpha v) - \lambda \int_u^v F\left(v\left(\alpha - \frac{1}{t}\right)\right)\left(1 - \frac{u}{t}\right)\frac{dt}{t}$$

$$= \frac{e^\gamma}{2}\left\{\log 3 - \lambda\frac{4}{3}\log 3 - \lambda\frac{1}{3}\log 9\right\} = \frac{e^\gamma}{2}\left(1 - \frac{2}{3}\lambda\right)\log 3.$$

Accordingly Theorem 9.1, with the above choice of the parameters, leads us to the estimate

$$W\left(\mathscr{X}; \mathfrak{P}_h, 8, \frac{8}{3}, \lambda\right) \geqslant XW(X^{1/8})\left\{\frac{e^\gamma}{2}\left(1 - \frac{2}{3}\lambda\right)\log 3 - \frac{B_{24}}{(\log X)^{1/15}}\right\}. \quad (2.3)$$

With the help of (5.2.5) (suitably interpreted—we need to remember that h is even, that $\omega(p) = 0$ if $p \mid h$ and that $\kappa = 1$) we can put $W(X^{1/8})$ into its final form

$$W(X^{1/8})$$

$$= \prod_{p \nmid h} \frac{1 - 1/(p-1)}{1 - 1/p} \prod_{p \mid h} \frac{p}{p-1} \frac{8e^{-\gamma}}{\log X} \left\{1 + O\left(\frac{\log\log 3N}{\log X}\right)\right\}$$

$$= \prod_{p>2} \frac{p(p-2)}{(p-1)^2} \prod_{2<p \mid h} \frac{(p-1)^2}{p(p-2)} 2 \prod_{2<p \mid h} \frac{p}{p-1} \frac{8e^{-\gamma}}{\log X} \left\{1 + O\left(\frac{\log\log 3N}{\log X}\right)\right\}$$

$$= 2 \prod_{p>2} \frac{p(p-2)}{(p-1)^2} \prod_{2<p \mid h} \frac{p-1}{p-2} \frac{8e^{-\gamma}}{\log X} \left\{1 + O\left(\frac{\log\log 3N}{\log X}\right)\right\}.$$

Since $X = li\, N$, (2.3) now implies, for $N \geqslant N_1$, that

$$W\left(\mathscr{X}; \mathfrak{P}_h, 8, \frac{8}{3}, \lambda\right) \geqslant 8 \prod_{p>2} \left(1 - \frac{1}{(p-1)^2}\right) \prod_{2<p \mid h} \frac{p-1}{p-2} \cdot \frac{N}{\log^2 N}$$

$$\times \left\{\left(1 - \frac{2}{3}\lambda\right) \log 3 - \frac{B_{25}}{(\log N)^{1/15}}\right\}. \tag{2.4}$$

We have now to "interpret" (2.4). We consider the elements $a = p + h$, counted in $W(\mathscr{X}, \mathfrak{P}_h, 8, \frac{8}{3}, \lambda)$. First of all we may disregard those a's for which $(a, h) > 1$; for then necessarily $(a, h) = p$, so that the number of such elements a is at most $\nu(h) = O(\log N)$, and can be absorbed into the error term. Next, we may set aside from further consideration also those numbers a that are divisible by the square of a prime p_1 satisfying

$$X^{1/8} \leqslant p_1 < X^{3/8}, \qquad p_1 \in \mathfrak{P}_h; \tag{2.5}$$

for their number is at most

$$\sum_{\substack{X^{1/8} \leqslant p_1 < X^{3/8} \\ p_1 \in \mathfrak{P}_h}} \sum_{\substack{p \leqslant N \\ p \equiv -h \bmod p_1^2}} 1 \leqslant \sum_{X^{1/8} \leqslant p_1 < X^{3/8}} \left(\frac{N}{p_1^2} + 1\right)$$

$$\ll NX^{-1/8} + X^{3/8} \ll N^{7/8} \log N, \tag{2.6}$$

and so may also be merged with the error term. For the remaining, non-excluded members a of $\mathscr{X}$ we may assume therefore that each is coprime with h *and* that it is "squarefree" with respect to the primes in (2.5).

Let a then be a non-excluded member of $\mathscr{X}$ whose weight

$$1 - \lambda \sum_{\substack{X^{1/8} \leqslant p_1 < X^{3/8} \\ p_1 | a,\ p_1 \in \mathfrak{P}_h}} \left(1 - \frac{8}{3}\frac{\log p_1}{\log X}\right)$$

makes a *positive* contribution to the left-hand side of (2.4). Remember that a has no prime factor less than $X^{1/8}$. Clearly a may have prime divisors $\geqslant X^{3/8}$, but for each of these

$$1 - \frac{8}{3}\frac{\log p_1}{\log X} \leqslant 0;$$

hence if we throw into the weight of a these terms corresponding to the large primes we can only increase the weight of a. Thus the weight of a is at most

$$1 - \lambda\left(\Omega(a) - \frac{8}{3}\frac{\log |a|}{\log X}\right).$$

Now take λ to be a number satisfying

$$\lambda > \tfrac{3}{4}. \tag{2.7}$$

Then if there were left an a in our counting function W such that $\Omega(a) \geqslant 4$, the weight of a would be at most

$$1 - \lambda\left(4 - \frac{8}{3}\frac{\log 2N}{\log X}\right) \leqslant 0$$

provided N is chosen large enough, and this is a contradiction since the weight of a is positive (we have used the fact that $\log |a| \leqslant \log (N + |h|) \leqslant \log (2N)$). Hence each (non-excluded) a with positive weight has $\Omega(a) \leqslant 3$; but then, since the weight is obviously $\leqslant 1$, we obtain at once

$$|\{p\colon p \leqslant N,\ p + h = P_3\}| \geqslant W\left(\mathscr{X}; \mathfrak{P}_h, 8, \tfrac{8}{3}, \lambda\right) + O(N^{7/8} \log N),$$
$$N \geqslant N_2, \tag{2.8}$$

for any λ satisfying (2.7).

Finally, we combine (2.4) with (2.8). For $\lambda = 3/4$,

$$8(1 - \tfrac{2}{3}\lambda) \log 3 = 4 \log 3 > \tfrac{13}{3},$$

so that we can evidently find a $\lambda > \frac{3}{4}$ such that the theorem is true.

Both the applications given in Theorem 9.2 are instances of a linearized sieve problem. In common with all such problems, the remainder terms are here a source of genuine difficulty, because they require information about the distribution of primes in arithmetic progressions. The choice

$$\alpha = \tfrac{1}{2}$$

is forced upon us by Bombieri's theorem, according to which

$$\sum_{d < N^{\alpha}/\log^{C_1} N} \mu^2(d)\, 3^{\nu(d)}\, E(N, d) \ll \frac{N}{\log^3 N}, \tag{2.9}$$

with a suitable constant C_1, is true if $\alpha = \frac{1}{2}$. It has been conjectured that (2.9) remains true even if $\frac{1}{2} < \alpha < 1$, and it is worth remarking that quite a small increase in the range of α's admissible for (2.9) would lead to the solvability of

$$N = p + P_2 \quad \textit{if} \quad 2 \mid N \ \textit{and}\ N\ \textit{is sufficiently large.} \tag{2.10}$$

If we took $\alpha v = 4, \alpha u = \frac{4}{3}$ in the above proof, (2.10) would follow if (2.9) were true with an $\alpha > \frac{4}{7}$; and a more careful calculation† with a better choice of αu shows that α needs to be only larger than 0·546. Further improvements are possible, and Buchstab has stated that (2.10) holds even if $\alpha = 0{\cdot}531$ is admissible in (2.9). It is possible that improvements of (2.9) lie very deep, but we now see that existing sieve methods are good enough to take advantage of even quite a small gain in the range of admissible α's.

3. The Weighted Sieve in Applicable Form

In the last section we gave two very special applications of Theorem 9.1. We shall now derive from the same theorem a very general result which is, nevertheless, remarkably simple to apply.

Naturally such a general result can be true only if we impose the same conditions as obtained in Theorem 9.1; but, with concrete applications in mind, we shall strengthen one of these, namely $(\Omega_2(1, L))$, to $(\Omega_2^*(1))$, where $(\Omega_2^*(\kappa))$ is given by

$$(\Omega_2^*(\kappa)) \qquad -A_2 \log\log 3X \leqslant \sum_{v \leqslant p < w} \frac{\omega(p)}{p} \log p - \kappa \log \frac{w}{v} \leqslant A_2$$

$$\text{if} \quad 2 \leqslant v \leqslant w.$$

† based on the proof of Theorem 9.2.

The effect of this change is to require that† the parameter L in $(\Omega_2(\kappa, L))$ satisfies

$$L \leqslant A_2 \log\log 3X. \tag{3.1}$$

We have also to introduce a further condition, (Ω_3) below, which is of minor importance but necessary for the arithmetical interpretation of the fundamental inequality (1.5), and will play a part in the proof of Theorem 9.3 analogous to that of (2.8) in the proof of Theorem 9.2. The new condition is‡

$$(\Omega_3) \qquad \sum_{\substack{z \leqslant p < y \\ p \in \mathfrak{P}}} |\mathscr{A}_{p^2}| \leqslant A_3 \left(\frac{X \log X}{z} + y\right) \quad \text{if} \quad 2 \leqslant z \leqslant y.$$

To interpret (1.5) we also need to know about the *size* of the elements a of $\mathscr{A}$; this was already evident in the proof of Theorem 9.2, but there, of course, such information was readily available. As a kind of exponential measure for the magnitude of the a's we introduce, for each positive integer r, the function

$$\Lambda_r = r + 1 - \frac{\log 4/(1 + 3^{-r})}{\log 3}; \tag{3.2}$$

and we remark that $\Lambda_1 = r$ and that

$$r + 1 - \frac{\log 4}{\log 3} \leqslant \Lambda_r \leqslant r + 1 - \frac{\log 3{\cdot}6}{\log 3} \quad \text{for} \quad r \geqslant 2. \tag{3.3}$$

THEOREM 9.3. (Ω_1), $(\Omega_2{}^*(1))$, (Ω_3), $(R(1, \alpha))$: *Suppose that*

$$(a, \mathfrak{P}) = 1 \quad \textit{for all} \quad a \in \mathscr{A}. \tag{3.4}$$

Let δ be a real number satisfying

$$0 < \delta \leqslant \tfrac{2}{3}, \tag{3.5}$$

and let r ($\geqslant 2$) be so large that

$$|a| \leqslant X^{\alpha(\Lambda_r - \delta)} \quad \textit{for all } a \in \mathscr{A}. \tag{3.6}$$

†Actually, (3.1) is more restrictive than our method requires. It will be clear from the proof of Theorem 9.3 that it would suffice to have $L = o((\log X)^{1/14})$.

‡ An even weaker condition would do. We could replace $\log X$ by $(\log X)^{A_6}$.

Then we have

$$|\{P_r : P_r \in \mathscr{A}\}| \geqslant \frac{\delta}{\alpha} \prod_p \frac{1 - \omega(p)/p}{1 - 1/p} \frac{X}{\log X} \quad \textit{for} \quad X \geqslant X_0 = X_0(r, \delta). \tag{3.7}$$

Moreover, if $q = q(P_r)$ *denotes the least prime factor of* P_r, *then for each* P_r *counted on the left of* (3.7) *we have*

$$q(P_r) \geqslant X^{\alpha/4}. \tag{3.8}$$

Remark 1. One cannot but be struck by the simplicity of this theorem. In effect it says: given any sieve problem satisfying the given conditions—and in practice this usually means any linear or linearized problem—all one has to do is to measure the largest element of $\mathscr{A}$ in terms of Λ_r (as in (3.6)), and if r is the least positive integer ($\geqslant 2$) for which (3.6) holds then, according to (3.7), $\mathscr{A}$ contains infinitely many almost-primes of order r. The theorem applies, for example, to the special sequences of Theorem 9.2: bearing in mind that there $\alpha = \frac{1}{2}$, (3.6) can obviously be satisfied with $r = 3$, and the essence of Theorem 9.2 follows at once. However, we proved Theorem 9.2 separately because, whereas in Theorem 9.3 the emphasis is on finding a smallest possible r for which δ can be chosen positive, in Theorem 9.2 we were concerned also to find as large an admissible δ as possible.

Remark 2. Note that the product on the right of (3.7) is uniformly positive since, by (5.2.6),

$$\prod_p \frac{1 - \omega(p)/p}{1 - 1/p} \geqslant \exp\{-A_1 A_2 (2 + A_2)\} > 0. \tag{3.9}$$

Proof of Theorem 9.3. We begin by choosing

$$v = \frac{4}{\alpha}, \qquad u = \frac{1 + 3^{-r}}{\alpha} \tag{3.10}$$

in conformity with requirement (1.3) of Theorem 9.1 (and the requirement for u and v in Lemma 9.1), and we take λ to be given by

$$\frac{1}{\lambda} = r + 1 - u\alpha(\Lambda_r - \delta). \tag{3.11}$$

By (3.3) we have $\Lambda_r \leqslant r$; hence, using $\delta > 0$, we find that

$$\frac{1}{\lambda} \geqslant r + 1 - (1 + 3^{-r})r = 1 - \frac{r}{3^r} \geqslant 1 - \frac{2}{9},$$

so that (1.4) is satisfied with $A_7 = 9/7$. With these choices we apply Lemma 9.1. We obtain

$$f(\alpha v) - \lambda \int_u^v F\left(v\left(\alpha - \frac{1}{t}\right)\right)\left(1 - \frac{u}{t}\right)\frac{dt}{t} = \frac{e^\gamma}{2}\log 3$$

$$\times \left\{1 - \lambda(1 + 3^{-r})\frac{\log 4/(1 + 3^{-r})}{\log 3} + \lambda 3^{-r}(r + 1)\right\}$$

$$= (r + 1 - (1 + 3^{-r})(\Lambda_r - \delta))^{-1}\frac{e^\gamma}{2}\log 3\,\{r + 1 - (1 + 3^{-r})$$

$$\times (\Lambda_r - \delta) - (1 + 3^{-r})(r + 1 - \Lambda_r) + 3^{-r}(r + 1)\}$$

$$= (r + 1 - (1 + 3^{-r})(\Lambda_r - \delta))^{-1}\frac{e^\gamma}{2}(\log 3)\,\delta\,(1 + 3^{-r})$$

$$\geqslant (r + 1 - \Lambda_r + \delta)^{-1}\frac{e^\gamma}{2}(\log 3)\,\delta \geqslant \delta\frac{e^\gamma}{4}\frac{\log 9}{\log 8} \tag{3.12}$$

since, by (3.3) and (3.5),

$$r + 1 - \Lambda_r + \delta \leqslant \frac{\log 4}{\log 3} + \frac{2}{3} \leqslant \log 4 + \log 2.$$

Further, by (5.2.5) we have

$$W(X^{1/v}) = \prod_p \frac{1 - \omega(p)/p}{1 - 1/p}\frac{4e^{-\gamma}}{\alpha \log X}\left\{1 + O\left(\frac{\log\log 3X}{\log X}\right)\right\},$$

so that by (1.5) and (3.12) we have

$$W(\mathscr{A}; \mathfrak{P}, v, u, \lambda) \geqslant \prod_p \frac{1 - \omega(p)/p}{1 - 1/p}\frac{X}{\log X}\left\{\frac{\delta}{\alpha}\frac{\log 9}{\log 8} - \frac{B_{26}(r)}{(\log X)^{1/15}}\right\}, \tag{3.13}$$

where v, u and λ are given by (3.10) and (3.11).

We have now to interpret (3.13). First of all, the a's counted on the left of (3.13) have, in view of (3.4), no prime divisor less than $X^{1/v}$ (note that 0 (if in $\mathscr{A}$) is not counted since we may assume that $X > 2^v$). Next, we may set aside those a's which are not squarefree with respect to the primes p satisfying

$$X^{1/v} \leqslant p < X^{1/u}, \qquad p \in \mathfrak{P}. \tag{3.14}$$

To see this we have only to observe that, by virtue of (Ω_3), their number is at most, using (3.10),

$$\sum_{\substack{X^{1/v} \leqslant p < X^{1/u} \\ p \in \mathfrak{P}}} \sum_{\substack{a \in \mathscr{A} \\ a \equiv 0 \bmod p^2}} 1 \leqslant A_3 (X^{1-1/v} \log X + X^{1/u}) = O\left(\frac{X}{\log^2 X}\right). \qquad (3.15)$$

We come finally to the remaining a's. For such an a, p can be a repeated prime factor only if $p \geqslant X^{1/u}$, and for such a p

$$1 - u\frac{\log p}{\log X} \leqslant 0.$$

Accordingly we only increase the weight of a if we extend the sum over the prime divisors of a satisfying (3.14) (cf. (1.2)) to *all* prime divisors of a, taking account even of their multiplicities. Hence the weight of a (remaining) a is at most

$$1 - \lambda \left\{\Omega(a) - u\frac{\log |a|}{\log X}\right\} \leqslant 1 - \lambda \{\Omega(a) - u\alpha(\Lambda_r - \delta)\} \qquad (3.16)$$

by (3.6). If $\Omega(a) \geqslant r + 1$ this quantity is, by (3.11), $\leqslant 0$, so that an a with positive weight necessarily has $\Omega(a) \leqslant r$. Since the weight of each a is at most 1, we may conclude from (3.4) and (3.15) that

$$W(\mathscr{A}; \mathfrak{P}, v, u, \lambda) \leqslant |\{P_r : P_r \in \mathscr{A},\ q(P_r) \geqslant X^{\alpha/4}\}| + O\left(\frac{X}{\log^2 X}\right).$$

If we now combine this inequality with (3.13), the theorem follows at once.

4. Almost-Primes in Intervals and Arithmetic Progressions

We first apply Theorem 9.3 to the existence of almost-primes in intervals and in arithmetic progressions. An approach to both problems can be derived from the following more general result; it will be convenient to have before us the formula

$$\Lambda_r = r + 1 - \frac{\log 4/(1+3^{-r})}{\log 3}$$

for the number Λ_r introduced in (3.2).

Theorem 9.4. *Let r be a positive integer $\geqslant 2$, and let k and l be a pair of coprime natural numbers. Then for every δ satisfying*

$$0 < \delta \leqslant \tfrac{2}{3}$$

there is a number $x_0 = x_0\ (r, \delta)$ *such that if*

$$x^{1-1/(\Lambda_r-\delta)} \geqslant k, \qquad x \geqslant x_0, \tag{4.1}$$

we have

$$|\{P_r\colon x - k\,x^{1/(\Lambda_r-\delta)} < P_r \leqslant x,\ P_r \equiv l \bmod k\}| \geqslant \delta\,(\Lambda_r - \delta)\frac{k}{\phi(k)}\frac{x^{1/(\Lambda_r-\delta)}}{\log x}. \tag{4.2}$$

Proof. We put

$$y = k\,x^{1/(\Lambda_r-\delta)}$$

and consider the sequence $\mathscr{A}$ given by

$$\mathscr{A} = \{n\colon x - y < n \leqslant x,\ n \equiv l \bmod k\},$$

in association with the sifting set $\mathfrak{P} = \mathfrak{P}_k$, where, we recall,

$$\mathfrak{P}_k = \{p\colon p \nmid k\}.$$

In view of (4.1), $\mathscr{A}$ consists entirely of positive integers. The sequence $\mathscr{A}$ was analysed in Example 2 (see Chapter 1), and in accordance with that analysis (cf. 1.3.8)) we now choose

$$X = \frac{y}{k} = x^{1/(\Lambda_r-\delta)}, \qquad \omega(p) = 1 \quad \text{if} \quad p \nmid k. \tag{4.3}$$

Then (Ω_1) is obviously satisfied (with $A_1 = 2$), and so is $(\Omega_2{}^*(1))$ as we can easily check from the elementary relations (5.1.1), (5.1.5) and from (4.1). Furthermore, by (1.3.9) and (1.4.9) we see that

$$|R_d| \leqslant 1 \quad \text{if} \quad \mu(d) \neq 0 \quad \text{and} \quad (d, k) = 1,$$

so that by Lemma 3.4 (with $h = 3$)

$$\sum_{\substack{d < X/\log^5 X \\ (d,k)=1}} \mu^2(d)\,3^{\nu(d)}\,|R_d| \ll \frac{X}{\log^2 X};$$

thus $(R(1, \alpha))$ is satisfied with

$$\alpha = 1. \tag{4.4}$$

Finally, (Ω_3) is valid, as can easily be seen from (1.3.6), condition (3.4) holds because $(k, l) = 1$, and (3.6) follows immediately from the choice (4.3) of X.

We may therefore apply Theorem 9.3. The conclusion of Theorem 9.4 then follows directly from (3.7) on making use of (4.3) and (4.1).

If we take $k = 1$ in Theorem 9.4 we obtain a statement about the existence of almost-primes of order r in short intervals or, what amounts to the same thing, about gaps between consecutive P_r's. If we take $k = [x^{1-1/(\Lambda_r-\delta)}]$ on the other hand, we obtain an upper estimate for the least P_r in the arithmetic progression l mod k. The latter result is perhaps not immediately obvious, but we have only to note that, with this choice of k, there is, according to (4.2), an almost-prime P_r satisfying

$$0 < P_r \leqslant (k+1)^{(\Lambda_r-\delta)/(\Lambda_r-\delta-1)} \leqslant k^{(\Lambda_r-2\delta)/(\Lambda_r-2\delta-1)}$$

provided that δ is sufficiently small and k is large enough.

The small values of r lead to especially interesting results, and we therefore place on record that

$$\Lambda_2 = 3 - \frac{\log \frac{18}{5}}{\log 3} > \frac{11}{6},$$

$$\Lambda_3 = 4 - \frac{\log \frac{27}{7}}{\log 3} > \frac{11}{4},$$

and

$$\Lambda_4 = 5 - \frac{\log \frac{162}{41}}{\log 3} > \frac{37}{10} > \frac{11}{3};$$

using rather a crude procedure, we may also deduce from (3.3) that†

$$r - \tfrac{2}{7} < \Lambda_r < r - \tfrac{1}{7} \quad \text{for all} \quad r \geqslant 2. \tag{4.5}$$

Using these facts, we arrive at‡

THEOREM 9.5. *There exist numbers*

P_2 in the interval $x - x^{6/11} < P_2 \leqslant x$ if $x \geqslant x_2$,

P_3 in the interval $x - x^{4/11} < P_3 \leqslant x$ if $x \geqslant x_3$,

P_4 in the interval $x - x^{3/11} < P_4 \leqslant x$ if $x \geqslant x_4$,

and P_r in the interval $x - x^{1/(r-2/7)} < P_r \leqslant x$ if $x \geqslant x_r$, $r \geqslant 2$.

THEOREM 9.6. *For any pair of coprime natural numbers k, l there exist numbers*

$$P_2 \leqslant k^{11/5}, \quad P_2 \equiv l \bmod k \quad \text{if} \quad k \geqslant k_2,$$

$$P_3 \leqslant k^{11/7}, \quad P_3 \equiv l \bmod k \quad \text{if} \quad k \geqslant k_3,$$

$$P_4 \leqslant k^{11/8}, \quad P_4 \equiv l \bmod k \quad \text{if} \quad k \geqslant k_4,$$

and $$P_r \leqslant k^{1+1/(r-9/7)}, \quad P_r \equiv l \bmod k \quad \text{if} \quad k \geqslant k_r, \quad r \geqslant 2.$$

† Actually, $\Lambda_r > r - \frac{4}{15} > r - \frac{3}{11} > r - \frac{2}{7}$ for all $r \geqslant 2$.

‡ Here x_r, k_r are sufficiently large constants which depend on r only.

The two theorems constitute natural analogues for almost-primes of classical problems relating to the distribution of primes in intervals and in arithmetic progressions. The first result in each theorem provides the closest point of comparison with the corresponding prime cases: thus we have at present the exponent $\frac{7}{12} + \varepsilon$ in place of $\frac{6}{11}$ for primes in intervals and the exponent 550 in place of $\frac{11}{5}$ for primes in arithmetic progressions.

5. Almost-Primes Representable by Irreducible Polynomials $F(n)$

The theme of this section is the classical problem about the existence of infinitely many primes in irreducible polynomial sequences. Theorem 9.7 below makes a contribution to the corresponding problem for almost-primes; it is the best result of its kind known at the present time.

Theorem 9.7. *Let $F(n)$ be an irreducible polynomial of degree $g(\geqslant 1)$ with integer coefficients. Let $\rho(p)$ denote the number of solutions of the congruence*

$$F(m) \equiv 0 \bmod p,$$

and suppose that

$$\rho(p) < p \quad \textit{for all } p. \tag{5.1}$$

Then we have

$$|\{n:\ 1 \leqslant n \leqslant x,\ F(n) = P_{g+1}\}| \geqslant \frac{2}{3}\prod_p \frac{1 - \rho(p)/p}{1 - 1/p}\frac{x}{\log x}$$

$$\textit{for} \quad x \geqslant x_0 = x_0(F). \tag{5.2}$$

In particular, there exist† *infinitely many integers n such that $F(n)$ consists of at most $g + 1$ prime factors.*

Remark 1. We note that the product on the right of (5.2) is, by (3.9), greater than some positive constant depending only on F.

Remark 2. The same argument will yield (cf. also proof of Theorem 9.5) the following gap theorem: *If F is as in Theorem 9.7, then there exist integers n in the interval $x - y < n \leqslant x$ such that $F(n) = P_{g+1}$ whenever $y \geqslant x^{g/(g+5/7)}$* $(x \geqslant x_1(F))$.

Proof. With Theorem 9.3 at our disposal, the proof is largely a matter of checking that Theorem 9.3 is applicable.

We take as $\mathscr{A}$, the sequence to be sifted, the sequence

$$\mathscr{A} = \{F(n):\ 1 \leqslant n \leqslant x\}$$

† To draw this last conclusion we must have $x_0(F)/x \to 0$ as $x \to \infty$.

which was analysed in Example 3, and as the sifting set $\mathfrak{P}$ the set $\mathfrak{P}_1$ of all primes. Thus, in accordance with (1.3.13) and (1.4.15), we choose

$$X = x, \qquad \omega(p) = \rho(p) \quad \text{for all } p; \tag{5.3}$$

and by (1.4.15), (1.3.14) and (1.3.16) we then have

$$|R_d| \leqslant g^{\nu(d)} \quad \text{if} \quad \mu(d) \neq 0. \tag{5.4}$$

We now check one by one that the conditions of Theorem 9.3 are satisfied. Certainly relations (5.1) and (1.3.16) together imply (Ω_1). Next, $(\Omega^*{}_2(1))$ is a consequence of Nagel's result (cf. (1.3.17))

$$\sum_{p<w} \frac{\rho(p)}{p} \log p = \log w + O_F(1). \tag{5.5}$$

As for $(R(1, \alpha))$, we see from (5.4) and Lemma 3.4 that

$$\sum_{d<X/\log^{3g+2}X} \mu^2(d)\, 3^{\nu(d)}\, |R_d| \ll \frac{X}{\log^2 X},$$

so that $(R(1, \alpha))$ is satisfied with

$$\alpha = 1. \tag{5.6}$$

Finally there is (Ω_3); but here we have

$$\rho(p^2) \leqslant gD^2 \tag{5.7}$$

if D denotes the discriminant of F, and therefore, by (1.3.12), (Ω_3) is satisfied. This settles the major conditions; there is also (3.4) to remember, but this holds trivially by virtue of our choice of $\mathfrak{P}$.

We may now apply Theorem 9.3. Taking

$$\delta = \tfrac{2}{3}, \qquad r = g + 1 \tag{5.8}$$

we find, by (4.5), that

$$\Lambda_r - \delta > g + 1 - \tfrac{2}{7} - \tfrac{2}{3} = g + \tfrac{1}{3} - \tfrac{2}{7}.$$

Therefore, using

$$F(n) = a_g n^g + O\,(n^{g-1}),$$

we see that (3.6) is satisfied if $x \geqslant x_0(F)$, where x_0 is a sufficiently large number depending only on F. Hence, if we use the values of our parameters as set out in (5.3), (5.6) and (5.8), Theorem 9.3 now yields (5.2).

6. Almost-Primes Representable by Irreducible Polynomials $F(p)$

The prime twins conjecture is best known and simplest instance of the question where we ask whether a given irreducible polynomial with a prime argument assumes infinitely many prime values. In this section we prove an analogous result for almost-primes which generalizes the first part of Theorem 9.2, and is the best result of its kind known at the present time.

Theorem 9.8. *Let $F(n)$ $(\neq \pm n)$ be an irreducible polynomial of degree g $(\geqslant 1)$ with integer coefficients. Let $\rho(p)$ denote the number of solutions of the congruence*

$$F(m) \equiv 0 \bmod p. \tag{6.1}$$

We shall suppose that

$$\rho(p) < p \quad \textit{for all } p, \tag{6.2}$$

and also that

$$\rho(p) < p - 1 \quad \textit{if} \quad p \nmid F(0) \quad \textit{and} \quad p \leqslant g + 1. \tag{6.3}$$

Then we have

$$|\{p : p \leqslant x, F(p) = P_{2g+1}\}| \geqslant \frac{4}{3} \prod_{p \nmid F(0)} \frac{1 - \rho(p)/(p-1)}{1 - 1/p}$$

$$\times \prod_{p \mid F(0)} \frac{1 - (\rho(p) - 1)/(p-1)}{1 - 1/p} \frac{x}{\log^2 x} \quad \textit{for} \quad x \geqslant x_0 = x_0(F). \tag{6.4}$$

In particular, there exist† *infinitely many primes p such that $F(p)$ consists of at most $2g + 1$ prime factors.*

Remark. Theorem 9.8 is similar in form to Theorem 9.7, but there are two additional conditions: $F(n) \neq \pm n$ and (6.3). Both are natural restrictions; indeed, if $F(n)$ were $\pm n$, our problem would be trivial. Let us look more closely at (6.3). Since F is irreducible, $\neq \pm n$ and has no fixed prime divisor (cf. (6.2)), we see that

$$F(0) \neq 0. \tag{6.5}$$

Suppose (6.3) is false. Then there exists a prime p_0 such that $\rho(p_0) = p_0 - 1$ and $p_0 \nmid F(0)$. This means that for all integers $m \not\equiv 0 \bmod p_0$, $F(m) \equiv 0 \bmod p_0$; in particular, $F(p)$ is divisible by p_0 for all primes $p \neq p_0$.

† cf. footnote attached to Theorem 9.7.

Proof. Once again the proof amounts to little more than checking that Theorem 9.3 is applicable. We consider the sequence

$$\mathscr{A} = \{F(p): p \leqslant x\},$$

and we take $\mathfrak{P}$ to be again the set of all primes, $\mathfrak{P}_1$. We refer back to Example 6 (with $k = 1$) where $\mathscr{A}$ was analysed and choose, in accordance with (1.3.48), (1.4.15) and (1.3.51),

$$X = li\, x, \qquad \omega(p) = \frac{\rho_1(p)}{p-1}p \quad \text{for all} \quad p, \tag{6.6}$$

where (cf. (1.3.53))

$$\rho_1(p) = \begin{cases} \rho(p) & \text{if} \quad p \nmid F(0) \\ \rho(p) - 1 & \text{if} \quad p \mid F(0). \end{cases} \tag{6.7}$$

We have from (1.4.15), (1.3.52) and (1.3.55) that

$$|R_d| \leqslant g^{\nu(d)}\,(E(N, d) + 1) \quad \text{if} \quad \mu(d) \neq 0. \tag{6.8}$$

It is now a matter of confirming the conditions under which Theorem 9.3 is valid. First consider (Ω_1). Here we see that, for $p \leqslant g + 1$, (6.7), (6.3) and (6.2) imply that

$$\rho_1(p) \leqslant p - 2,$$

and hence that

$$\omega(p) \leqslant \frac{p-2}{p-1}p \leqslant \left(1 - \frac{1}{g}\right)p \quad \text{if} \quad p \leqslant g + 1;$$

if, on the other hand, $p \geqslant g + 2$, then $\rho_1(p) \leqslant \rho(p) \leqslant g$ by (1.3.54) and we find that

$$\omega(p) \leqslant \frac{g}{p-1}p \leqslant \frac{g}{g+1}p = \left(1 - \frac{1}{g+1}\right)p,$$

thus verifying (Ω_1) with $A_1 = g + 1$. Condition $(\Omega_2{}^*(1))$ is, as before, a consequence of (5.5). As for $(R(1, \alpha))$, we have from (6.8), Lemma 3.5 (with $U_1 = 3$, $h = 3g$, $k = 1$, $C_1' \geqslant C_1\,(3, 3g, 1)$ and also $C_1' \geqslant \frac{5}{2}$) and Lemma 3.4 that, for sufficiently large values of x,

$$\sum_{d < X^{1/2}/\log^{C_1'} X} \mu^2(d)3^{\nu(d)}|R_d| \ll \frac{x}{\log^3 x} \ll \frac{X}{\log^2 X}.$$

Thus $(R(1, \alpha))$ holds with

$$\alpha = \tfrac{1}{2}. \tag{6.9}$$

We come to (Ω_3), but here we find, if we refer back to the corresponding stage in the proof of Theorem 9.7, that

$$|\{p' \colon p' \leqslant x, F(p') \equiv 0 \bmod p^2\}|$$
$$\leqslant |\{n \colon n \leqslant x, F(n) \equiv 0 \bmod p^2\}| \ll \frac{x}{p^2} + 1 \ll \frac{X \log X}{p^2} + 1,$$

so that (Ω_3) is satisfied. Finally, because of our choice of $\mathfrak{P}$, (3.4) is trivially true.

We may now apply Theorem 9.3. We take

$$\delta = \tfrac{2}{3} \quad \text{and} \quad r = 2g + 1$$

and find that, by (6.9) and (4.5),

$$\alpha(\Lambda_r - \delta) > \tfrac{1}{2}(2g + 1 - \tfrac{2}{7} - \tfrac{2}{3}) = g + \tfrac{5}{14} - \tfrac{1}{3},$$

so that (3.6) is satisfied if $x \geqslant x_1 = x_1(F)$. Hence, by (6.6), Theorem 9.3 yields

$$|\{p \colon p \leqslant x, F(p) = P_{2g+1}\}| \geqslant \frac{4}{3} \prod_p \frac{1 - \rho_1(p)/(p-1)}{1 - 1/p} \frac{li\, x}{\log(li\, x)}.$$

From this we arrive easily at (6.4), using (6.7) and the observation that

$$\frac{li\, x}{\log(li\, x)} \geqslant \frac{x}{\log^2 x}$$

for all sufficiently large x.

The simplest polynomial of the kind considered in Theorem 9.8 is

$$F(n) = n + h, \qquad h \neq 0.$$

Here $\rho(p) = 1$ for all primes p and (6.3) requires that h be even. Under these circumstances we can derive from Theorem 9.8 as we know already from Theorem 9.2, that there exist infinitely many primes p such that

$$p + h = P_3.$$

In this special case Theorem 9.2 has, however, the advantage of providing a better lower estimate than (6.4).

Nevertheless, Theorem 9.8 deals also with the general linear case

$$F(n) = an + b,$$

where

$$ab \neq 0, \quad (a, b) = 1 \ \text{and} \ 2 \mid ab, \tag{6.10}$$

(this condition is an amalgam of (6.2), (6.3) and the requirement $b \neq 0$) and yields in these circumstances the inequality

$$|\{p\colon p \leqslant x,\ ap + b = P_3\}| \geqslant \frac{8}{3} \prod_{p>2} \left(1 - \frac{1}{(p-1)^2}\right) \prod_{2<p|ab} \frac{p-1}{p-2} \frac{x}{\log^2 x} \quad \text{for} \quad x \geqslant x_0 = x_0(a, b); \tag{6.11}$$

in particular (for fixed a, b satisfying (6.10)) there are infinitely many primes p such that

$$ap + b = P_3.$$

If we take

$$F(n) = n^2 + n + 1,$$

we obtain from Theorem 9.8 the lower estimate

$$|\{p\colon p \leqslant x,\ p^2 + p + 1 = P_5\}| \geqslant \frac{8}{3} \prod_{p>2} \left(1 - \frac{1}{(p-1)^2}\right) \prod_{p>3} \left(1 - \frac{\chi(p)}{p-2}\right) \frac{x}{\log^2 x} \quad \text{for} \quad x \geqslant x_0,$$

where x_0 is an absolute constant and

$$\chi(p) = \begin{cases} 1 & \text{if} \quad p \equiv 1 \bmod 3 \\ -1 & \text{if} \quad p \equiv -1 \bmod 3. \end{cases}$$

This result should be compared with Corollary 5.9.2 where we obtained the expression on the right with 8 in place of 8/3 (together with an error term) as an upper bound for the number of primes $\leqslant x$ for which $p^2 + p + 1$ is a prime. Heuristically one would expect 1 in place of 8 in this last result.

Notes

9.1. Kuhn's weighting method occurs for the first time in Kuhn [1]. Other descriptions are given in Wang [8], Levin [5], Halberstam–Roth [1] (Chapter 4), to mention only a few. His contribution to sieve theory has not received the credit it deserves; as Chapters 9, 10 and 11 show, a weighting procedure enhances any of the methods we have described so far, and that to quite an extraordinary extent, and should probably be regarded as indicating a vital component of an optimal sieve method.

The Littlewood weights $\exp(-p \log x/x)$ $(p < x)$ were introduced in this context by Rényi [1, 2]. Implicitly, their use amounts to working with fewer primes in the sifting set (primes less than $2N/\log 2N$) and arriving at a number of unsifted elements that is smaller than the expected result by a factor of order of magnitude $1/\log 2N$.

Of the two Selberg weighted sieves referred to here, the first was described in an unpublished ms (Selberg [2]) which he kindly made available to us; we give a sketch in Section 10.6. The second is described in Miech [1, 3]; Sections 10.4–5 contain a sharper and more general version.

The logarithmic weight function was introduced in Ankeny–Onishi [1], Ankeny [1] and elaborated in Richert [1]. Our account is based on Richert [1], and all the principal results of this chapter are due to him.

Ankeny–Onishi [1] hint, on p.38, at a "more sophisticated" sieve.

Buchstab's weighted sieve was sketched in Buchstab [9], and Buchstab [10] contains a full version. One might say that Buchstab's method is the natural generalization of Kuhn's idea.

Theorem 9.1: cf. Richert [1], Theorem 1.

9.2. Theorem 9.2: cf. Halberstam–Jurkat–Richert [1]; Richert [1], Theorem 8; Buchstab [10]. The last of these appeared first; it gives a result similar to ours but differs in respect of the prime factors of the P_3, reflecting the difference between the methods used. Buchstab proves (in his Theorem 1) that every sufficiently large even number $2N$ can be put in the form $2N = p + n$ where n is either a P_2 or has three prime factors of which one lies in the interval $(N^{1/19}, N^{2/15})$. These were, until recently (cf. Chapter 11), the best available approximations to the prime twin and binary Goldbach problems. Earlier approximations have been mentioned in the Notes for Chapters 2 and 7, but we do not claim to have assembled an exhaustive list. Chen [2] claims (1, 2) in place of our (1, 3) but no detailed calculations are given there; however, his claim is fully justified in his later paper (see Chen [3]), on which our Chapter 11 is based.

With 2 in place of 13/3, the expressions on the right of (2.1) and (2.2) are the asymptotic estimates conjectured by Hardy–Littlewood [2] (pp. 32 and 42) for the counting numbers $|\{p: p \leqslant N, p + h = p'\}|$ and $|\{p: p \leqslant N, N - p = p'\}|$ (with h, N even). For conjectured asymptotic formulae for the number of representations of N as $N = P_r + P_s$ see Miech (*J. r. ang. Math.* **233** (1968), 1–27).

(2.6): The exclusion of numbers that are not squarefree with respect to primes of intermediate size is present already in Kuhn [1], but some expositions of Kuhn's method, e.g. Levin [5] and Halberstam–Roth [1] mistakenly omit this step. Selberg [2] has another way of disposing of such numbers.

(2.9): see Notes for Chapter 3 for conjecture that (2.9) holds with some $\alpha > \frac{1}{2}$. Buchstab's claim cited here occurs in Buchstab [10], but we are not sure that it is justified.

Jutila [1] proves an important variation of Bombieri's theorem which relates to $\pi(x;k,l) - \pi(x-y;k,l)$ in place of $\pi(x;k,l)$ and he states that he can derive from this that every large enough even number is representable as the sum of two almost equal integers one of which is prime and the other a P_8.

9.3. Condition (Ω_3): originally we worked with a stronger condition (cf. Richert [1], (A_4)); Greaves [1] pointed out that the weaker condition (Ω_3) fits equally well (in fact, better) into our theory. As Greaves found, this change is important, for in the application that interested him (Ω_3) could be shown to hold whereas (A_4) could not.

Theorem 9.3: cf. Richert [1], Theorem 2. Jurkat–Richert [1] have a less general result which is also not as sharp because they worked with a worse "exponential measure" (in place of our Λ_r).

9.4. Theorem 9.4: cf. Richert [1], Theorem 3; Jurkat–Richert [1] for an earlier, weaker, version.

Ramachandra pointed out, in correspondence, that from Ramachandra [1] and Theorem 9.4 (quoted from Richert [1]) it follows that there exists a constant $\alpha < \frac{1}{2}$ such that, if x is large enough, there exists a number n in $(x - x^{\alpha}, x)$ with largest prime factor greater than $n^{5/8}$.

Theorem 9.5: cf. Richert [1], Theorem 4. The problem was first studied by Brun [7], who showed that there is a P_{11} in $(x - x^{\frac{1}{2}}, x)$, $x \geqslant x_0$; then by Wang [5]; Mientka [1]; Kuhn [1, 2]; Linnik [2]; Uchiyama [3]; Jurkat–Richert [1]. Theorem 9.5 should be viewed against the background of current information about the distribution of primes; from the recent work of Huxley, *Invent. Math.* **15** (1972), 164–170 we know that there is a prime in $(x - x^{7/12+\varepsilon}, x)$, $x \geqslant x_0(\varepsilon)$ (note that $6/11 < 7/12$!). On the basis of *RH* there is a prime in $(x - x^{\frac{1}{2}+\varepsilon}, x)$, and it is conjectured that there is a prime even in $(x - x^{\varepsilon}, x)$, both for large enough x. (See Schinzel [1] for numerical evidence relating to this question.) The last result in Theorem 9.5 clearly constitutes an approximation to this latter conjecture, in terms of almost-primes.

Theorem 9.6: cf. Richert [1], Theorem 5. The problem was first studied by Fluch [1], who showed that there exists a $P_2 \leqslant k^{15+\varepsilon}$, $P_2 \equiv l \bmod k$, $k \geqslant k_0(\varepsilon)$; then by Levin [8]; Uchiyama [4]; Jurkat–Richert [1]. Theorem 9.6 relates to the famous result of Linnik (*Mat. Sbornik* **15** (1947), 139–178) according to which $p(k,l)$, the least prime in the arithmetic progression $l \bmod k$, satisfies $p(k,l) < k^C$ for some absolute constant C and k sufficiently large; recently Jutila has shown that $C \leqslant 550$ (*Ann. Acad. Sci. Fennicae* **A471**, 1970). On the basis of *GRH*, $p(k,l) < \phi(k)\log^2 k$ for each sufficiently large k and almost all l mod k (Turán, *Acta Sci. Math. Szeged* **8** (1937), 226–235); also, it is known unconditionally (Elliott–Halberstam, Studies in Pure Mathematics, ed. L. Mirsky (1971), Academic Press, 59–61) that

$p(k, l) < \phi(k) \log k . \psi(k)$ for almost all k, and for almost all $l \bmod k$, where $\psi(k)$ is any function tending arbitrarily slowly to ∞. It is conjectured that the latter result is always true, and Theorem 9.6, with r large, offers some support in terms of an almost-prime analogue.

Fluch [2] has proved that there exists a squarefree $P_4 \ll k^{3/2}$ in $l \bmod k$, which sharpens old results of Prachar, *Monatsh. Math.* **62** (1958) 173–176, Erdös, *Monatsh. Math.* **64** (1960), 314–316 on the least squarefree number in an arithmetic progression. Using Fluch's idea and the method of Roth, *J. London Math. Soc.* **26** (1951), 263–268, Halberstam–Roth, *ibid* 268–273, one can show that the assertions of Theorem 9.5 hold with P_r squarefree for $r = 2, 3, 4$ and $[\frac{1}{2}(r+1)]$—free for $r \geqslant 5$; similarly, the assertions of Theorem 9.6 hold with P_r squarefree for $r = 2, 3$. The result of Erdös cited above seems overdue for improvement, and there may be an approach via the sieve. However, one would then need to know *either* how to isolate the squarefree numbers in the sequence $\{n : n \leqslant x, n \equiv l \bmod k\}$, in the sense of establishing an $R(1, \alpha)$—condition with $\alpha > \frac{2}{3}$ for the sequence

$$\mathscr{A} = \{n : n \leqslant x, n \equiv l \bmod k, \mu(n) \neq 0\}$$

(cf. Orr, *J. Number Th.* **3** (1971), 474–497); *or* how to sift with a set $\mathfrak{P}$ consisting of primes $< z$ and of *all* prime squares p^2.

9.5. Theorem 9.7: cf. Richert [1], Theorem 6. The fact that $F(n) = P_{g+1}$ infinitely often was first proved for $g = 1$ by Ricci [6], for $g = 2$ by Kuhn [2], for $g = 3, 4, 5$ by Wang [9] and for $g = 6, 7$ by Levin [10]. For general g the result was first stated by Buchstab [10] (Theorem 4) (but without proof) to be a consequence of his new weighted sieve. The problem was first investigated by Rademacher [1], who obtained $4g - 1$ in place of $g + 1$; Ricci [6] then obtained $3g - 1$; Kuhn [2], Wang [9] obtained expressions of the form $g + c \log g$ by combining Selberg's sieve with Kuhn weights; and Levin [10] arrived at other improvements for g of intermediate size. From a quantitative point of view, (5.2) should be compared with the heuristically expected result (Bateman–Stemmler [1], p.149)

$$|\{n:\ 1 \leqslant n \leqslant x,\ F(n) = p\}| \sim \frac{1}{g} \prod_p \frac{1 - \rho(p)/p}{1 - 1/p} \frac{x}{\log x} \qquad (x \to \infty).$$

(5.7): see Nagel [1].

Ricci [6] (p.77 *et seq.*), [7] (p.109 *et seq.*) are among the best sources for arithmetical aids needed when sifting polynomial sequences. See also Ricci [2, 6] for other remarkable sieve applications which might well be reviewed in the light of the improved sieve techniques now available.

While Theorem 9.7 tells us that $n^2 + 1 = P_3$ infinitely often, it is known that $m^2 + n^2 + 1 = p$ infinitely often (see Notes on Section 3.5). Thus we may hope to do relatively better with polynomials in two variables, and Greaves [1] has confirmed this with the following interesting theorem (proved on the basis of Theorem 9.3, quoted from Richert [1]): "Let f be an irreducible binary form with integer coefficients, degree $g \geqslant 3$, having no fixed prime divisors. Then there is a $\delta > 0$ such that

$$|\{m, n: 1 \leqslant m, n \leqslant x, f(m, n) = P_k\}| \geqslant \delta x^2/\log x^2, \qquad x \geqslant x_0(f),$$

where $k = [\frac{1}{2}g] + 1$; moreover, for each pair m, n counted, $q(P_k) > x^{\frac{1}{2}-\varepsilon}$." Note that $k = 2$ when $g = 3$.

9.6. Theorem 9.8: cf. Richert [1], Theorem 7. Using Brun's sieve, Miech [2] proved such a result with cg in place of $2g + 1$ for some sufficiently large c. In Levin [5] it is stated that $p^2 - p + 1 = P_{11}$ (P_9 on *GRH*), and $\frac{1}{2}(p^2 + 1) = P_{11}$, each for infinitely many primes p. Levin [10] also gives $p^2 - 2 = P_7$ (P_5 under *GRH*), and $p^2 - 2 = P_5$ is proved unconditionally in Rieger [17]. Theorem 9.8 can be extended to polynomials in several variables. Greaves [2] shows in a recent paper that under the usual appropriate conditions on the irreducible polynomial $F(m, n)$ in two variables, $F(p, p') = P_{g+1}$ for more than $\delta' x^2/\log^3 x$ pairs of primes p, p' not exceeding $x(\delta' > 0, x \geqslant x_1(F))$. In his proof, the rôle of Bombieri's theorem is taken by a result close to the theorem of Barban described in the Notes for Section 3.5.

(6.11): this is an approximation to conjecture D (with $a = 1$) of Hardy–Littlewood [2]. Take $a = 8, b = 1$: Vaughan [2] has noted that *either* there are infinitely many primes p for which $8p + 1 = P_2$, a special result better than (6.11) (with $a = 8, b = 1$); *or* there are infinitely many primes p such that $8p + 1 = p_1 p_2 p_3$ where p_1, p_2, p_3 are distinct primes, in which case $\tau(8p + 1) = \tau(8p)$ infinitely often in confirmation of an old conjecture (Erdös–Mirsky *Proc. London Math. Soc.* (3) **2** (1952), 257–271) to the effect that $\tau(n + 1) = \tau(n)$ for infinitely many n. (Vaughan quotes Richert [1], Theorem 7).

In the last result of Section 9.6, the heuristic evidence derives from Bateman–Horn [1] (see also Bateman–Horn [2] and Shanks–Wrench, *Math. Comp.* **17** (1963), 136–54).

Chapter 10

Weighted Sieves: The General Case

1. The First Method

In this chapter we shall begin by extending the weighted sieve method of Chapter 9 to the class of κ-dimensional sieve problems, where $\kappa \neq 1$. We shall also describe, later on in the chapter, several other methods of attacking these problems.

In Chapter 9 we were able to build on the sharp Theorems 8.3 and 8.4, and still we found that the introduction of (carefully chosen) weights enabled us to go, in the nature and quality of the final results, appreciably beyond what would have followed from Theorem 8.4 alone. When $\kappa > 1$ we have at our disposal no result that compares in quality with the main results of Chapter 8; the most we have in this case is Theorem 6.3 (for the upper bound problem) and Theorem 7.4 (for the lower bound). Not only do these results fall short of the precision of Chapter 8, but they involve auxiliary functions (such as σ_κ and η_κ) about which we know little, and which enter the results in various awkward ways. Nevertheless, we have considered it worthwhile to include the results below, and this for two reasons: First, we have no doubt that Theorems 6.3 and 7.4 will be improved in time—indeed, we should not be surprised if better results were not even at this moment on the way—and our results are so formulated that they can take advantage of any such improvement; and second, we demonstrate in several of the theorems in later sections how powerful the weighted sieve is even when based on results that are far from the best possible. To explain this second point, we refer the reader to Theorems 10.4 and 10.5 below where, without having recourse to numerical integration and using only the simplest (although not always obvious) properties of the functions σ_κ and η_κ, we nevertheless achieve remarkably good concrete results about the occurrence of P_r's in various polynomial sequences. These results would be better still if more were known about the numbers ν_κ ($\kappa > 1$). So far as integer values of κ are concerned, only $\nu_2, \ldots, \nu_{16}$ are known; so that when we come to apply Theorem 10.3 we are able to take full advantage of the quality of this result only when $\kappa = 2, \ldots, 16$ (or when κ is very large). In all other cases we have to fall back on the inequality (7.4.7) for ν_κ (or on the sharper inequalities described in Chapter 7, Notes). In this respect the elegant, technically simpler method of Section 10.4 retains some interest, although it is, as it stands, less powerful.

To sum up: we hope that the results and methods of this chapter will be of value to research workers even when many other results in this book have been overtaken by new developments.

The reader should have in mind the preceding discussion when he looks at the (formidable) statement of Theorem 10.1. Here we take a sieve problem of dimension κ and we postulate for it just enough basic information (which may or may not be available at present—it usually is, but in less than best possible form) to allow us to deduce (non-trivially) that the sequence $\mathscr{A}$ of the problem contains many almost-primes P_r (cf. (1.11)) with r relative small (cf. (1.10)).

THEOREM 10.1. (Ω_1), $(\Omega_2(\kappa, L))$, (Ω_3), $(R(\kappa, \alpha))$:

Suppose that

$$(a, \mathfrak{P}) = 1 \quad \textit{for all} \quad a \in \mathscr{A}, \tag{1.1}$$

and suppose also that, with a suitable constant μ, we have

$$|a| \leqslant X^{\alpha\mu} \quad \textit{for all} \quad a \in \mathscr{A}. \tag{1.2}$$

Let u and v be two real numbers (independent of X) such that

$$\frac{1}{\alpha} < u < v, \qquad \frac{2}{\alpha} < v. \tag{1.3}$$

Suppose further that for some constant δ satisfying

$$0 < \delta \leqslant 1, \tag{1.4}$$

we have

$$S(\mathscr{A}; \mathfrak{P}, X^{1/v}) \geqslant XW(X^{1/v}) \left\{ f_\kappa(\alpha v) - C_{11} \frac{L}{(\log X)^\delta} \right\} \tag{1.5}$$

where

$$f_\kappa(\alpha v) > 0; \tag{1.6}$$

and that for any real number $\xi > 1$ satisfying

$$\frac{1}{C_{12}} \leqslant v \frac{\log \xi^2}{\log X} \leqslant \alpha v - 1, \tag{1.7}$$

we have for all primes $X^{1/v} \leqslant p < X^{1/u}$, $p \in \mathfrak{P}$, that†

$$S(\mathscr{A}_p; \mathfrak{P}, X^{1/v}) \leqslant \frac{\omega(p)}{p} XW(X^{1/v}) \left\{F_\kappa\left(v\frac{\log \xi^2}{\log X}\right) + C_{13}\frac{L}{(\log \xi)^\delta}\right\}$$

$$+ C_{14} \sum_{\substack{d < \xi^2 \\ d | P(X^{1/v})}} 3^{\nu(d)} |R_{pd}| \tag{1.8}$$

where, for $t > 0$, $F_\kappa(t)$ is a non-negative, decreasing and continuous function satisfying

$$F_\kappa(t_1) - F_\kappa(t_2) \leqslant C_{15}\frac{t_2 - t_1}{t_1} \quad \textit{for} \quad \frac{1}{C_{12}} \leqslant t_1 < t_2. \tag{1.9}$$

Let r be a positive integer such that

$$r > u\alpha\mu - 1 + \frac{\kappa \displaystyle\int_u^v F_\kappa\left(v\left(\alpha - \frac{1}{t}\right)\right)\left(1 - \frac{u}{t}\right)\frac{dt}{t}}{f_\kappa(\alpha v)}. \tag{1.10}$$

Then we have

$$|\{P_r : P_r \in \mathscr{A}\}| \geqslant |\{P_r : P_r \in \mathscr{A}, q(P_r) \geqslant X^{1/v}\}|$$

$$\geqslant XW(X^{1/v}) \left\{f_\kappa(\alpha v) - \frac{\kappa}{r + 1 - u\alpha\mu}\int_u^v F_\kappa\left(v\left(\alpha - \frac{1}{t}\right)\right)\right.$$

$$\left. \times \left(1 - \frac{u}{t}\right)\frac{dt}{t} - C_{16}\frac{L}{(\log X)^\delta}\right\}. \tag{1.11}$$

Remark 1. By (5.2.5) we have

$$W(X^{1/v}) = \prod_p \frac{1 - \omega(p)/p}{(1 - 1/p)^\kappa} \frac{v^\kappa e^{-\gamma\kappa}}{\log^\kappa X} \left\{1 + O\left(\frac{vL}{\log X}\right)\right\}. \tag{1.12}$$

By imposing the condition (1.10) we have arranged for the difference of the first two terms inside the curly brackets on the right of (1.11), to be positive. Therefore, if this difference exceeds a positive constant δ_0 which is independent of X, and if $L = o((\log X)^\delta)$, we infer from (1.11) that

$$|\{P_r : P_r \in \mathscr{A}\}| \geqslant \delta_0 v^\kappa e^{-\gamma\kappa} \prod_p \frac{1 - \omega(p)/p}{(1 - 1/p)^\kappa} \frac{X}{\log^\kappa X} \quad \text{for} \quad X \geqslant X_0. \tag{1.13}$$

† The result remains true, if, in the remainder term on the right of (1.8), 3 is replaced by some constant A_8 along with the corresponding change in $(R(\kappa, \alpha))$.

Moreover, by (5.2.6) our infinite product is uniformly positive and so, by making $X \to \infty$, we obtain the result that there are infinitely many P_r's, of the type described, in our set $\mathscr{A}$.

Remark 2. When one comes to apply Theorem 10.1, the most important problem is to find a pair of numbers u and v such that, while (1.3) remains true, the expression on the right of (1.10) is as small as possible. If then r is chosen to be the least integer satisfying (1.10), statement (1.11) represents the best that can be achieved (so far as r is concerned) by the method of Theorem 10.1. In seeking to make the right choice of u and v in (1.10), the main difficulty comes from the expression

$$\frac{\kappa\int_u^v F_\kappa\left(v\left(\alpha - \frac{1}{t}\right)\right)\left(1 - \frac{u}{t}\right)\frac{dt}{t}}{f_\kappa(\alpha v)} = I(\alpha, u, v),$$

say. It is easy to verify that

$$I\left(\alpha, \frac{u}{\alpha}, \frac{v}{\alpha}\right) = I(1, u, v); \tag{1.14}$$

and this observation could be useful from a practical point of view for, once we have determined or estimated this expression in the case $\alpha = 1$ with a "good" pair, u, v, we could use the *same* value in the general α-case by employing u/α and v/α in place of u and v. On the other hand, we must stress again that u and v have to be chosen so as to minimize the whole expression on the right of (1.10), and this choice might well not be consistent—usually it is not—with the use of (1.14).

Remark 3. The constants $C_{11}, \ldots, C_{16}$ occurring in the statement of the theorem, and also the constants $C_{17}, \ldots, C_{26}$ that occur in the proof below, may depend, like the B_i's, on the structure constants A_i, κ and α, but they may depend in addition also on u, v, μ, r and δ. The reader should be clear that, in any given realization of Theorem 10.1, δ is a fixed positive number, μ is in the nature of a structure constant, u and v take numerical values (consistent with (1.3)) and the positive integer r is chosen, again numerically, in optimal fashion on the basis of (1.10) (cf. Remark 1).

Proof. We define λ by

$$\frac{1}{\lambda} = r + 1 - u\alpha\mu, \tag{1.15}$$

so that (1.10) ensures that

$$\lambda > 0. \tag{1.16}$$

As we did in Theorem 9.1, we now introduce the weighted sum

$$W(\mathscr{A}; \mathfrak{P}, v, u, \lambda) = \sum_{\substack{a \in \mathscr{A} \\ (a, P(X^{1/v}))=1}} \left\{1 - \lambda \sum_{\substack{X^{1/v} \leqslant p < X^{1/u} \\ p|a, p \in \mathfrak{P}}} \left(1 - u\frac{\log p}{\log X}\right)\right\}. \tag{1.17}$$

We may clearly assume that

$$X > C_{17}$$

where C_{17} is sufficiently large, and we shall use the abbreviations

$$z = X^{1/v}, \; y = X^{1/u}.$$

Then we obtain a connection between the weight function W and our more familiar S functions, in the form of the relation

$$W(\mathscr{A}; \mathfrak{P}, v, u, \lambda) = S(\mathscr{A}; \mathfrak{P}, z) - \lambda \sum_{\substack{z \leqslant p < y \\ p \in \mathfrak{P}}} \left(1 - \frac{\log p}{\log y}\right) S(\mathscr{A}_p; \mathfrak{P}, z). \tag{1.18}$$

We shall deal separately with each of the expressions on the right. For the first we have at our disposal (1.5), according to which

$$S(\mathscr{A}; \mathfrak{P}, z) \geqslant XW(z)\left\{f_\kappa(\alpha v) - C_{11}\frac{L}{(\log X)^\delta}\right\}. \tag{1.19}$$

It will take a little longer to deal with the second expression on the right of (1.18). We begin by putting

$$\xi^2 = \frac{X^\alpha}{(\log X)^{A_4}},$$

and we apply (1.8), with ξ^2/p instead of ξ^2, to each $S(\mathscr{A}_p; \mathfrak{P}, z)$; we then obtain

$$\begin{aligned}\sum_{\substack{z \leqslant p < y \\ p \in \mathfrak{P}}} &\left(1 - \frac{\log p}{\log y}\right) S(\mathscr{A}_p; \mathfrak{P}, z) \\ &\leqslant \sum_{\substack{z \leqslant p < y \\ p \in \mathfrak{P}}} \left(1 - \frac{\log p}{\log y}\right) \left\{\frac{\omega(p)}{p} XW(z)\left(F_\kappa\left(\frac{\log \xi^2/p}{\log z}\right)\right.\right. \\ &\quad \left.\left. + C_{13}\frac{L}{(\log \xi/\sqrt{p})^\delta}\right) + C_{14}\sum_{\substack{d < \xi^2/p \\ d|P(z)}} 3^{\nu(d)}|R_{pd}|\right\}.\end{aligned} \tag{1.20}$$

We have to check that it was in order to use (1.8) with ξ^2/p in place of ξ^2, and this amounts to confirming that the left-hand inequality in (1.7) holds for ξ^2/p instead of ξ^2; but if $p < y$,

$$\frac{\xi^2}{p} \geqslant \frac{X^{\alpha - 1/u}}{(\log X)^{A_4}} \tag{1.21}$$

and, since $u > 1/\alpha$, the inequality is indeed true. We need (1.21) again almost at once, in order to replace ξ^2 by X^α in the argument of F. We also need (1.9) for this purpose; by (1.9) and (1.21)

$$F_\kappa\left(\frac{\log \xi^2/p}{\log z}\right) - F_\kappa\left(\frac{\log X^\alpha/p}{\log z}\right) \leqslant C_{15}A_4 \frac{\log\log X}{\log \xi^2/p} \leqslant C_{18}\frac{\log\log X}{\log X},$$

and therefore, using (1.21) yet a third time, we see that the right-hand side of (1.20) is at most

$$XW(z)\sum_{z\leqslant p<y}\left\{\left(1-\frac{\log p}{\log y}\right)\frac{\omega(p)}{p}F_\kappa\left(\frac{\log X^\alpha/p}{\log z}\right) + C_{19}\frac{\omega(p)}{p}\frac{L}{(\log X)^\delta}\right\}$$

$$+ C_{14}\sum_{\substack{z\leqslant p<y\\ p\in\mathfrak{P}}}\ \sum_{\substack{d<\xi^2/p\\ d|P(z)}} 3^{\nu(d)}|R_{pd}|$$

$$\leqslant XW(z)\left\{\sum_{z\leqslant p<y}\left(1-\frac{\log p}{\log y}\right)\frac{\omega(p)}{p}F_\kappa\left(\frac{\log X^\alpha/p}{\log z}\right) + C_{20}\frac{L}{(\log X)^\delta}\right\}$$

$$+ C_{14}\sum_{\substack{n<\xi^2\\ (n,\overline{\mathfrak{P}})=1}} \mu^2(n)\,3^{\nu(n)}|R_n|,$$

since, by (2.3.3),

$$\sum_{z\leqslant p<y}\frac{\omega(p)}{p} \leqslant \kappa\log\frac{v}{u} + \frac{A_2}{\log 2}.$$

If now we appeal to ($R(\kappa,\alpha)$) and (2.3.6) we are able to conclude that

$$\sum_{\substack{z\leqslant p<y\\ p\in\mathfrak{P}}}\left(1-\frac{\log p}{\log y}\right)S(\mathscr{A}_p;\mathfrak{P},z)$$

$$\leqslant XW(z)\left\{\sum_{z\leqslant p<y}\left(1-\frac{\log p}{\log y}\right)\frac{\omega(p)}{p}F_\kappa\left(\frac{\log X^\alpha/p}{\log z}\right) + C_{21}\frac{L}{(\log X)^\delta}\right\}. \tag{1.22}$$

We are still not ready to substitute back into the right-hand side of (1.18). We still have to approximate the sum on the right of (1.22) by an integral. By (5.2.1) and ($\Omega_2(\kappa, L)$) we have

$$\sum_{z\leqslant p<w}\left(1-\frac{\log p}{\log y}\right)\frac{\omega(p)}{p} = \kappa\left\{\log\frac{\log w}{\log z} - \frac{\log w}{\log y} + \frac{u}{v}\right\} + O\left(\frac{C_{22}L}{\log X}\right) \quad \text{for} \quad z \leqslant w \leqslant y, \tag{1.23}$$

and it will be convenient to write, if $z \leqslant w \leqslant y$,

$$D(w) = \sum_{z\leqslant p<w}\left(1-\frac{\log p}{\log y}\right)\frac{\omega(p)}{p}$$

and

$$E(w) = F_\kappa\left(\frac{\log X^\alpha/w}{\log z}\right).$$

Then, using (1.23), we obtain, in exactly the same way as we did at the corresponding stage in the proof of Theorem 9.1,

$$\sum_{z\leqslant p<y}\left(1-\frac{\log p}{\log y}\right)\frac{\omega(p)}{p}F_\kappa\left(\frac{\log X^\alpha/p}{\log z}\right) = \sum_{z\leqslant p<y}\left(1-\frac{\log p}{\log y}\right)\frac{\omega(p)}{p}E(p)$$

$$= \kappa\left\{\log\frac{v}{u} - 1 + \frac{u}{v} + O\left(\frac{C_{23}L}{\log X}\right)\right\}E(y) - \int_z^y D(w)\,dE(w)$$

$$= \kappa\left\{\log\frac{v}{u} - 1 + \frac{u}{v}\right\}E(y) + O\left(\frac{C_{24}L}{\log X}\right)$$

$$- \kappa\int_z^y\left\{\log\frac{\log w}{\log z} - \frac{\log w}{\log y} + \frac{u}{v}\right\}dE(w)$$

$$= \kappa\int_{X^{1/v}}^{X^{1/u}} E(w)\left(1-\frac{u\log w}{\log X}\right)\frac{dw}{w\log w} + O\left(\frac{C_{24}L}{\log X}\right)$$

$$= \kappa\int_u^v F_\kappa\left(v\left(\alpha-\frac{1}{t}\right)\right)\left(1-\frac{u}{t}\right)\frac{dt}{t} + O\left(\frac{C_{24}L}{\log X}\right)$$

on using in the last step the substitution $w = X^{1/t}$. If now we combine this with (1.22) and substitute back in (1.18) together with (1.19), we arrive finally at

$$W(\mathscr{A};\mathfrak{P},v,u,\lambda) \geqslant XW(X^{1/v})$$
$$\times \left\{ f_\kappa(\alpha v) - \lambda\kappa \int_u^v F_\kappa\left(v\left(\alpha - \frac{1}{t}\right)\right)\left(1 - \frac{u}{t}\right)\frac{dt}{t} - C_{25}\frac{L}{(\log X)^\delta}\right\}. \quad (1.24)$$

To prove the theorem we now require an upper estimate for $W(\mathscr{A};\mathfrak{P},v,u,\lambda)$. To begin with, (1.1) ensures that the a's counted in (1.17) have no prime divisor $< X^{1/v}$; we might note also that any zero in $\mathscr{A}$ is not counted if we assume that $X > 2^v$, which is implied by the fact that $X > C_{17}$ (for (1.1) implies that if $\mathscr{A}$ contains 0 then $\bar{\mathfrak{P}}$ is empty and so $2 \in \mathfrak{P}$). Next we may assume that any a that is counted is squarefree with respect to the primes p in the interval

$$X^{1/v} \leqslant p < X^{1/u}, \qquad p \in \mathfrak{P}.$$

For by (Ω_3), (2.3.6) and (1.3) the number of a's that are not squarefree in this sense is at most

$$\sum_{\substack{X^{1/v} \leqslant p < X^{1/u} \\ p \in \mathfrak{P}}} \sum_{\substack{a \in \mathscr{A} \\ a \equiv 0 \bmod p^2}} 1 \leqslant A_3(X^{1-1/v}\log X + X^{1/u}) \leqslant XW(X^{1/v})\frac{C_{26}}{\log X}, \quad (1.25)$$

and so may, eventually, be absorbed into the error term.

We may therefore concentrate on the remaining a's. Any one of these may have a prime divisor with multiplicity greater than 1, but clearly only if that prime is $\geqslant X^{1/u}$; however, for such a prime divisor p

$$1 - u\frac{\log p}{\log X} \leqslant 0.$$

Hence, if we extend the inner summation on the right of (1.17) over *all* primes dividing a, and count them according to their multiplicities, we see that the weight of each a in (1.17) is, in view of (1.2), at most

$$1 - \lambda\left\{\Omega(a) - u\frac{\log |a|}{\log X}\right\} \leqslant \lambda\left\{\frac{1}{\lambda} - \Omega(a) + u\alpha\mu\right\}.$$

Thus, by (1.15), the weight of each of the a's in (1.17) for which $\Omega(a) \geqslant r+1$ is negative or zero. Since the weight is always at most 1 we obtain, using (1.1) and (1.25), that

$$|\{P_r\colon P_r \in \mathscr{A}\}| \geqslant |\{P_r\colon P_r \in \mathscr{A}, q(P_r) \geqslant X^{1/v}\}|$$

$$= \sum_{\substack{a \in \mathscr{A} \\ (a, P(X^{1/v}))=1 \\ \Omega(a) \leqslant r}} 1 \geqslant W(\mathscr{A}; \mathfrak{P}, v, u, \lambda) - \frac{C_{26}}{\log X} XW(X^{1/v}). \tag{1.26}$$

If now we combine (1.24) and (1.26), the proof of the theorem is complete.

2. The First Method in Applicable Form

It is now clear that if we were to take $\kappa = 1$, Theorems 8.3 and 8.4 would provide us with suitable results of type (1.8) and (1.5), and if we were then to apply Theorem 10.1 we should obtain once again the essence of Chapter 9. If $\kappa \neq 1$ we lack results of comparable precision, but we shall now investigate how far the weaker Theorems 7.4 and 6.3 can take us.

We shall concentrate on κ-dimensional problems with $\kappa > 1$. With $\kappa = 1$ we have dealt already, and we could, in principle, obtain results also for the dimensions $\kappa < 1$. However, when $\kappa < 1$ the sieve problems involve "thin" sifting sets $\mathfrak{P}$ (at any rate, so far as problems involving polynomial sequences are concerned), and here even non-trivial lower estimates do not (usually) lead to results of genuine arithmetical interest; we stress that we are talking about lower estimates, which are the subject of this chapter—we saw in earlier chapters that sieve methods are able to give some interesting information of the upper estimate variety even when $\mathfrak{P}$ is "thin". (It seems that what one should look for in a successful lower bound "thin" sieve is compensation between the density of $\mathfrak{P}$ and the size of the admissible "cut-off" point z – the thinner $\mathfrak{P}$ is, the larger one ought to be able to take z; we know of no such method. In the usual sieve the only kind of compensation that seems possible is between the density of $\mathfrak{P}$ and the dimension κ; thus to sift a reducible polynomial sequence of degree g by a $\mathfrak{P}$ of density $1/g$ might give a one-dimensional sieve problem. But here the arithmetical applications will tend to be rather artificial.)

Let us suppose then that $\kappa > 1$. We see that (1.3) implies the condition (7.6.1) of Theorem 7.4, and then Theorem 7.4 yields (1.5) with

$$f_\kappa(\alpha v) = 1 - \eta_\kappa(\alpha v).$$

From (7.4.6) we infer that requiring (1.6) amounts to the condition

$$\alpha v > \nu_\kappa$$

on v, and, by (7.4.11), this implies that $v > 2/\alpha$. The other crucial requirement in Theorem 10.1 is (1.8), and this is supplied by Theorem 6.3 with

$$F_\kappa(t) = \frac{1}{\sigma_\kappa(t)}.$$

Then (1.9) and the other properties of $F_\kappa(t)$ required by Theorem 10.1 are valid by virtue of (6.3.9), (6.3.4) and (6.3.5). It is also convenient again to replace $(\Omega_2(\kappa, L))$ by condition $(\Omega_2{}^*(\kappa))$; this amounts, as we know (cf. (9.3.1) and footnote †), to requiring that $L \leqslant A_2 \log\log 3X$, which, in fact, could be replaced by a much weaker condition.

Since, according to Theorems 7.4 and 6.3, δ can be any positive constant < 1, we now obtain from Theorem 10.1:

THEOREM 10.2. (Ω_1), $(\Omega_2{}^*(\kappa))$, (Ω_3), $(R(\kappa, \alpha))$:

Suppose that $\kappa > 1$ *and that*

$$(a, \mathfrak{P}) = 1 \quad \textit{for all} \quad a \in \mathscr{A}; \tag{2.1}$$

and suppose also that, with a suitable constant μ, *we have*

$$|a| \leqslant X^{\alpha\mu} \quad \textit{for all} \quad a \in \mathscr{A}. \tag{2.2}$$

Let u *and* v *be two real numbers (independent of* X*) such that*

$$\frac{1}{\alpha} < u < v, \qquad \frac{\nu_\kappa}{\alpha} < v. \tag{2.3}$$

Let r *be a positive integer such that*

$$r > u\alpha\mu - 1 + \frac{\kappa \displaystyle\int_u^v \frac{1}{\sigma_\kappa(v(\alpha - 1/t))}\left(1 - \frac{u}{t}\right)\frac{dt}{t}}{1 - \eta_\kappa(\alpha v)}. \tag{2.4}$$

Then we have

$$|\{P_r : P_r \in \mathscr{A}\}| \geqslant XW(X^{1/v})\left\{1 - \eta_\kappa(\alpha v) - \frac{\kappa}{r + 1 - u\alpha\mu}\int_u^v \frac{1}{\sigma_\kappa(v(\alpha - 1/t))}\left(1 - \frac{u}{t}\right)\frac{dt}{t} - \frac{C_{27}}{\sqrt{(\log X)}}\right\}. \tag{2.5}$$

Remark 1. This result will enable us to obtain, with $\kappa > 1$, much better lower estimates than those provided by Theorem 7.4 alone. The situation is just as it was in Chapter 9 in relation to Theorem 8.4. The constant C_{27} in (2.5) may depend, apart from the A_i's, κ and α, also on u, v, μ and r. Note (cf. Remark 3 following Theorem 10.1) that here $\delta = \frac{1}{2}$.

Remark 2. It is clear from (2.4) that the chief difficulty in applying Theorem 10.2 rests with the quotient

$$\frac{\displaystyle\int_u^v \frac{1}{\sigma_\kappa(v(\alpha - 1/t))}\left(1 - \frac{u}{t}\right)\frac{dt}{t}}{1 - \eta_\kappa(\alpha v)}.$$

We have already stressed several times that very little is known about the functions σ_κ and η_κ, as well as about the critical number ν_κ; and, indeed, all previous investigations using this approach have turned at this stage to numerical integration—which, in practice, usually meant restricting oneself to $\kappa = 1$ or 2. To evaluate this expression precisely is still, so far as we know, beyond the present state of knowledge; but the following lemma gives an estimate which, as we shall see, is surprisingly effective.

LEMMA 10.1. *Let* $0 < \alpha \leqslant 1$ *and* $\kappa > 1$. *Then for any* $\zeta > 0$, *putting*

$$\alpha u = 1 + \zeta - \frac{\zeta}{\nu_\kappa} \quad \textit{and} \quad \alpha v = \frac{\nu_\kappa}{\zeta} + \nu_\kappa - 1, \tag{2.6}$$

we have the estimate

$$\frac{\displaystyle\kappa\int_u^v \frac{1}{\sigma_\kappa(v(\alpha - 1/t))}\left(1 - \frac{u}{t}\right)\frac{dt}{t}}{1 - \eta_\kappa(\alpha v)} \leqslant (\kappa + \zeta)\log\frac{\nu_\kappa}{\zeta} - \kappa + \zeta\frac{\kappa}{\nu_\kappa}. \tag{2.7}$$

Proof. We begin by noting that

$$\frac{u}{v} = \frac{\zeta}{\nu_\kappa}. \tag{2.8}$$

If we put $v(\alpha - 1/t) = w - 1$ in the integral, so that w ranges from

$$\alpha v - \frac{v}{u} + 1 = \alpha v - \frac{\nu_\kappa}{\zeta} + 1 = \nu_\kappa$$

(at $t = u$, cf. (2.6)), to αv (at $t = v$), the left-hand side of (2.7) becomes

$$\frac{\zeta\kappa}{1 - \eta_\kappa(\alpha v)} \int_{\nu_\kappa}^{\alpha v} \frac{w/\nu_\kappa - 1}{\sigma_\kappa(w-1)} \frac{dw}{\alpha v + 1 - w}.$$

We next integrate by parts, to obtain

$$\frac{\zeta\kappa}{1 - \eta_\kappa(\alpha v)} \left\{ \left[-\log(\alpha v + 1 - w) \frac{w/\nu_\kappa - 1}{\sigma_\kappa(w-1)} \right]_{w=\nu_\kappa}^{w=\alpha v} \right.$$

$$\left. + \int_{\nu_\kappa}^{\alpha v} \log(\alpha v + 1 - w) \left(\frac{1/\nu_\kappa}{\sigma_\kappa(w-1)} - \left(\frac{w}{\nu_\kappa} - 1 \right) \frac{\sigma_\kappa'(w-1)}{\sigma_\kappa^2(w-1)} \right) dw \right\}.$$

The term inside the square brackets vanishes both at $w = \alpha v$ and at $w = \nu_\kappa$. Moreover, since we are interested only in an upper bound, we may omit the term under the integral sign involving $\sigma_\kappa'/\sigma_\kappa^2$ because, by (6.3.5), $\sigma_\kappa'(w-1) > 0$ throughout the range of integration. Thus the expression to be estimated is at most

$$\frac{\zeta/\nu_\kappa}{1 - \eta_\kappa(\alpha v)} \int_{\nu_\kappa}^{\alpha v} \frac{\kappa}{\sigma_\kappa(w-1)} \log(\alpha v + 1 - w)\, dw.$$

We integrate by parts a second time, this time using (7.4.4), and we find that the latter expression equals

$$\frac{\zeta/\nu_\kappa}{1 - \eta_\kappa(\alpha v)} \int_{\nu_\kappa}^{\alpha v} (w^\kappa(1 - \eta_\kappa(w)))'\, w^{1-\kappa} \log(\alpha v + 1 - w)\, dw$$

$$= \frac{\zeta/\nu_\kappa}{1 - \eta_\kappa(\alpha v)} \int_{\nu_\kappa}^{\alpha v} w^\kappa(1 - \eta_\kappa(w)) \left\{ \frac{w^{1-\kappa}}{\alpha v + 1 - w} \right.$$

$$\left. + (\kappa - 1) w^{-\kappa} \log(\alpha v + 1 - w) \right\} dw$$

$$\leqslant \frac{\zeta}{\nu_\kappa} \int_{\nu_\kappa}^{\alpha v} \left\{ \frac{w}{\alpha v + 1 - w} + (\kappa - 1) \log(\alpha v + 1 - w) \right\} dw,$$

where at the very last step we have used the fact that $1 - \eta_\kappa(w)$ is increasing with w. We are now left with an elementary integral which equals

$$\frac{\zeta}{\nu_\kappa}\{-\alpha v + \nu_\kappa + (\alpha v + 1)\log(\alpha v + 1 - \nu_\kappa)$$

$$+ (\kappa - 1)((\alpha v + 1 - \nu_\kappa)\log(\alpha v + 1 - \nu_\kappa) - (\alpha v + 1 - \nu_\kappa) + 1)\}$$

$$= \frac{\zeta}{\nu_\kappa}\left\{1 - \frac{\nu_\kappa}{\zeta} + \left(\frac{\nu_\kappa}{\zeta} + \nu_\kappa\right)\log\frac{\nu_\kappa}{\zeta} + (\kappa - 1)\frac{\nu_\kappa}{\zeta}\log\frac{\nu_\kappa}{\zeta}\right.$$

$$\left. - (\kappa - 1)\frac{\nu_\kappa}{\zeta} + \kappa - 1\right\}$$

$$= (\kappa + \zeta)\log\frac{\nu_\kappa}{\zeta} - \kappa + \zeta\frac{\kappa}{\nu_\kappa},$$

after repeated use of the second relation in (2.6). This completes the proof of our lemma.

Although one cannot but be pleased with this simple argument, it is only fair to draw attention to its shortcomings. First, and most important, is the choice of ν_κ as the lower limit of integration (after the first integration by parts); it is hard to see what other limit could be chosen here without knowing a good deal more about the function η_κ. As a consequence we are forced to choose u larger than it probably should be, and this in turn affects the size of the expression on the right of (2.4).

A more obvious defect is the omission of the term involving $\sigma_\kappa'/\sigma_\kappa^2$. Actually, we could estimate this term from below very crudely by using (6.3.7) (to get rid of σ_κ'), the obvious—and highly wasteful—inequalities $\sigma_\kappa(w) \leqslant 1$ and $1 - \eta_\kappa(\alpha v) \leqslant 1$ and a shortened range of integration; but the resulting gain is so small that we have not been able to take advantage of it in any interesting application, and we have therefore been content to estimate the term from below by zero.

One other comment: it follows from the second remark following Theorem 10.1 that it might have been enough to prove Lemma 10.1 with $\alpha = 1$ (cf. (1.14)). However, by choosing u and v differently for each α we shall be able to reach slightly better results.

With the help of Lemma 10.1 we shall now derive from Theorem 10.2 a result which, while slightly weaker, is more explicit and leads readily to a wide range of applications.

THEOREM 10.3. (Ω_1), $(\Omega_2{}^*(\kappa))$, (Ω_3), $(R(\kappa, \alpha))$:

Suppose that $\kappa > 1$ *and that*

$$(a, \mathfrak{P}) = 1 \qquad \textit{for all} \quad a \in \mathscr{A}; \tag{2.9}$$

and suppose also that for some suitable constant μ *we have*

$$|a| \leqslant X^{\alpha\mu} \quad \textit{for all} \quad a \in \mathscr{A}. \tag{2.10}$$

Let ζ *be a real number satisfying*

$$0 < \zeta < \nu_\kappa \tag{2.11}$$

and let r *be a positive integer such that*

$$r > (1 + \zeta)\,\mu - 1 + (\kappa + \zeta)\log\frac{\nu_\kappa}{\zeta} - \kappa - \zeta\,\frac{\mu - \kappa}{\nu_\kappa}. \tag{2.12}$$

Then there exists a positive number $\delta = \delta(r, \mu, \kappa, \zeta)$ *such that*

$$|\{P_r\colon P_r \in \mathscr{A}\}| \geqslant \delta\,\frac{X}{\log^\kappa X}\left(1 - \frac{C_{28}}{\sqrt{(\log X)}}\right), \tag{2.13}$$

where C_{28} *depends at most on* r, μ *and* ζ *(as well as on the* A_i*'s,* κ *and* α*).*

Proof. In order to apply Lemma 10.1 to Theorem 10.2 we have to check that the conditions (2.3) on u and v in the statement of Theorem 10.2 are satisfied by the choice (2.6). Since $\alpha u = 1 + \zeta - \zeta/\nu_\kappa$, it follows at once from (7.4.11) (which states that $\nu_\kappa > 2$ for all $\kappa \geqslant 1$) that $u > 1/\alpha$. Moreover, if u and v are given by (2.6) then

$$\frac{v}{u} = \frac{\nu_\kappa}{\zeta}$$

(cf. (2.8) and $\alpha v = \nu_\kappa + (\nu_\kappa/\zeta - 1)$), so that our condition (2.11) implies both that $v > u$ and $\nu_\kappa/\alpha < v$. We may now apply Lemma 10.1. We see at once that (2.4) is a consequence of (2.12), so that (2.5) is true; and (2.13) then follows from (1.12) and the associated discussion.

3. ALMOST-PRIMES REPRESENTABLE BY POLYNOMIALS

We are now able to derive some applications from Theorem 10.3.

THEOREM 10.4. *Let* $F_1(n), \ldots, F_g(n)$ $(g > 1)$ *be distinct irreducible polynomials with integral coefficients, and write* $F(n)$ *for the product* $F_1(n) \ldots$

$F_g(n)$; let G denote the degree of F. Further let $\rho(p)$ denote the number of solutions of

$$F(n) \equiv 0 \bmod p,$$

and suppose that

$$\rho(p) < p \quad \textit{for all } p. \tag{3.1}$$

Then, for any natural number r satisfying

$$r > G - 1 + \left(g + \frac{g}{G}\right) \log\left(\frac{G}{g} v_g\right) - \frac{g}{G}\frac{G-g}{v_g}, \tag{3.2}$$

there is a positive number $\delta = \delta(r, F)$ such that, as $x \to \infty$,

$$|\{n: 1 \leqslant n \leqslant x, F(n) = P_r\}| \geqslant \delta \frac{x}{\log^g x}\left(1 + O_F\left(\frac{1}{\sqrt{\log x}}\right)\right). \tag{3.3}$$

Proof. We take

$$\mathscr{A} = \{F(n): 1 \leqslant n \leqslant x\},$$

and for $\mathfrak{P}$ we choose the set $\mathfrak{P}_1$ of all primes. The reader might wish at this point to refer back to Example 3 of Chapter 1, where the relevant arithmetical properties of F are analysed. We choose

$$X = x, \qquad \omega(d) = \rho(d),$$

where $\rho(d)$ denotes the number of solutions of the congruence $F(n) \equiv 0 \bmod d$. Then, as in the proof of Theorem 5.3 (with $y = x$), we find (cf. (1.3.16)) that

$$|R_d| \leqslant \omega(d) \leqslant G^{\nu(d)} \quad \text{if} \quad \mu(d) \neq 0, \tag{3.4}$$

and that (Ω_1) and $(\Omega_2(\kappa, L))$ are satisfied with

$$A_1 = O_F(1), \qquad \kappa = g, \qquad A_2 = O_F(1), \qquad L = O_F(1);$$

and this implies, incidentally, that $(\Omega_2{}^*(\kappa))$ is fulfilled. Condition (Ω_3) is satisfied by virtue of the fact (cf. (9.5.7)) that

$$\rho(p^2) = O_F(1)$$

and (1.3.12). Finally, by (3.4) and Lemma 3.4 we obtain that

$$\sum_{d < X/\log^{3G+g+1} X} \mu^2(d)\, 3^{\nu(d)}\, |R_d| \ll \frac{X}{\log^{g+1} X}\,.$$

which shows that $(R(\kappa, \alpha))$ is satisfied with

$$\alpha = 1. \tag{3.5}$$

We are now in a position to apply Theorem 10.3. Since $\mathfrak{P} = \mathfrak{P}_1$, (2.9) represents no restriction whatever. As to condition (2.10), we note that

$$F(n) = O_F(x^G),$$

so that, in view of (3.5), (2.10) holds with any constant μ satisfying $\mu > G$ if x is sufficiently large. Finally, choosing $\zeta = g/G$ which satisfies (2.11) because of (7.4.11), we see that (2.12) can be written in the form of our condition (3.2).

A particularly interesting special case of Theorem 10.4 relates to the problem of prime k-tuplets, a natural extension of the prime twins problem. Here the appropriate polynomials F_i are all linear, so that $G = g$, and we derive at once from Theorem 10.4 the following analogue for almost-primes.

THEOREM 10.5. *Let g be a natural number > 1, and let a_i, b_i $(i = 1, \ldots, g)$ be integers satisfying*

$$\prod_{i=1}^{g} a_i \prod_{1 \leqslant t < s \leqslant g} (a_t b_s - a_s b_t) \neq 0.$$

Suppose also that

$$\prod_{i=1}^{g} (a_i n + b_i)$$

has no fixed prime divisor. Then, for any natural number r satisfying

$$r > (g + 1) \log \nu_g + g - 1, \tag{3.6}$$

there is a positive number δ such that, as $x \to \infty$,

$$\left| \left\{ n : 1 \leqslant n \leqslant x, \prod_{i=1}^{g} (a_i n + b_i) = P_r \right\} \right| \geqslant \delta \frac{x}{\log^g x} \left(1 + O\left(\frac{1}{\sqrt{(\log x)}} \right) \right), \tag{3.7}$$

where δ and the O-constant depend only on r, g and on the a_i's and b_i's.

We might pause for a moment to consider the quality of this result, with special reference, of course, to the values of r indicated by (3.6). If g is a large (but fixed) number, so that, in view of (7.4.8), we may regard the right-hand side of (3.6) as being essentially $g \log g + g(1 + \log 2{\cdot}44..)$, Theorem 10.5 represents a surprisingly good result. For g of intermediate magnitude—say $g = 40$ or 50—it is hard to have any feeling for what a reasonable result would be. So let us look at the small values of g, the cases of main interest at the present time.

(i) $g = 2$: We deduce from (7.4.9) that

$$3 \log \nu_2 + 1 < 5{\cdot}47,$$

so that $r = 6$ is an admissible choice in (3.6).

(ii) $g = 3$: Here (7.4.9) tells us that

$$4 \log \nu_3 + 2 < 9{\cdot}72,$$

so that $r = 10$ is an admissible choice of r in (3.6).

For values of g between 4 and 16 similar calculations lead to the following results:

TABLE 3

g	4	5	6	7	8	9	10	11	12	13	14	15	16
r	15	19	24	29	34	39	45	50	55	61	66	72	78

Coming back to the case $g = 2$, the result we obtained above, namely that $(a_1 n + b_1)(a_2 n + b_2) = P_6$ infinitely often (subject to the cited conditions on the constants a_1, a_2, b_1, b_2), could be improved by replacing Lemma 10.1 with a piece of numerical integration. From the work of Ankeny–Onishi, as corrected by Porter, it appears, as a consequence of certain numerical computations, that if $\alpha = 1$, $u = 1{\cdot}5$, $v = 9$, then (using our notation)

$$\frac{2\displaystyle\int_u^v \frac{1}{\sigma_2(v(1 - 1/t))}\left(1 - \frac{u}{t}\right)\frac{dt}{t}}{1 - \eta_2(v)} < 2{\cdot}3;$$

turning back to Theorem 10.2 this result makes the right-hand side of (2.4) less than

$$\tfrac{3}{2}\mu - 1 + 2{\cdot}3 = \tfrac{3}{2}\mu + 1{\cdot}3.$$

This would imply that we may actually take

$$r = \tfrac{3}{2}G + \begin{cases} 2, & G \text{ even} \\ \tfrac{3}{2}, & G \text{ odd}, \end{cases} \quad \text{when } g = 2 \tag{3.8}$$

in Theorem 10.4; and consequently that

$$r = 5 \quad \text{when} \quad g = 2$$

in Theorem 10.5. We therefore derive a better result for two linear forms than that cited above; in particular, we obtain the well-known result that, for infinitely many n,

$$n(n + 2) = P_5.$$

There are, of course, many other special applications of Theorem 10.4 that one could give; the possibilities are almost limitless. To make a selection becomes a matter of taste rather than judgement. One application that appears to us, after Theorem 10.5, to possess some intrinsic interest is the case of the g polynomials F_i all having the same degree, call it h; so that $G = gh$. (This is clearly a generalization of Theorem 10.5.) Here the critical number r (cf. (3.2)) has to satisfy the inequality

$$r > hg - 1 + \left(g + \frac{1}{h}\right) \log (h\nu_g) - \frac{h-1}{h} \cdot \frac{g}{\nu_g}.$$

The numerical calculations for $2 \leqslant g \leqslant 15$ (where ν_g is known) and $2 \leqslant h \leqslant 7$ have been carried out. The result—given a product $F(n)$ of g polynomials (satisfying the requirements of Theorem 10.4) of equal degree h, there exist infinitely many integers n such that

$$F(n) = P_r$$

—for each of these special cases is tabulated below. The number in the $g-h$ square is the least number r admissible by Theorem 10.4.

TABLE 4

h / g	2	3	4	5	6	7
2	9	11	14	16	18	20
3	14	18	22	25	29	32
4	20	26	31	35	40	44
5	27	33	39	45	51	57
6	33	41	48	56	63	69
7	40	49	58	66	74	82
8	46	57	67	77	86	95
9	53	65	76	87	98	108
10	60	74	86	98	110	121
11	67	82	96	109	122	134
12	74	90	106	120	134	148
13	81	99	115	131	146	161
14	89	108	125	142	159	175
15	96	116	135	153	171	188

A second application of Theorem 10.4 which we have studied a little more closely is the case $g = 2$, $G \geqslant 3$, when $F(n) = F_1(n) F_2(n)$ and G is the degree of F. Since $\nu_2 < 4{\cdot}43$, any number r satisfying the inequality

$$r > G + 2 \log G + 0{\cdot}14 + \frac{1}{G}(2 \log G + 2{\cdot}5). \tag{3.9}$$

is admissible. The following table shows the least admissible values of r when $3 \leqslant G \leqslant 8$:

TABLE 5

G	3	4	5	6	7	8	
							$(g = 2)$
r	7	9	10	11	12	14	

Thus, for example, the product of two polynomials which satisfy the requirements of Theorem 10.4 and have combined degree 5, is infinitely often the product of at most ten primes.

If $G \geqslant 5$, any number r satisfying

$$r \geqslant G + 2 \log G + 1 \tag{3.10}$$

is admissible. This inequality is not as sharp as (3.9), but may serve as a convenient "rule of thumb" for anyone wishing to have information for larger G.

The reader should have no difficulty in compiling similar information for other cases.

Another class of problems relates to the existence of almost-primes in sequences generated by polynomials with *prime* arguments. For these we have

THEOREM 10.6. *Let $F_i(n)$ ($\neq \pm n$), $i = 1, \ldots, g (g > 1)$ be distinct irreducible polynomials with integral coefficients, and write $F(n)$ for the product $F_1(n) \ldots F_g(n)$; let G be the degree of F. Further, let $\rho(p)$ denote the number of solutions of*

$$F(n) \equiv 0 \bmod p,$$

and suppose that

$$\rho(p) < p \quad \textit{for all } p$$

as well as that

$$\rho(p) < p - 1 \quad \textit{if} \quad p \nmid F(0).$$

Then, for any natural number r satisfying

$$r > 2G - 1 + \left(g + \frac{g}{2G}\right)\log\left(\frac{2G}{g}\nu_g\right) - \frac{g}{2G}\frac{2G-g}{\nu_g} \tag{3.11}$$

there is a positive number $\delta = \delta(r, F)$ such that, as $x \to \infty$,

$$|\{p : p \leqslant x, F(p) = P_r\}| \geqslant \delta \frac{x}{\log^{g+1} x}\left(1 + O_F\left(\frac{1}{\sqrt{(\log x)}}\right)\right).$$

Proof. We consider the sequence

$$\mathscr{A} = \{F(p) : p \leqslant x\}$$

of Example 6 (with $k = 1$; see Chapter 1), and again take $\mathfrak{P}$ to be the set $\mathfrak{P}_1$ of all primes. In accordance with (1.3.48), (1.4.15) and (1.3.51), we choose

$$X = li\, x, \qquad \omega(p) = \frac{\rho_1(p)}{p-1}p \quad \text{for all} \quad p,$$

where, by (1.3.53),

$$\rho_1(p) = \begin{cases} \rho(p) & \text{if} \quad p \nmid F(0) \\ \rho(p) - 1 & \text{if} \quad p \mid F(0); \end{cases}$$

and by (1.4.15), (1.3.52) and (1.3.55) we have

$$|R_d| \leqslant G^{\nu(d)}(E(x, d) + 1) \quad \text{if} \quad \mu(d) \neq 0. \tag{3.12}$$

Then, as in the proofs of Theorems 4.3 and 10.4, we see that conditions (Ω_1), $(\Omega_2{}^*(\kappa))$ and (Ω_3) are satisfied with

$$A_1 = O_F(1), \qquad \kappa = g, \qquad A_2 = O_F(1), \qquad A_3 = O_F(1).$$

As for condition $(R(\kappa, \alpha))$, we deduce from (3.12), Lemma 3.5 (with $U_1 = g + 3$, $h = 3G$, $k = 1$, $C = C_1(g + 2, 3G, 1)$) and Lemma 3.4 that

$$\sum_{d < X^{1/2}/\log^{C_1} X} \mu^2(d)\, 3^{\nu(d)} |R_d| \ll \frac{x}{\log^{g+2} x} \ll \frac{X}{\log^{g+1} X};$$

thus $(R(\kappa, \alpha))$ holds with

$$\alpha = \tfrac{1}{2},$$

and it follows that this time (cf. the corresponding stage of the argument in the proof of Theorem 10.4) (2.10) is satisfied with any constant

$$\mu > 2G$$

if X is sufficiently large.

We choose ζ next, in accord with (2.11) (cf. (7.4.11)), to be

$$\zeta = \frac{g}{2G}.$$

Then the condition (2.12) can be written in the form of (3.11), and our result now follows from Theorem 10.3.

There is again the special case when each $F_i(n)$ is a linear polynomial. Here G $= g$ and we derive at once from Theorem 10.6

THEOREM 10.7. *Let g be a natural number > 1, and let a_i, b_i $(i = 1, \ldots g)$ be integers satisfying*

$$\prod_{i=1}^{g} a_i b_i \prod_{1 \leqslant t < s \leqslant g} (a_t b_s - a_s b_t) \neq 0.$$

Let $\rho(p)$ denote the number of solutions of

$$\prod_{i=1}^{g} (a_i n + b_i) \equiv 0 \bmod p,$$

and suppose that

$$\rho(p) < p \quad \textit{for all } p$$

as well as that

$$\rho(p) < p - 1 \quad \textit{if} \quad p \nmid b_1 \ldots b_g.$$

Then, for any natural number r satisfying

$$r > (g + \tfrac{1}{2}) \log (2\nu_g) + 2g - 1 - \tfrac{1}{2}\frac{g}{\nu_g}, \tag{3.13}$$

there is a positive number δ such that, as $x \to \infty$,

$$\left|\left\{p : p \leqslant x, \prod_{i=1}^{g} (a_i p + b_i) = P_r\right\}\right| \geqslant \delta \frac{x}{\log^{g+1} x}\left(1 + O\left(\frac{1}{\sqrt{(\log x)}}\right)\right) \tag{3.14}$$

where δ and the O-constant depend only on r, g and on the a_i's and b_i's.

Let us check the quality of this result as we did in the case of Theorem 10.5. For a large (but fixed) number g, the right-hand side of (3.13) is dominated by $g \log g + (2 + \log 4{\cdot}88..)g$. In the special cases $g = 2$ and $g = 3$ we have, from (7.4.9), that the expression on the right of (3.13) is

$$\tfrac{5}{2}\log(2\nu_2) + 3 - \frac{1}{\nu_2} < 8{\cdot}23$$

if $g = 2$, and

$$\tfrac{7}{2}\log(2\nu_3) + 5 - \frac{3}{2\nu_3} < 13{\cdot}96$$

if $g = 3$. Hence we may take

$$r = 9 \text{ when } g = 2, \quad \text{and} \quad r = 14 \text{ when } g = 3.$$

To state these results explicitly: For any constants a_i, b_i satisfying the requirements of Theorem 10.7, there are infinitely many primes p such that

$$(a_1 p + b_1)(a_2 p + b_2) = P_9,$$

and infinitely many primes p such that

$$(a_1 p + b_1)(a_2 p + b_2)(a_3 p + b_3) = P_{14}.$$

It is easy to check in the same way that $r = 20$ is an admissible choice when $g = 4$. According to Porter one can show, using numerical integration, that when $g = 2$, P_9 may even be replaced by P_7.

Once again there are many special instances we might give, apart from Theorem 10.7, of the general problem dealt with in Theorem 10.6; but we prefer instead to leave such a selection to the taste of each reader. We shall give here only the analogues of the two applications discussed in connection with Theorem 10.4. First we list a few results concerning the special case when the polynomials F_i all have the same degree, call it h as before, and $g = 2$ or 3:

Given g irreducible polynomials F_i of the kind considered in Theorem 10.6, each of degree h, there exist infinitely many primes p such that

$$F_1(p)\ldots F_g(p) = P_r, \qquad \deg F_i = h \quad (1 \leqslant i \leqslant g),$$

where r is the least integer satisfying

$$r > 2gh - 1 + \left(g + \frac{1}{2h}\right)\log(2h\nu_g) - \frac{g}{\nu_g}\cdot\frac{2h-1}{2h}.$$

TABLE 6

h / g	2	3	4	5	6	7
2	14	18	23	27	31	36
3	22	29	36	42	49	55
4	31	40	49	58	66	75

Next (and last), let us consider the case when $g = 2$ and $G \geqslant 3$ in Theorem 10.6. Given two polynomials F_1 and F_2 (of the kind considered in Theorem 10.6) of combined degree G, there exist infinitely many primes p such that

$$F_1(p)\,F_2(p) = P_r, \qquad \deg F_1 + \deg F_2 = G,$$

where r is the least integer satisfying

$$r > 2G + 2\log G + 1{\cdot}53 + \frac{1}{G}(\log G + 2). \tag{3.15}$$

(We have used $\nu_2 < 4{\cdot}43$.) The results for $3 \leqslant G \leqslant 8$ are as follows:

TABLE 7

G	3	4	5	6	7	8	
r	11	14	16	18	20	23	$(g = 2)$.

The weaker inequality

$$r \geqslant 2(G + \log G + 1) \qquad (G > 5)$$

provides a simple means of finding results for large G, although these will not always be as good as those derived from (3.15).

Similar tables may easily be compiled for other cases.

4. THE SECOND METHOD

The work of Chapter 9 and of the first three sections of this chapter leaves us in no doubt that a highly effective procedure for deriving lower sieve estimates is to work with weighted sums. Up to now we have used only the sum $W(\mathscr{A};\mathfrak{P},v,u,\lambda)$, but it is natural to ask whether there may not be some advantage to working with the slightly more general function

$$W_\vartheta(\mathscr{A};\mathfrak{P},v,u,\lambda) = \sum_{\substack{a\in\mathscr{A}\\ (a,P(X^{1/v}))=1}} \left\{1-\lambda\sum_{\substack{X^{1/v}\leqslant p<X^{1/u}\\ p\mid a,\ p\in\mathfrak{P}}}\left(1-\vartheta u\frac{\log p}{\log X}\right)\right\}, \tag{4.1}$$

where, in order that we should be able to derive information from the inequality $W_\vartheta > 0$ along the same lines as before, ϑ should be some real number satisfying

$$0 \leqslant \vartheta \leqslant 1. \tag{4.2}$$

The case $\vartheta = 1$ corresponds to our old sifting function W. If $\vartheta = 0$, $W_\vartheta = W_0$ counts with a non-negative weight those numbers in $\mathscr{A}$ which have no prime divisor in $\mathfrak{P}$ less than $X^{1/v}$ and at most λ^{-1} prime divisors in the interval $X^{1/v} \leqslant p < X^{1/u}$; since each weight is at most 1, we have, in other words, the inequality

$$W_0(\mathscr{A};\mathfrak{P},v,u,\lambda) \leqslant \left|\left\{a: a\in\mathscr{A}, (a,P(X^{1/v}))=1, \sum_{\substack{X^{1/v}\leqslant p<X^{1/u}\\ p\mid a,\ p\in\mathfrak{P}}} 1 \leqslant \frac{1}{\lambda}\right\}\right|. \tag{4.3}$$

The sum W_0 was introduced into the subject by Kuhn and is the basis of most applications of lower sieve estimates in the literature. The idea was a most fruitful one; Buchstab's powerful new method is a direct extension of Kuhn's—where Kuhn attached the constant weight λ^{-1} to all the primes of a certain interval, Buchstab attaches different (but still constant) weights to (many) different sub-intervals—and, to judge by the results, it is always an improvement where comparisons are possible.

The last remark applies equally well to the comparison between W_0 and $W_1 = W$, the latter being, so far as one can tell, about as powerful as Buchstab's method (but much simpler to apply—cf. Theorem 9.3 or Theorem 10.3). What is quite certain is that if W_0 is used in the problem area of Chapter 9 it always gives weaker results. The situation with respect to the problems being dealt with in this chapter is not quite so clear. Indeed, we cannot get, using W_0, as good a result as Theorem 10.3, but the chief reason is that we cannot obtain an analogue for W_0 of Lemma 10.1 that is of comparable quality. We suspect that here too W_1 will prove to be superior to W_0 when enough information becomes available about the functions σ_κ and η_κ for a comparison to be made, at any rate most of the time. In the lower bound sieve method we are about to describe, we can also distinguish between weight functions closely similar to W_0 and W_1, and there we shall find that, for very good reasons, the analogue of W_0 sometimes does give better results.

Before we embark on a significantly different method, we draw the reader's attention—if he has not already made this observation—to the fact that, in principle, all sorts of other variations and generalizations of (4.1) may lead to good lower estimates, or even to improvements; for example, the function

$$\sum_{\substack{a\in\mathscr{A}\\(a,P(X^{1/v}))=1}}\left\{1-\lambda\sum_{\substack{X^{1/v}\leqslant p<X^{1/u}\\p|a,\ p\in\mathfrak{P}}}\left(1-\vartheta u\frac{\log p}{\log X}\right)^k\right\}$$

has been suggested, and there are even reasons for expecting success with it. At the moment we are able to say only that some preliminary investigations have proved disappointing; it is quite clear that this is an area where more work needs to be done.

The advantage of weight functions such as W_ϑ is that they can be expressed as linear combinations of functions of type $S(\mathscr{A};\mathfrak{P},z)$, and are then estimated in terms of these. The method we shall describe now uses a weight function involving many parameters which have to be chosen in such a way as to make the function as large as possible—the procedure is close to that of Chapter 3, and indeed the idea of the method derives from Selberg. In principle the method appears to be sounder than that based on W_0 or W_1 —since a weight function is necessary, let us find the right one by an optimizing process—but in practice the associated maximum problem is too difficult (at any rate, it has proved too difficult so far) and the approximate choice of parameters may fall far short of the truth. Nevertheless, Selberg's method is simple and leads to results that are elegant, easy to apply and involve only elementary functions. The results are somewhat weaker than the earlier theorems in this chapter, but have the advantage of not depending on knowledge of the behaviour of ν_κ; while the numbers ν_κ for $\kappa > 16$ are undetermined, the results of this section retain some interest for problems of dimension higher than 16.

We may take (4.1) as our point of departure. Roughly speaking, we shall remove the condition $(a, P(X^{1/v})) = 1$ from the outer summation and instead we shall introduce into each term an expression

$$\left(\sum_{d|a,\ d|P(X^{1/v})}\lambda_d\right)^2$$

(cf. Chapter 3) which is 1 if a and $P(X^{1/v})$ are coprime and is small otherwise. In theory the problem is to choose the λ_d's optimally, but we cannot solve this problem; fortunately the numbers λ_d introduced in Chapter 3 provide a fairly successful approximate solution. The following result forms the basis of the method.

THEOREM 10.8. (Ω_1): *Let z_1, z, y, λ and ϑ be real numbers satisfying*

$$2 \leqslant z_1 < z < y, \qquad \lambda > 0, \qquad 0 \leqslant \vartheta \leqslant 1. \tag{4.4}$$

Using the numbers λ_d defined in (3.1.4), *form the weighted sum*

$$W_\vartheta^* (\mathscr{A}; \mathfrak{P}, z_1, z, y, \lambda)$$

$$= \sum_{a \in \mathscr{A}} \left\{ 1 - \sum_{\substack{p < z_1 \\ p \mid a,\ p \in \mathfrak{P}}} 1 - \lambda \sum_{\substack{p < y \\ p \mid a,\ p \in \mathfrak{P}}} \left(1 - \vartheta \frac{\log p}{\log y} \right) \right\} \left(\sum_{\substack{d \mid a \\ d \mid P(z)}} \lambda_d \right)^2. \tag{4.5}$$

Then we have

$$W_\vartheta^* (\mathscr{A}; \mathfrak{P}, z_1, z, y, \lambda)$$

$$\geqslant \frac{X}{G(z)} \left\{ 1 - \sum_{p < z_1} \frac{\omega(p)}{p} \left(1 - \frac{G(z/p)}{G(z)} \right) - \lambda \sum_{p < z} \left(1 - \vartheta \frac{\log p}{\log y} \right) \frac{\omega(p)}{p} \right.$$

$$\left. \times \left(1 - \frac{G(z/p)}{G(z)} \right) - \lambda \sum_{z \leqslant p < y} \left(1 - \vartheta \frac{\log p}{\log y} \right) \frac{\omega(p)}{p} \right\}$$

$$- (1 + \lambda) \sum_{\substack{n < z^2 y \\ (n, \overline{\mathfrak{P}}) = 1}} \mu^2(n)\, 3^{\nu(n)}\, |R_n|\, (1 + \nu(n)), \tag{4.6}$$

where $G(x)$ is the function defined in (1.4.14).

Remark 1. Definition (4.5) needs a little explanation. A number a of $\mathscr{A}$ cannot have attached to it (in W_ϑ) a positive weight if it is divisible by a prime of $\mathfrak{P}$ less than z_1. Nevertheless, the sum over the primes p (of $\mathfrak{P}$) $< z_1$ is present in the weight not so much to remove a's divisible by small primes as to get rid later on, when it comes to interpreting (4.6), of those a's that are not squarefree with respect to the primes up to y.

The job of "removing" those a's that have prime factors in $\mathfrak{P}$ less than z belongs rather to the factor $(\sum_d \lambda_d)^2$, as we have already pointed out; putting $\vartheta = 0$ in (4.5), just for the moment, we then see that the function of the second inner sum on the right of (4.5)—the sum over primes less than y—is, as in Kuhn's weights, to ensure than an a with a positive weight has at most λ^{-1} prime factors between z and y.

Remark 2. Note that we have imposed only one general condition, namely (Ω_1). This is needed in order that $G(z)$ should be well defined, and hence also the functions $G_p(x)$ (cf. (3.1.3)) and the numbers λ_d.

Proof of Theorem 10.8. Let us write

$$D = [d_1, d_2]$$

as we did in Chapter 3. We then obtain

$$W_\vartheta^* (\mathscr{A}; \mathfrak{P}, z_1, z, y, \lambda) = \sum_{\substack{d_\nu | P(z) \\ \nu = 1,2}} \lambda_{d_1} \lambda_{d_2} \left\{ |\mathscr{A}_D| - \sum_{\substack{p < z_1 \\ p \in \mathfrak{P}}} |\mathscr{A}_{[p,D]}| \right.$$

$$\left. - \lambda \sum_{\substack{p < y \\ p \in \mathfrak{P}}} \left(1 - \vartheta \frac{\log p}{\log y}\right) |\mathscr{A}_{[p,D]}| \right\},$$

and with the notation (1.4.14) this becomes

$$W_\vartheta^* (\mathscr{A}; \mathfrak{P}, z_1, z, y, \lambda) \geqslant X \sum_{\substack{d_\nu | P(z) \\ \nu = 1,2}} \lambda_{d_1} \lambda_{d_2}$$

$$\times \left\{ \frac{\omega(D)}{D} - \sum_{p < z_1} \frac{\omega([p,D])}{[p, D]} - \lambda \sum_{p < y} \left(1 - \vartheta \frac{\log p}{\log y}\right) \frac{\omega([p,D])}{[p,D]} \right\}$$

$$- \sum_{\substack{d_\nu | P(z) \\ \nu = 1,2}} |\lambda_{d_1} \lambda_{d_2}| \left\{ |R_D| + (1 + \lambda) \sum_{\substack{p < y \\ p \in \mathfrak{P}}} |R_{[p,D]}| \right\} = X \Sigma_0 - \Sigma_R, \tag{4.7}$$

say; note that the change from equality to inequality derives solely from the remainder terms, and that no precision has been lost so far as the leading terms are concerned.

First consider Σ_0. With our choice (3.1.4) of the λ_d's, the first sum in Σ_0 is precisely the sum Σ_1 of (3.1.1), and we saw in (3.1.7) that $\Sigma_1 = 1/G(z)$. We may therefore write

$$\Sigma_0 = \frac{1}{G(z)} - \sum_{p < z_1} \Sigma_p - \lambda \sum_{p < y} \left(1 - \vartheta \frac{\log p}{\log y}\right) \Sigma_p \tag{4.8}$$

where

$$\Sigma_p = \sum_{\substack{d_\nu | P(z) \\ \nu = 1,2}} \lambda_{d_1} \lambda_{d_2} \frac{\omega([p,D])}{[p,D]}. \tag{4.9}$$

To estimate W_ϑ^* and therefore Σ_0 from below, we require upper bounds for the sums Σ_p. Let us concentrate on a typical sum Σ_p; a simple upper bound for Σ_p is given in (4.15), but to reach (4.15) requires some careful analysis. We have (remember that ω is multiplicative)

$$\Sigma_p = \sum_{\substack{d_\nu | P(z) \\ \nu=1,2 \\ D \equiv 0 \bmod p}} \lambda_{d_1}\lambda_{d_2}\frac{\omega(D)}{D} + \frac{\omega(p)}{p} \sum_{\substack{d_\nu | P(z) \\ \nu=1,2 \\ D \not\equiv 0 \bmod p}} \lambda_{d_1}\lambda_{d_2}\frac{\omega(D)}{D}$$

$$= \sum_{\substack{d_\nu | P(z) \\ \nu=1,2}} \lambda_{d_1}\lambda_{d_2}\frac{\omega(D)}{D} - \left(1 - \frac{\omega(p)}{p}\right) \sum_{\substack{d_\nu | P(z) \\ (d_\nu, p)=1 \\ \nu=1,2}} \lambda_{d_1}\lambda_{d_2}\frac{\omega(D)}{D}. \tag{4.10}$$

We now refer back to the proof of (3.1.5.). The first sum on the right of (4.10) is again Σ_1, while the only difference between the second sum and Σ_1 is the pair of conditions $(d_\nu, p) = 1$, which are equivalent to removing the prime p from the set $\mathfrak{P}$. Therefore, if we write

$$y_l = \sum_{\substack{d | P(z) \\ d \equiv 0 \bmod l}} \lambda_d \frac{\omega(d)}{d}, \tag{4.11}$$

we obtain, by (3.1.5),

$$\Sigma_p = \sum_{\substack{l<z \\ l | P(z)}} \frac{1}{g(l)} y_l^2 - \left(1 - \frac{\omega(p)}{p}\right) \sideset{}{'}\sum_{\substack{l<z \\ l | P(z) \\ (l,p)=1}} \frac{1}{g(l)} (y_l - y_{lp})^2, \tag{4.12}$$

where the dash again indicates that the summation extends over those numbers l only for which $\omega(l) \neq 0$ (and therefore also $g(l) \neq 0$).

Let us assume first of all that also $\omega(p) \neq 0$, so that

$$g(p) = \frac{\omega(p)/p}{1 - \omega(p)/p} \neq 0. \tag{4.13}$$

Then, since $\lambda_d = 0$ for $d \geqslant z$ implies that $y_{lp} = 0$ for $l \geqslant z/p$, we can write (4.12) in the form

$$\Sigma_p = \sideset{}{'}\sum_{\substack{l<z \\ l | P(z) \\ (l,p)=1}} \frac{1}{g(l)} y_l^2 + \frac{1}{g(p)} \sideset{}{'}\sum_{\substack{m<z/p \\ m | P(z) \\ (m,p)=1}} \frac{1}{g(m)} y_{mp}^2 - \left(1 - \frac{\omega(p)}{p}\right) \sideset{}{'}\sum_{\substack{l<z/p \\ l | P(z) \\ (l,p)=1}} \frac{1}{g(l)} (y_l - y_{lp})^2$$

$$- \left(1 - \frac{\omega(p)}{p}\right) \sum_{\substack{z/p \leqslant l<z \\ l | P(z) \\ (l,p)=1}} \frac{1}{g(l)} y_l^2 = \frac{\omega(p)}{p} \sideset{}{'}\sum_{\substack{z/p \leqslant l<z \\ l | P(z) \\ (l,p)=1}} \frac{1}{g(l)} y_l^2$$

$$+ \sideset{}{'}\sum_{\substack{l<z/p \\ l | P(z) \\ (l,p)=1}} \frac{1}{g(l)} \left\{ y_l^2 + \frac{1}{g(p)} y_{lp}^2 - \left(1 - \frac{\omega(p)}{p}\right)(y_l - y_{lp})^2 \right\},$$

and if we apply (4.13) in the last expression, we obtain

$$\Sigma_p = \frac{\omega(p)}{p} \sum_{\substack{z/p \leqslant l < z \\ l|P(z) \\ (l,p)=1}}' \frac{1}{g(l)} y_l^2 + \frac{\omega(p)}{p} \sum_{\substack{l < z/p \\ l|P(z) \\ (l,p)=1}}' \frac{1}{g(l)} \left(y_l + \frac{1}{g(p)} y_{lp} \right)^2. \tag{4.14}$$

In order to make the lower bound for $W_\vartheta{}^*$ and therefore for Σ_0 as large as possible, we require the sums Σ_p to be as small as possible. We have remarked already that we do not know how to solve the underlying extremal problem, but in (4.14) we see the virtue of using the λ_d's of Chapter 3: it turns out that the second sum on the right of (4.14) vanishes. The summation conditions in this sum ensure that (3.1.6) is applicable, and from (3.1.6) together with the fact that $(l,p) = 1$ it follows that if $l < z/p$ and $l|P(z)$, then

$$y_l + \frac{1}{g(p)} y_{lp} = \frac{\mu(l)\,g(l)}{G(z)} + \frac{1}{g(p)} \frac{\mu(lp)\,g(lp)}{G(z)} = 0.$$

Using (3.1.6) also in the first sum on the right of (4.14) we therefore arrive at the estimate

$$\Sigma_p = \frac{\omega(p)}{p} \sum_{\substack{z/p \leqslant l < z \\ l|P(z) \\ (l,p)=1}}' \frac{1}{g(l)} \frac{\mu^2(l)\,g^2(l)}{G^2(z)}$$

$$= \frac{\omega(p)/p}{G^2(z)} \sum_{\substack{z/p \leqslant l < z \\ (l,p)=1}} \mu^2(l)\,g(l) \leqslant \frac{\omega(p)/p}{G^2(z)} \left\{ G(z) - G\left(\frac{z}{p}\right) \right\}. \tag{4.15}$$

In the last step we introduced the inequality sign by dropping the condition $(l,p) = 1$; we were able to do this because the function $g(l)$ is non-negative.

In the course of the argument leading up to (4.15) we assumed that $\omega(p) \neq 0$. Suppose now that $\omega(p) = 0$. Using (4.11), it is then clear that $y_m = 0$ whenever $p \mid m$. We may therefore add the condition $(l,p) = 1$ in the first sum on the right of (4.12), whilst in the second each term $y_{lp} = 0$. Hence (4.12) shows that $\Sigma_p = 0$ if $\omega(p) = 0$, and therefore (4.15) holds also in this case.

We need one last comment on the range of validity of (4.15). We have not used the condition $p \leqslant z$ at all, so that (4.15) holds even if $p \geqslant z$. Here it takes a somewhat simpler form because $G(z/p) = 0$ if $p \geqslant z$.

Thus (4.15) provides an upper bound for each sum Σ_p occurring in (4.8), and hence we arrive at a lower bound for Σ_0 (cf. the coefficient of X in (4.6)).

It remains to deal with Σ_R. By (3.1.2) and (3.1.8) we have

$$\Sigma_R \leqslant \sum_{\substack{d_\nu < z \\ d_\nu|P(z) \\ \nu=1,2}} |R_D| + (1+\lambda) \sum_{\substack{p<y \\ p \in \mathfrak{P}}} \sum_{\substack{d_\nu < z \\ d_\nu|P(z) \\ \nu=1,2}} |R_{[p,D]}|,$$

and (3.1.11) then brings us to

$$\Sigma_R \leqslant \sum_{\substack{d<z^2 \\ d|P(z)}} 3^{\nu(d)} |R_d| + (1 + \lambda) \sum_{\substack{d<z^2 \\ d|P(z)}} 3^{\nu(d)} \sum_{\substack{p<y \\ p\in\mathfrak{P}}} |R_{[p,d]}|$$

$$\leqslant \sum_{\substack{n<z^2y \\ (n,\overline{\mathfrak{P}})=1}} \mu^2(n)\, 3^{\nu(n)} |R_n| \{1 + (1 + \lambda)\, \nu(n)\}. \tag{4.16}$$

In view of (4.7), this completes the proof of our theorem.

The leading term on the right of (4.6) looks rather complicated, but we shall now simplify it with the help of the information in Chapter 5 concerning the asymptotic behaviour of $G(x)$. We shall need an Ω_2-condition—the simpler $\Omega_2{}^*(\kappa)$ will do—and express it in a form suitable for the applications we have in mind.

LEMMA 10.2. (Ω_1), $(\Omega_2{}^*(\kappa))$: *Let κ be a natural number. Let u, v, λ and ϑ be real numbers (independent of X) satisfying*

$$1 < u < v, \qquad \lambda \geqslant 0, \qquad 0 \leqslant \vartheta \leqslant 1, \tag{4.17}$$

and write

$$z_1 = \log^{\kappa+2} X, \qquad z = X^{1/v}, \qquad y = X^{1/u}. \tag{4.18}$$

Then for $X \geqslant X_0 = X_0(u, v)$ we have that

$$1 - \sum_{p<z_1} \frac{\omega(p)}{p}\left(1 - \frac{G(z/p)}{G(z)}\right) - \lambda \sum_{p<z}\left(1 - \vartheta\frac{\log p}{\log y}\right)\frac{\omega(p)}{p}\left(1 - \frac{G(z/p)}{G(z)}\right)$$

$$- \lambda \sum_{z\leqslant p<y}\left(1 - \vartheta\frac{\log p}{\log y}\right) \geqslant 1 - \lambda\left\{\kappa\sum_{j=1}^{\kappa}\frac{1}{j} + \kappa\log\frac{v}{u} - \vartheta\kappa + \vartheta\frac{u}{v}\frac{\kappa}{\kappa+1}\right\}$$

$$- (1 + \lambda)C_{29}\frac{(\log\log 3X)^2}{\log X} \tag{4.19}$$

where C_{29} depends only on u, v (as well as on the A_i's and κ).

Remark. The expression estimated from below in (4.19) is the leading term on the right of (4.6) apart from the factor $X/G(z)$. We have introduced the requirement that κ be a natural number to simplify calculations and because all applications will be to polynomial sequences. It will be clear from the proof that if anything were to be gained by it, this restriction on κ could easily be removed. Incidentally, the constants $C_{30}, \ldots, C_{35}$ occurring below have the same dependency as C_{29}.

Proof. The quotients $G(z/p)/G(z)$ on the left of (4.19) occur with non-negative weights only, and therefore have to be estimated from below. For this we shall use the formulae

$$G\left(\frac{z}{p}\right) \geqslant c_3 \log^{\kappa}\left(\frac{z}{p}\right) + O(c_3 \log^{\kappa-1} z) \tag{4.20}$$

and

$$\frac{1}{G(z)} = \frac{1}{c_3}\log^{-\kappa} z + O\left(\frac{1}{c_3} L \log^{-\kappa-1} z\right) \tag{4.21}$$

where c_3 is given by (5.3.13). To verify these results we note first of all that the conditions of the lemma imply those of Lemma 5.4 with

$$L \leqslant A_2 \log\log 3X. \tag{4.22}$$

Hence (4.21) is an immediate consequence of (5.3.1), (5.2.5) and the definition of c_3; the only requirement is that $X \geqslant 2^{v}$ (which merely restates the condition $z \geqslant 2$). Coming to (4.20), we have at our disposal the asymptotic formula (5.3.10) which is valid subject to (5.3.3), so that (4.20) is certainly true (with equality even) if (cf. (5.3.3))

$$\log\frac{z}{p} \geqslant B_8 L.$$

If, on the other hand, this condition is not satisfied, that is, if

$$\log\frac{z}{p} < B_8 L,$$

then (4.20) is still true by virtue of the trivial estimate

$$G\left(\frac{z}{p}\right) - c_3 \log^{\kappa}\frac{z}{p} \geqslant -c_3 \log^{\kappa-1}\frac{z}{p} \,.\, B_8 L \geqslant -B_8 c_3 L \log^{\kappa-1} z.$$

We may now apply (4.20) and (4.21). From these relations and (4.22) it follows, using the choice of z in (4.18), that

$$\frac{G(z/p)}{G(z)} \geqslant \frac{\log^{\kappa} z/p}{\log^{\kappa} z} - C_{30}\frac{\log\log 3X}{\log X}. \tag{4.23}$$

We substitute this inequality on the left of (4.19), and first estimate the total contribution of all the error terms introduced in (4.23). By (5.2.1) (or Lemma 2.3) this is at most

$$C_{30}\frac{\log\log 3X}{\log X}\left\{\sum_{p<z_1}\frac{\omega(p)}{p}+\lambda\sum_{p<z}\frac{\omega(p)}{p}\right\}$$

$$\leqslant C_{31}(1+\lambda)\frac{(\log\log 3X)^2}{\log X}. \tag{4.24}$$

Hence the expression the left of (4.19) is at least

$$1-\sum_{p<z_1}\frac{\omega(p)}{p}\left(1-\left(\frac{\log z/p}{\log z}\right)^{\kappa}\right)$$

$$-\lambda\sum_{p<z}\left(1-\vartheta\frac{\log p}{\log y}\right)\frac{\omega(p)}{p}\left(1-\left(\frac{\log z/p}{\log z}\right)^{\kappa}\right)$$

$$-\lambda\sum_{z\leqslant p<y}\left(1-\vartheta\frac{\log p}{\log y}\right)\frac{\omega(p)}{p}-C_{31}(1+\lambda)\frac{(\log\log 3X)^2}{\log X}. \tag{4.25}$$

In the first sum of (4.25) we merely apply Bernoulli's inequality

$$1-(1-x)^{\kappa}\leqslant \kappa x \quad (0<x\leqslant 1) \tag{4.26}$$

to give

$$1-\left(\frac{\log z/p}{\log z}\right)^{\kappa}=1-\left(1-\frac{\log p}{\log z}\right)^{\kappa}\leqslant\kappa\frac{\log p}{\log z}, \tag{4.27}$$

so that, using our choice of z_1 and z in (4.18) as well as $(\Omega_2{}^*(\kappa))$,

$$\sum_{p<z_1}\frac{\omega(p)}{p}\left(1-\left(\frac{\log z/p}{\log z}\right)^{\kappa}\right)\leqslant\frac{\kappa}{\log z}\sum_{p<z_1}\frac{\omega(p)}{p}\log p\leqslant C_{32}\frac{\log\log 3X}{\log X}. \tag{4.28}$$

For the remaining terms of (4.25) we need to observe that, by (5.2.1) and invoking $(\Omega_2{}^*(\kappa))$,

$$\sum_{p<w}\left(1-\vartheta\frac{\log p}{\log y}\right)\frac{\omega(p)}{p}=\kappa\log\log w+O(L)-\frac{\vartheta}{\log y}(\kappa\log w+O(L)),$$

$$w\geqslant 2. \tag{4.29}$$

Hence

$$\sum_{p<z}\left(1-\vartheta\frac{\log p}{\log y}\right)\frac{\omega(p)}{p}\left(1-\left(\frac{\log z/p}{\log z}\right)^{\kappa}\right)$$

$$=\int_2^z\left\{\kappa\log\log w+O(L)-\frac{\vartheta}{\log y}(\kappa\log w+O(L))\right\}$$

$$\times\, d\left\{1-\left(1-\frac{\log w}{\log z}\right)^{\kappa}\right\}, \tag{4.30}$$

and we see at once that, by (4.18), (4.22) and (4.26), the contribution from the error terms in the integral is at most

$$C_{33}\frac{\log\log 3X}{\log X}. \tag{4.31}$$

If we integrate by parts now, we obtain (from the lower limit of integration) another term which, by the same sort of argument, is seen to be of the same order of magnitude as (4.31), so that altogether now we have from (4.30)

$$\sum_{p<z}\left(1-\vartheta\frac{\log p}{\log y}\right)\frac{\omega(p)}{p}\left(1-\left(\frac{\log z/p}{\log z}\right)^{\kappa}\right)$$

$$\leqslant\kappa\int_2^z\left(\frac{1}{w\log w}-\frac{\vartheta}{\log y}\frac{1}{w}\right)\left(1-\left(1-\frac{\log w}{\log z}\right)^{\kappa}\right)dw+C_{34}\frac{\log\log 3X}{\log X}. \tag{4.32}$$

The integral on the right can be evaluated. On writing

$$\frac{\log w}{\log z}=1-t$$

and substituting from (4.18), and using the fact that κ is a natural number, the integral is at most

$$\kappa\int_0^1\left(\frac{1}{1-t}-\vartheta\frac{u}{v}\right)(1-t^{\kappa})\,dt=\kappa\sum_{j=1}^{\kappa}\int_0^1 t^{j-1}\,dt-\vartheta\frac{u}{v}\kappa\int_0^1(1-t^{\kappa})\,dt$$

$$=\kappa\sum_{j=1}^{\kappa}\frac{1}{j}-\vartheta\frac{u}{v}\kappa\left(1-\frac{1}{\kappa+1}\right), \tag{4.33}$$

and this allows us to deduce from (4.32) that

$$\sum_{p<z}\left(1-\vartheta\frac{\log p}{\log y}\right)\frac{\omega(p)}{p}\left(1-\left(\frac{\log z/p}{\log z}\right)^{\kappa}\right)$$

$$\leqslant \kappa\sum_{j=1}^{\kappa}\frac{1}{j}-\vartheta\frac{u}{v}\kappa\left(1-\frac{1}{\kappa+1}\right)+C_{34}\frac{\log\log 3X}{\log X}. \tag{4.34}$$

Turning finally to the last sum in (4.25), we derive at once from (4.29) and (4.18) that

$$\sum_{z\leqslant p<y}\left(1-\vartheta\frac{\log p}{\log y}\right)\frac{\omega(p)}{p}=\kappa\log\frac{\log y}{\log z}+O(L)-\frac{\vartheta}{\log y}\left(\kappa\log\frac{y}{z}+O(L)\right)$$

$$\leqslant \kappa\log\frac{v}{u}-\vartheta\kappa\left(1-\frac{u}{v}\right)+C_{35}\frac{\log\log 3X}{\log X}. \tag{4.35}$$

We have only to collect up (4.25), (4.28), (4.34) and (4.35) to see that the proof of the lemma is now complete.

A combination of Theorem 10.8 and Lemma 10.2 gives a simple lower bound for the weight function $W_\vartheta{}^*$. Rather than state this as a separate result, we proceed directly to apply it to the problem of finding a lower estimate for the number of almost-primes P_r in a general sequence $\mathscr{A}$. Since $W_\vartheta{}^*$ is linear in ϑ we may expect the best results to come from the end-points $\vartheta=0$ or $\vartheta=1$. This is actually true, and we deal in Theorem 10.9 below with the case $\vartheta=1$ which is slightly simpler and also gives in most applications the better result. In order to apply the case $\vartheta=0$ to greatest effect we shall need to introduce a further basic condition (see (4.58) in Theorem 10.10 below).

THEOREM 10.9. (Ω_1), $(\Omega_2{}^*(\kappa))$, (Ω_3), $(R(\kappa,\alpha))$: *Let κ be a natural number*†. *Suppose that*

$$(a,\mathfrak{P})=1 \quad \text{for all} \quad a\in\mathscr{A}, \tag{4.36}$$

† This condition is actually not necessary. It was used only once, in evaluating the integral on the left of (4.33), in the course of proving Lemma 10.2. For any $\kappa>0$ one can readily verify that the part of the integral not involving ϑ (which could be isolated by putting $\vartheta=0$) is equal to

$$\kappa\sum_{j=1}^{\infty}\left(\frac{1}{j}-\frac{1}{j+k}\right)=\kappa\frac{\Gamma'}{\Gamma}(\kappa)+\gamma\kappa+1; \tag{4.41}$$

so that we could now easily state Theorem 10.9 for an arbitrary positive κ merely by replacing the term $\kappa\sum_1^{\kappa}1/j$ in (4.39) by (4.41).

and suppose also that for some constant μ satisfying

$$\mu > \frac{\kappa^2}{2(\kappa + 1)}, \tag{4.37}$$

we have

$$|a| < X^{\alpha\mu} \quad \textit{for all } a \in \mathscr{A}. \tag{4.38}$$

Let r be a positive integer such that

$$r > \mu - 1 + \kappa \sum_{j=1}^{\kappa} \frac{1}{j} + \kappa \log\left(\frac{2\mu}{\kappa} + \frac{1}{\kappa + 1}\right). \tag{4.39}$$

Then there exists a positive number $\delta = \delta(r, \mu, \kappa, \alpha)$ such that

$$|\{P_r : P_r \in \mathscr{A}\}| \geqslant \delta \frac{X}{\log^{\kappa} X}\left(1 - \frac{C_{36}}{\log\log 3X}\right), \tag{4.40}$$

where C_{36} depends at most on r and μ (as well as on the A_i's, κ and α).

Remark. Theorem 10.9 should be compared with Theorem 10.3. Both furnish the same kind of result under similar conditions. It is now known that Theorem 10.3 is the sharper of the two; however Theorem 10.9 holds some advantage for $\kappa > 16$ (and integral) by virtue of its explicit character. The additional condition (4.37) is rather curious, but appears to be necessary for the method to succeed. In the case of sequences $\mathscr{A}$ generated by polynomials it is satisfied trivially.

Another feature of Theorem 10.9 that requires comment is the unexpectedly large error term in (4.40). In fact, this apparent weakness has a literary rather than mathematical origin, and is, in fact, due solely to our reluctance to introduce a new R-condition at this late stage, one more appropriate to the method being used here than the familiar $(R(\kappa, \alpha))$. As will be evident from the proof of (4.48) below, a slight modification of $(R(\kappa, \alpha))$ would have allowed us to replace the term

$$\frac{1}{\log\log 3X}$$

in (4.40) by

$$\frac{(\log\log 3X)^2}{\log X}. \tag{4.41}$$

The constants $C_{37}, \ldots, C_{42}$ have the same dependency as C_{36}.

Proof. We apply Theorem 10.8 and Lemma 10.2 with $\vartheta = 1$, and we choose, as we did in (4.18),

$$z_1 = \log^{\kappa+2} X, \qquad z = X^{1/v}, \qquad y = X^{1/u}; \tag{4.42}$$

now, however, we also choose u and v according to

$$\alpha v = 2 + \left(\frac{2\mu}{\kappa} + \frac{1}{\kappa+1}\right)(1+\varepsilon), \quad \alpha u = \frac{2}{2\mu/\kappa + 1/(\kappa+1)} + 1 + \varepsilon, \tag{4.43}$$

where $\varepsilon > 0$ is sufficiently small, so that

$$\frac{v}{u} = \frac{2\mu}{\kappa} + \frac{1}{\kappa+1} > 1 \tag{4.44}$$

by (4.37). We further choose λ by putting

$$\frac{1}{\lambda} = r + 1 - u\alpha\mu = r + 1 - \frac{2\mu}{2\mu/\kappa + 1/(\kappa+1)} - \mu - \varepsilon\mu > 0 \tag{4.45}$$

for $0 < \varepsilon \leqslant \varepsilon_0(\mu, \kappa)$, since, by (4.39), $r + 1 > \mu + \kappa$. The form of (4.40) permits us to assume that

$$X \geqslant C_{37},$$

where C_{37} is sufficiently large. With all these choices of our parameters we see that the conditions of Theorem 10.8 and Lemma 10.2 are satisfied. It follows from a combination of these results, together with (4.45) and (4.44), that

$$\begin{aligned} W_1^*(\mathscr{A}; \mathfrak{P}, z_1, z, y, \lambda) \geqslant \frac{X}{G(z)} \lambda \Bigg\{ & r + 1 - \frac{2\mu}{2\mu/\kappa + 1/(\kappa+1)} - \mu - \varepsilon\mu \\ & - \kappa \sum_{j=1}^{\kappa} \frac{1}{j} - \kappa \log\left(\frac{2\mu}{\kappa} + \frac{1}{\kappa+1}\right) + \kappa - \frac{\kappa/(\kappa+1)}{2\mu/\kappa + 1/(\kappa+1)} \\ & - \left(\frac{1}{\lambda} + 1\right) C_{38} \frac{(\log\log 3X)^2}{\log X} \Bigg\} \\ & - (1+\lambda) \sum_{\substack{n < z^2 y \\ (n, \overline{\mathfrak{P}}) = 1}} \mu^2(n)\, 3^{\nu(n)} |R_n| (1 + \nu(n)). \end{aligned} \tag{4.46}$$

It follows from (4.39), if $0 < \varepsilon \leqslant \varepsilon_1(r, \mu, \kappa)$ and if we then require that $X \geqslant C_{39}$, that there exists a positive number $\delta_1 = \delta_1(r, \mu, \kappa, \alpha)$ making the leading term on the right of (4.46) at least as large as $\delta_1 X/G(z)$. By (4.21), (5.3.13), (5.2.6) and (4.22) this latter expression is at least

$$\delta_1 \frac{X}{c_3 \log^\kappa z}\left\{1 + O\left(\frac{L}{\log z}\right)\right\} \geqslant \delta_2 v^\kappa \frac{X}{\log^\kappa X}\left\{1 + O\left(\frac{v \log\log 3X}{\log X}\right)\right\}, \quad (4.47)$$

where δ_2 is positive and has the same dependence as δ_1.

Having dealt with the leading term on the right of (4.46), we turn to the remainder term. By (4.42) and (4.43)

$$z^2 y = X^{2/v + 1/u}$$

and

$$\frac{2}{v} + \frac{1}{u} = \alpha \cdot \frac{1}{\alpha v}\left(2 + \frac{v}{u}\right) = \alpha \frac{2 + v/u}{2 + (v/u)(1+\varepsilon)} < \alpha;$$

hence, using the well-known estimate

$$\nu(n) \ll \frac{\log n}{\log\log 3n},$$

it follows from $(R(\kappa, \alpha))$ that

$$(1 + \lambda) \sum_{\substack{n < z^2 y \\ (n, \overline{\mathfrak{P}}) = 1}} \mu^2(n)\, 3^{\nu(n)} |R_n| (1 + \nu(n)) \ll (1 + \lambda) \frac{X}{\log^\kappa X} \frac{1}{\log\log 3X}. \quad (4.48)$$

We can now put these results together and conclude from (4.46), (4.47) and (4.48) that

$$W_1^* (\mathscr{A}; \mathfrak{P}, z_1, z, y, \lambda) \geqslant \delta_2 \frac{X}{\log^\kappa X}\left(1 - \frac{C_{40}}{\log\log 3X}\right). \quad (4.49)$$

Finally, we have to interpret (4.49) arithmetically, that is, to find a suitable upper bound of W_1^*. We begin by recalling from (3.1.4) that $\lambda_d = 0$ if $\mu(d) = 0$, so that, by (3.1.8),

$$\left(\sum_{\substack{d \mid a \\ d \mid P(z)}} \lambda_d\right)^2 \leqslant \left(\sum_{d \mid a} \mu^2(d)\right)^2 = 2^{2\nu(a)}. \quad (4.50)$$

We are now interested only in those members a of $\mathscr{A}$ whose weights

$$1 - \sum_{\substack{p < z_1 \\ p \mid a}} 1 - \lambda \sum_{\substack{p < y \\ p \mid a}} \left(1 - \frac{\log p}{\log y}\right) \quad (4.51)$$

are positive (in view of (4.36), the condition $p \in \mathfrak{P}$ in these summations is redundant and has been dropped). In fact, even among these we specialize further and concentrate on those numbers a that are squarefree with respect to the primes $p < y$. By so doing we exclude a negligible number of elements of $\mathscr{A}$; for if a has positive weight it cannot have a prime factor $< z_1$, and therefore the numbers of excluded a's is, by (Ω_3), at most

$$A_3\left(\frac{X \log X}{z_1} + y\right) \leqslant C_{41}\frac{X}{\log^{\kappa+1} X}. \tag{4.52}$$

Moreover, for each excluded a we have by (4.45), (4.51) and (4.38) that

$$r + 1 - u\alpha\mu = \frac{1}{\lambda} > \sum_{\substack{p<y\\ p|a}}\left(1 - \frac{\log p}{\log y}\right) \geqslant \sum_{p|a}\left(1 - \frac{\log p}{\log y}\right)$$

$$\geqslant \nu(a) - \frac{\log |a|}{\log y} \geqslant \nu(a) - u\alpha\mu, \tag{4.53}$$

so that $\nu(a) \leqslant r$; and therefore the contribution of all excluded a's to $W_1{}^*$ is, by (4.50) and (4.52), at most

$$C_{42}\frac{X}{\log^{\kappa+1} X}. \tag{4.54}$$

For each a that remains, i.e. is not excluded, we know that it is squarefree with respect to the primes $p < y$ and therefore

$$\sum_{\substack{p<y\\ p|a}}\left(1 - \frac{\log p}{\log y}\right)$$

extends over *all* prime divisors (counted according to multiplicity) of a. If we interpret the sum in this extended sense and drop the condition $p < y$ we do not increase the sum because

$$1 - \frac{\log p}{\log y} \leqslant 0 \text{ for each prime } p \geqslant y.$$

Hence we have for each remaining a, in place of (4.53),

$$r + 1 - u\alpha\mu = \frac{1}{\lambda} > \sum_{\substack{p<y\\ p|a}}\left(1 - \frac{\log p}{\log y}\right) \geqslant \sum_{\substack{p,m\\ p^m|a}}\left(1 - \frac{\log p}{\log y}\right)$$

$$= \Omega(a) - \frac{\log |a|}{\log y} \geqslant \Omega(a) - u\alpha\mu, \tag{4.55}$$

so that for it even $\Omega(a) \leqslant r$ is true. It therefore follows, again using (4.50), that

$$W_1^* (\mathscr{A}; \mathfrak{P}_1, z_1, z, y, \lambda) \leqslant 4^r \sum_{P_r \in \mathscr{A}} 1 + C_{42} \frac{X}{\log^{\kappa+1} X}, \tag{4.56}$$

and a combination of (4.56) with (4.49) now proves the theorem.

We now take $\vartheta = 0$ in Theorem 10.8 and Lemma 10.2, and set about proving Theorem 10.10 below. Theorem 10.10 has an interesting feature in condition (4.58) (which replaces the usual (4.38)); considered in the light of a polynomial sequence, we see that the result of Theorem 10.10 takes some account of the number and degrees of the irreducible factors of the generating polynomial. The details of this application will be found in Theorems 10.11 and 10.12.

THEOREM 10.10. (Ω_1), $(\Omega_2^*(\kappa))$, (Ω_3), $(R(\kappa, \alpha))$: *Let κ be a natural number, and suppose that*

$$(a, \mathfrak{P}) = 1 \quad \textit{for all } a \in \mathscr{A}. \tag{4.57}$$

Suppose further that there is a natural number g together with suitable constants $\mu_1, \ldots, \mu_g$, such that each a in $\mathscr{A}$ can be written as

$$a = a_1 \ldots a_g$$

with the integers a_i always satisfying

$$|a_i| \leqslant X^{\alpha\mu_i}, \qquad i = 1, \ldots, g. \tag{4.58}$$

Let u be a real number (independent of X) such that

$$1 < \alpha u \leqslant 3, \tag{4.59}$$

and let r be a positive integer satisfying

$$r > \sum_{i=1}^{g} [\alpha u \mu_i] - 1 + \kappa \sum_{j=1}^{\kappa} \frac{1}{j} + \kappa \log \frac{2}{\alpha u - 1}. \tag{4.60}$$

Then there exists a positive number $\delta = \delta(r, \mu_i, \kappa, u\alpha)$ such that

$$|\{P_r : P_r \in \mathscr{A}\}| \geqslant \delta \frac{X}{\log^{\kappa} X} \left(1 - \frac{C_{43}}{\log\log 3X}\right), \tag{4.61}$$

where C_{43} depends at most on r and the μ_i's (as well as on the A_i's, κ and α).

Remark. $C_{44}, \ldots, C_{50}$ below have the same dependency as C_{43}.

Proof. We begin by deriving a lower estimate for W_0^*, and here the details are similar to the corresponding part of the proof of Theorem 10.9. We take $\vartheta = 0$, of course, and (cf. (4.18)) we choose

$$z_1 = \log^{\kappa+2} X, \qquad z = X^{1/v}, \qquad y = X^{1/u} \tag{4.62}$$

with

$$v = u\left(\frac{2}{\alpha u - 1} + \varepsilon\right), \tag{4.63}$$

where $\varepsilon > 0$ is sufficiently small; this time we leave the parameter u arbitrary except of course that it must satisfy (4.59). We put

$$\frac{1}{\lambda} = r + 1 - \sum_{i=1}^{g} [\alpha u \mu_i] \geqslant 1 \tag{4.64}$$

in view of (4.60).

To prove (4.61) it is clearly in order to assume immediately that

$$X \geqslant C_{44}$$

where C_{44} is sufficiently large, and our choice of parameters now ensures that the conditions of Theorem 10.8 and Lemma 10.2 are satisfied. Hence, by (4.63) and (4.64), we deduce that

$$\begin{aligned} W_0^*(\mathscr{A}; \mathfrak{P}, z_1, z, y, \lambda) \geqslant \frac{X}{G(z)}\lambda \Bigg\{ r + 1 - \sum_{i=1}^{g} [\alpha u \mu_i] - \kappa \sum_{j=1}^{\kappa} \frac{1}{j} \\ - \kappa \log\left(\frac{2}{\alpha u - 1} + \varepsilon\right) - \left(\frac{1}{\lambda} + 1\right) C_{45} \frac{(\log\log 3X)^2}{\log X} \Bigg\} \\ - (1 + \lambda) \sum_{\substack{n < z^2 y \\ (n, \overline{\mathfrak{P}}) = 1}} \mu^2(n)\, 3^{\nu(n)} |R_n| (1 + \nu(n)). \end{aligned} \tag{4.65}$$

It follows from (4.60) that if $0 < \varepsilon \leqslant \varepsilon_1(r, \mu_i, \kappa, \alpha u)$ and then $X \geqslant C_{46}$, then there exists a positive number $\delta_1 = \delta_1(r, \mu_i, \kappa, \alpha u)$ such that the first term on the right of (4.65) is at least

$$\delta_1 \frac{X}{G(z)}.$$

We now substitute for $G(z)$; arguing as in the case of (4.47) we find that this expression, and hence the whole first term on the right of (4.65), is at least

$$\delta_2 \frac{X}{\log^\kappa X}\left(1 - C_{47}\frac{\log\log 3X}{\log X}\right),$$

where δ_2 is positive and has the same dependence as δ_1. As for the remainder term in (4.65), we have by (4.63) that

$$\frac{2}{v} + \frac{1}{u} = \frac{1}{u}\left(\frac{2}{2/(\alpha u - 1) + \varepsilon} + 1\right) < \alpha,$$

so that this term may be estimated in exactly the same way, using $(R(\kappa, \alpha))$, as in the previous theorem, by (4.48). Altogether then we arrive at the lower estimate

$$W_0^*(\mathscr{A}; \mathfrak{P}, z_1, z, y, \lambda) \geqslant \delta_2 \frac{X}{\log^\kappa X}\left(1 - \frac{C_{48}}{\log\log 3X}\right). \tag{4.66}$$

We now require an upper bound for W_0^*, and, as in Theorem 10.9, we concentrate on those a's whose weights ($\vartheta = 0$ now)

$$1 - \sum_{\substack{p<z_1\\ p|a}} 1 - \lambda \sum_{\substack{p<y\\ p|a}} 1 \tag{4.67}$$

are positive. Clearly none of these a's has a prime factor less than z_1. (As before, (4.57) has rendered the summation condition $p \in \mathfrak{P}$ in each of the sums of (4.67) unnecessary.)

We first deal with the a's which are not square-free with respect to the primes $< y$. By (Ω_3) their number is, as in (4.52), at most

$$C_{49}\frac{X}{\log^{\kappa+1} X}. \tag{4.68}$$

To estimate the contribution of each we need the new condition (4.58). If we denote by t_i the total number of prime factors of a_i which are $\geqslant y$, we obtain from (4.58), for each i, that

$$X^{t_i/u} = y^{t_i} \leqslant |a_i| \leqslant X^{\alpha\mu_i},$$

or $t_i \leqslant u\alpha\mu_i$, so that

$$t_i \leqslant [\alpha u\mu_i].$$

Hence if a has positive weight, it follows from (4.67) that

$$\nu(a) < \frac{1}{\lambda} + \sum_{i=1}^{g} [\alpha u\mu_i] = r + 1$$

by (4.64), so that $\nu(a) \leqslant r$. By (4.68) and (4.50), the total contribution to $W_0{}^*$ of the 'non-square-free' a's with positive weight is therefore at most

$$C_{50} \frac{X}{\log^{\kappa+1} X}. \tag{4.69}$$

We are left with the a's having positive weights which are square-free with respect to the primes $< y$. For each of these the preceding argument shows that even

$$\Omega(a) < \frac{1}{\lambda} + \sum_{i=1}^{g} [\alpha u \mu_i] = r + 1,$$

so that $\Omega(a) \leqslant r$. We conclude that

$$W_0{}^* (\mathscr{A}; \mathfrak{P}, z_1, z, y, \lambda) \leqslant 4^r \sum_{P_r \in \mathscr{A}} 1 + C_{50} \frac{X}{\log^{\kappa+1} X}, \tag{4.70}$$

and our theorem follows at once from (4.66) and (4.70).

5. Almost-Primes Representable by Polynomials

We shall now consider the applications of Theorems 10.9 and 10.10 of the preceding section, to polynomial sequences.

THEOREM 10.11. *Let $F_1(n), \ldots, F_g(n)$ be distinct irreducible polynomials with integral coefficients, and let $F(n)$ denote their product; also, let h_i denote the degree of F_i and G the degree of F. Let $\rho(p)$ be the number of solutions of the congruence*

$$F(n) \equiv 0 \bmod p,$$

and suppose that

$$\rho(p) < p \quad \textit{for all } p.$$

Then there exists a positive number $\delta = \delta(r, F)$ such that, as $x \to \infty$,

$$|\{n: 1 \leqslant n \leqslant x, F(n) = P_r\}| \geqslant \delta \frac{x}{\log^g x} \left\{1 + O_F\left(\frac{1}{\log\log x}\right)\right\} \tag{5.1}$$

for any natural number r which satisfies either one of the two conditions

$$r > G - 1 + g \sum_{j=1}^{g} \frac{1}{j} + g \log\left(\frac{2G}{g} + \frac{1}{g+1}\right) \tag{5.2}$$

or, for any number ζ *satisfying* $0 < \zeta < 1$,

$$r > G - 1 + g \sum_{j=1}^{g} \frac{1}{j} + g \log \frac{2}{\zeta} + \sum_{i=1}^{g} [\zeta h_i]. \tag{5.3}$$

Proof. Theorem 10.11 should be compared with Theorem 10.4. It stands in the same relation to Theorems 10.9 and 10.10 as Theorem 10.4 does to Theorem 10.3. If we take $\mathscr{A} = \{F(n): 1 \leqslant n \leqslant x\}$ and $X = x$, as we did in Theorem 10.4, all the conditions of Theorem 10.9 are satisfied with $\kappa = g$, $\alpha = 1$ and any $\mu > G$ if x is sufficiently large. The same applies to the conditions of Theorem 10.10, except that here (4.58) requires $\mu_i > h_i$ $(i = 1, \ldots, g)$. Hence if we choose $\mu = G + \varepsilon$ in Theorem 10.9 and $\mu_i = h_i + \varepsilon$ $(i = 1, \ldots, g)$, $u = 1 + \zeta$ in Theorem 10.10, with $\varepsilon > 0$ and sufficiently small (i.e. x sufficiently large) in both, (5.1) subject to (5.2) follows from Theorem 10.9, and (5.1) subject to (5.3) from Theorem 10.10. In the latter case the range $2 \leqslant \zeta \leqslant 3$ too is admissible, but we have omitted it because for these large values of ζ the bound for r in (5.3) is, as one can easily check, always worse than that given in (5.2).

An interesting special case of Theorem 10.11, which was already briefly mentioned in Section 3, is when all the polynomials F_i have the same degree, h, so that

$$h_i = h \quad \text{for} \quad i = 1, \ldots, g, \quad \text{and} \quad G = gh. \tag{5.4}$$

When we make this choice in Theorem 10.11 and use (5.3) with $\zeta = 1/h - \varepsilon$, where $\varepsilon > 0$ is small enough (so that $\sum_{i=1}^{g} [\zeta h_i] = 0$), we obtain

COROLLARY 10.11.1. *Under the assumptions of Theorem* 10.11, *if*

$$\deg F_i = h \quad \textit{for} \quad i = 1, \ldots, g, \tag{5.5}$$

then, for any natural number r satisfying

$$r > gh - 1 + g \sum_{j=1}^{g} \frac{1}{j} + g \log 2h, \tag{5.6}$$

there exists a positive number $\delta = \delta(r, F)$ *such that, as* $x \to \infty$,

$$|\{n: 1 \leqslant n \leqslant x, F(n) = P_r\}| \geqslant \delta \frac{x}{\log^g x} \left\{1 + O_F\left(\frac{1}{\log\log x}\right)\right\}. \tag{5.7}$$

The special case $h = 1$, when all the F_i's are linear polynomials, has already been considered in Theorem 10.5 as a special case of Theorem 10.4. Here

$$h_i = 1 \quad \text{or} \quad i = 1, \ldots, g \quad \text{and} \quad G = g. \tag{5.8}$$

If we put $h = 1$ in Corollary 10.11.1, we obtain the parallel result

COROLLARY 10.11.2. *Under the assumptions of Theorem* 10.5, *but now for any natural number r satisfying*

$$r > g - 1 + g\sum_{j=1}^{g}\frac{1}{j} + g\log 2, \tag{5.9}$$

there exists a positive number δ such that, as $x \to \infty$,

$$\left|\left\{n\colon 1 \leqslant n \leqslant x, \prod_{i=1}^{g}(a_i n + b_i) = P_r\right\}\right| \geqslant \delta\frac{x}{\log^g x}\left\{1 + O\left(\frac{1}{\log\log x}\right)\right\} \tag{5.10}$$

where δ and the O-constant depend at most on r, g, the a_i's and the b_i's.

The case $g = 1$, although strictly speaking it is of no interest since Chapter 9 gives much better results, is worth recording as an indication of the quality of Theorem 10.11. Here Corollary 10.11.1 with $h = G$ gives, as the smallest admissible value of r,

$$r = G + 1 + [\log(2G)],$$

as compared with $r = G + 1$ of Theorem 9.7.

If $g \geqslant 2$, let us look first at the simplest case, the product of g linear polynomials, where Corollary 10.11.2 applies. When $g = 2$ and 3, the expression in (5.9) is actually smaller than the corresponding expression in (3.6) (see Theorem 10.5), but both lead to the same results. When $g > 16$, Theorem 10.5 requires an upper estimate for ν_g, whereas Corollary 10.11.2 continues to yield results in a simple and straightforward manner; the same remark applies all the more in a comparison between Theorem 10.4 and Theorem 10.11 (and so Corollary 10.11.1).

As an illustration of Corollary 10.11.2, the reader may easily verify from (5.9) the following table of admissible choices of r (in the case of products of linear polynomials):

g	2	3	4	5	6
r	6	10	15	19	24

In particular, Corollary 10.11.2 shows that for infinitely many n

$$(a_1 n + b_1)(a_2 n + b_2)(a_3 n + b_3)(a_4 n + b_4)$$

has at most 15 prime divisors, provided only that

$$\prod_{i=1}^{4} a_i \prod_{1 \leqslant t < s \leqslant 4}(a_t b_s - a_s b_t) \neq 0$$

and that the product polynomial has no fixed prime factors.

In the case of Corollary 10.11.1, where we may now assume that $h > 1$, Theorem 10.4 always gives better results when $2 \leqslant g \leqslant 16$; and the same is true when g is large, even though Theorem 10.4 must then be used in conjunction with (7.4.7).

We have considered more closely also another application of Theorem 10.11, in which $G = g + 1$—this is the case when F is the product of $g - 1$ linear polynomials and one quadratic polynomial. Here it turns out that (5.3) is always better than (5.2); and that the most effective form of (5.3) (with the choice $\zeta = 1 - \varepsilon$) is

$$r > g\left\{\sum_{j=1}^{g}\frac{1}{j} + 1 + \log 2\right\} + 1.$$

We come next to the analogue of Theorem 10.6.

THEOREM 10.12. *Let $F_i(n)$ ($\neq \pm n$) $i = 1, \ldots, g$, be distinct irreducible polynomials with integral coefficients, and write $F(n)$ for the product $F_1(n) \ldots F_g(n)$; let h_i be the degree of F_i and denote by G the degree of F. Further, let $\rho(p)$ denote the number of solutions of*

$$F(n) \equiv 0 \bmod p,$$

and suppose that

$$\rho(p) < p \quad \textit{for all } p$$

as well as that

$$\rho(p) < p - 1 \quad \textit{if} \quad p \nmid F(0).$$

Then there exists a positive number $\delta = \delta(r, F)$ such that, as $x \to \infty$,

$$|\{p : p \leqslant x,\ F(p) = P_r\}| \geqslant \delta \frac{x}{\log^{g+1} x}\left\{1 + O_F\left(\frac{1}{\log\log x}\right)\right\} \tag{5.11}$$

for any natural number r satisfying either one of the conditions:

$$r > 2G - 1 + g\sum_{j=1}^{g}\frac{1}{j} + g\log\left(\frac{4G}{g} + \frac{1}{g+1}\right) \tag{5.12}$$

or, for any number ζ satisfying $0 < \zeta < 1$,

$$r > 2G - 1 + g\sum_{j=1}^{g}\frac{1}{j} + g\log\frac{4}{\zeta} + \sum_{i=1}^{g}[\zeta h_i]. \tag{5.13}$$

Proof. We derive this result from Theorems 10.9 and 10.10 in much the same way that Theorem 10.6 followed from Theorem 10.3; if we do now as we did then, and take $\mathscr{A} = \{F(p) : p \leqslant x\}$, $X = li\,x$, we may infer from the earlier proof that all the conditions of Theorem 10.9 are satisfied with $\kappa = g$, $\alpha = 1/2$ and any $\mu > 2G$ if x is sufficiently large. For the same reason the conditions of Theorem 10.10 are fulfilled with $\mu_i > 2h_i$, $i = 1, \ldots, g$ in (4.58). Hence if we choose $\mu = 2G + \varepsilon$, $\mu_i = 2h_i + \varepsilon$, $i = 1, \ldots, g$ and $u = 2 + \zeta$, and take $\varepsilon > 0$ sufficiently small (in other words, x sufficiently large) we see that (5.11) follows subject to (5.12) from Theorem 10.9, and subject to (5.13) from Theorem 10.10. Once again we have cut down the wider range $0 < \zeta \leqslant 4$ for ζ to $0 < \zeta < 1$, since for the remaining ζ's (5.12) will always give a better result than (5.13).

We follow the pattern of applications of Theorem 10.11. We begin with the special case when all the polynomials F_i are of the same degree, that is, when condition (5.4) is satisfied. With this added condition in Theorem 10.12, use of (5.13) with $\zeta = 1/h - \varepsilon$, where $\varepsilon > 0$ is sufficiently small, yields

COROLLARY 10.12.1. *Under the conditions of Theorem 10.12, and if we assume also that for some natural number h,*

$$\deg F_i = h \quad \textit{for} \quad i = 1, \ldots, g; \tag{5.14}$$

then for any natural number r satisfying

$$r > 2gh - 1 + g \sum_{j=1}^{g} \frac{1}{j} + g \log(4h) \tag{5.15}$$

there exists a positive number $\delta = \delta(r, F)$ such that, as $x \to \infty$,

$$|\{p : p \leqslant x, F(p) = P_r\}| \geqslant \delta \frac{x}{\log^{g+1} x} \left\{1 + O_F\left(\frac{1}{\log\log x}\right)\right\}. \tag{5.16}$$

The special case $h = 1$ of Corollary 10.12.1, when all the polynomials F_i are linear and therefore

$$h_i = 1 \quad \text{for} \quad i = 1, \ldots, g, \text{ and } G = g, \tag{5.17}$$

has already been considered in Theorem 10.7 as a special case of Theorem 10.6.

If we put $h = 1$ in Corollary 10.12.1 we obtain at once

COROLLARY 10.12.2. *Under the conditions of Theorem 10.7, but now for any natural number r satisfying*

$$r > 2g - 1 + g \sum_{j=1}^{g} \frac{1}{j} + 2g \log 2, \tag{5.18}$$

there exists a positive number δ such that, as $x \to \infty$,

$$\left|\left\{p: p \leqslant x, \prod_{i=1}^{g} (a_i p + b_i) = P_r\right\}\right| \geqslant \delta \frac{x}{\log^{g+1} x}\left\{1 + O\left(\frac{1}{\log\log x}\right)\right\}, \quad (5.19)$$

where δ and the O-constant depend only on r, g, the a_i's and the b_i's.

We shall not illustrate these results numerically. Their main function is to complement the results from the earlier sections of this chapter for $\kappa > 16$; the reader can easily compute actual illustrations for himself.

6. ANOTHER METHOD

The method of the last section may be viewed as a special case of a quite general sieve procedure. Let $D(n)$ be a positive arithmetic function which increases as $\Omega(n)$, the number of prime factors of n, increases. Let λ_d be a real function defined on the positive square-free integers d, and suppose that†

$$\lambda_d = 0 \quad \text{if} \quad d \geqslant z,$$

where $z \geqslant 2$ is a number to be chosen later. Then consider the sum

$$\sum_{a \in \mathscr{A}} \{1 - \lambda D(a)\}\left(\sum_{d|a} \lambda_d\right)^2, \quad (6.1)$$

where $\mathscr{A}$ is, as usual, a sequence in which we attempt to locate members with few prime factors, λ is a positive number and it is to be understood that d takes square-free values only.

If for some λ and some choice of the numbers λ_d the sum (6.1) is positive, we may infer that there are members a in $\mathscr{A}$ for which

$$D(a) < \frac{1}{\lambda}; \quad (6.2)$$

and in view of the nature of D, this implies a limitation on $\Omega(a)$. To give an example, take $D(a) = \mu^2(a)2^{\nu(a)}$: if (6.1) were positive with $1/\lambda = 16$ (for some choice of the λ_d's), we could say that there exists at least one a having at most 3 prime factors.

In the last section we took D to be an (additive) function of type

$$D(a) = \sum_{p|a} \gamma_p,$$

† The function of z (cf. Chapter 3) is to control the size of error terms.

where γ_p is equal to 1 for the small primes, $\gamma_p = 0$ for the large primes and γ_p is chosen in yet another way for primes of intermediate size; but there might be some advantage in working with other kinds of functions D.

We write the sum (6.1) as

$$Q_1 - \lambda Q_2,$$

where

$$Q_1 = \sum_{a \in \mathscr{A}} \Big(\sum_{d \mid a} \lambda_d \Big)^2$$

and

$$Q_2 = \sum_{a \in \mathscr{A}} \Big(\sum_{d \mid a} \lambda_d \Big)^2 D(a);$$

and the ideal prescription for success is to choose the λ_d's in such a way that Q_1, Q_2 are of the same order of magnitude and Q_2/Q_1 is as small as possible, $1/\lambda$ being then chosen as a good upper bound for this ratio. In practice this objective may well prove too difficult, and one might have to be satisfied with minimizing only the (by far) more complicated expression Q_2, or even just the "dominant" portion of Q_2.

To see a little more clearly what is involved, let us consider Q_2. We have

$$Q_2 = \sum_{d_1, d_2} \lambda_{d_1} \lambda_{d_2} \sum_{\substack{a \in \mathscr{A} \\ a \equiv 0 \bmod d}} D(a), \qquad d = [d_1, d_2],$$

which is an expression reminiscent of the original Selberg sieve. But whereas there is a good deal (though not enough) of information available about sums of type

$$\sum_{\substack{a \in \mathscr{A} \\ a \equiv 0 \bmod d}} 1,$$

arithmetic expressions of type

$$\sum_{\substack{a \in \mathscr{A} \\ a \equiv 0 \bmod d}} D(a) \tag{6.3}$$

are notoriously hard to deal with, even on average over d, and even for relatively simple functions D. Indeed, so little information of this kind is available that there is no specific application of this method in the literature. There is an unpublished manuscript of Selberg in which the method has been applied to the sequence

$$\mathscr{A} = \{n\,(n+2) \colon 1 \leqslant n \leqslant x,\ n \text{ odd}\}$$

with

$$D(a) = D(n\,(n+2)) = \tau(n) + \tau(n+2).$$

Here the expression (6.3) reduces to a combination of sums of the type

$$\sum_{\substack{1 \leqslant n \leqslant x \\ n \equiv h \bmod d}} \tau(n); \tag{6.4}$$

and Selberg succeeded in estimating this sum asymptotically with the remarkably good error term $O(\tau(d)x^{1/3})$. This circumstance made it possible for Selberg to prove, after considerable technical complications, that for infinitely many square-free integers n, $n(n+2) = P_5$ (even with $1/\lambda = 14$).

The method sketched here is, potentially, of great interest. The fact that it is not readily applicable now, is due not so much to a defective sieve technique as to our lack of knowledge of the deeper properties of the distributions of arithmetic functions over arithmetic progressions. We saw in earlier chapters that Bombieri's theorem, a result of an averaging type, was able to take the place of an unproved result as deep as the grand Riemann Hypothesis; and in the same way we may now hope that a Bombieri-type theorem can be found for the sum (6.4) (taking the place of $\pi(x; k, h)$)—or, more ambitiously still, for a whole class of sums (6.3). With effective results of this type, the method described above may become very important; though the technical difficulties are such that we may expect it to be, for some time at least, most successful with rather special problems, that is to say, with problems where the sequences $\mathscr{A}$ have a relatively simple structure.

NOTES

10.1. Sections 10.1-3 derive from Halberstam–Richert [2], and no further reference will be made to this fact; the basic tools are Theorems 6.3 and 7.4, and the logarithmic weights procedure introduced in Chapter 9, and so Sections 10.1-3 may be viewed as refinements and extensions of the methods of Ankeny–Onishi [1], Ankeny [1]. There is numerical evidence to suggest that, if $\kappa > 1$, Selberg's sieve leads to better results than Rosser's.

The possibility of improving Theorems 6.3, 7.4: we have already referred to this in the Notes for Chapter 7, in connection with Porter's thesis and the current research of Diamond–Jurkat. There appear to be grounds for optimism.

For earlier general weighted sieves applicable to non-linear problems, see Kuhn [2] who states that his method can be used in this context; Levin–Maksudov [1] who give a general lower bound sieve using Kuhn weights (numerical tables of results are given at the end of their paper; see also Levin [1, 5, 9]).

10.2. Sieve problems of dimension $\kappa < 1$: our remark here perhaps requires modification in the light of the recent work of Iwaniec [2], who applies

Rosser's sieve for dimension $\kappa = 1/2$ (here, and for $\kappa = 1$, Rosser's sieve is optimal—see Selberg [5]; Selberg's method would work too, along the lines of Chapter 8, but gives weaker results) to prove the following interesting result: "Let $\phi_0(m, n) = am^2 + bmn + cn^2$ with $a > 0$, $(a, b, c) = 1$, and $D = b^2 - 4ac$ not a perfect square; also let $A \neq 0$, $B \geqslant 1$ be coprime integers, and Γ_{ϕ_0} the genus of ϕ_0. Then

$$\frac{N}{\phi(B)\log^{3/2} N} + O(N\log^{-5} N) \ll |\{p: p \leqslant N, p = B\phi(m, n) + A, \phi \in \Gamma_{\phi_0}\}|$$

$$\ll \frac{N}{\phi(B)\log^{3/2} N}$$

with the O-constants depending at most on A and ϕ_0." He gives similar results for $|\{p: p \leqslant N, p = B\phi(m, n) + A, (m, n) = 1, \phi \in \Gamma_{\phi_0}\}|$ and $|\{p: p \leqslant N, p = B\phi_0(m, n) + A\}|$.

See James [1, 2, 3] for other significant $\kappa = 1/2$ results via Brun's method.

Theorems 10.2, 10.3: general theorems of the type of Theorem 9.3.

"Previous investigations": see e.g. Levin–Maksudov [1]; Levin [5].

Lemma 10.1: the effectiveness of this elementary estimate of such a complicated transcendental expression—indeed, the very existence of such an estimate—seems to us to demonstrate the power of the weighted sieve.

10.3. In all the applications given here we state that $g > 1$. We do so merely because the case $g = 1$ is covered by Chapter 9. In fact, all the results of sections 10.1-3 (including the critical Lemma 10.1) hold with $(\kappa =)\ g = 1$ and, in particular, the applications to Goldbach's and the prime twin problems of the methods of Chapter 10 (sections 1-3) or Chapter 11 are qualitatively as good as those of Chapter 9. This remark may be of use to anyone preparing a short lecture course on this subject.

Theorem 10.4: cf. Ricci [2, 3, 6]; Levin [1, 2] (for quadratic polynomials); Kuhn [3]; Ankeny-Onishi [1], Ankeny [1].

Theorem 10.5: cf. Schinzel–Wang [1], where applications are given in connection with various multiplicative functions; Fluch [1]; Kuhn [3]; Levin [5] gives applications with $g = 2, 3$; Hsieh [1, 2] with $g = 3$ (using a combination of Selberg's, Brun's and Kuhn's methods).

Reference to numerical integration: see Ankeny–Onishi [1], p. 57; Porter [2].

The classical result $n(n + 2) = P_5$ is due to Selberg [2] (first stated in Selberg [4]).

So far as we know, there are no general results in the literature like Theorems 10.6, 10.7 and the various applications following them other than the special results of Porter [2] to some of which we have referred in the text.

10.4. The work of this and the next section is a generalization and refinement of Miech [1, 3], and Porter (Ph.D. thesis, Nottingham 1973; unpublished) via the improved weighting procedure of Ankeny–Onishi–Richert. Miech quotes some Princeton lecture notes of A. Selberg for the main idea of his method. So far as we know, these notes are not generally available.

The methods of Kuhn and Buchstab referred to here have already been cited.

10.5. Hagedorn [1] has proved, by means of his new inequality for ν_κ, described in the Notes for Chapter 7, that the results of this section are always numerically inferior to the corresponding results of Section 10.3.

10.6. The method sketched here is that of Selberg [2]. In Selberg [5] there is mention of a still incomplete attempt by Hofmeister to refine this method with a view to improving the result (2, 3).

(6.3): There exist average estimates of such sums, of Bombieri type; cf. Wolke [1]; Siebert [2]. These have many striking applications.

Chapter 11

Chen's Theorem

1. Introduction

Our object in this chapter will be to prove the following remarkable result of Jing-run Chen, which came to our attention only after Chapters 1–10 had gone to press; it constitutes a splendid climax to any account of sieve theory:

THEOREM 11.1. *There exists an absolute constant N_0 such that, if N is even and $N > N_0$, then*

$$|\{p: p \leqslant N, N - p = P_2\}|$$
$$> 0{\cdot}67 \prod_{p>2} \left(1 - \frac{1}{(p-1)^2}\right) \prod_{2<p|N} \frac{p-1}{p-2} \cdot \frac{N}{\log^2 N};$$

in particular, every sufficiently large even number N can be represented in the form

$$N = p + P_2.$$

Remark 1. The line of argument is close to that suggested at the end of the Remark attached to Theorem 9.2: to get a good lower estimate of the number of representations of N in the form $N = p + P_3$ and then to remove those representations in which $N = p + p_1p_2p_3$. We return to the use of Kuhn's weights because these give one a better control over the location of the prime factors of P_3 and weight the representations $N = p + p_1p_2p_3$ with a factor $\frac{1}{2}$. The upper bound for the number of these latter representations involves, essentially, an averaging of Bombieri's inequality (cf. Chapter 3, Lemmas 3.3 and 3.5) and this is accomplished by means of the large sieve in a way that is similar to, but much easier than, Gallagher's proof of Bombieri's theorem. We shall explain this part of the argument later.

Remark 2. Chen's proof fits well into our development of Selberg theory, as we shall show. But the close interlocking of sieve procedures and analytic methods leads to additional complications—the basic idea is not unlike that used in recent times by Hooley and others to sharpen the Brun–Titchmarsh inequality—and we have therefore not removed Theorem 9.2. Chen's method will certainly have other applications, but it is special in a sense in which the approach of Chapter 9 is not; of course, the conjugate result that $p + h = P_2$ $(2|h)$ for infinitely many primes p can be proved in precisely the same way.

2. The Weighted Sieve

Let

$$\mathscr{G} = \{N - p\colon p \leqslant N\} \quad \text{and} \quad \mathfrak{P} = \mathfrak{P}_N.$$

Then

$$|\{N - p\colon p \leqslant N, N - p = P_2\}|$$

$$\geqslant \sum_{\substack{p \leqslant N \\ (N-p, P(N^{1/10}))=1}}{}' \left\{1 - \frac{1}{2} \sum_{\substack{N^{1/10} \leqslant p_1 < N^{1/3} \\ p_1 | N-p, p_1 \in \mathfrak{P}_N}} 1 - \frac{1}{2} \sum_{\substack{N^{1/10} \leqslant p_1 < N^{1/3} \\ p_1 | N-p, p_1 \in \mathfrak{P}_N}} \sum_{\substack{N^{1/3} \leqslant p_2 < (N/p_1)^{1/2} \\ p_2 | N-p, p_2 \in \mathfrak{P}_N \\ N-p = p_1 p_2 p_3}} 1\right\}, \tag{2.1}$$

where Σ' signifies that $(N - p, N) = 1$, i.e. that $p \nmid N$, and that $N - p$ is square-free relative to the primes p_1 and p_2 appearing in the inner sums on the right. The point is that only those primes $p \leqslant N$ for which $N - p = P_2$ have a weight attached to them (the expression in parentheses) that is positive. To see this, we observe first of all that the only positive values taken by the weight are 1 or $\frac{1}{2}$. Now if the weight is equal to 1, both the inner sums must be 0 so that, in fact,

$$(N - p, P(N^{1/3})) = 1;$$

and in this case $N - p$ can plainly be only a P_2.

If now the weight is $\frac{1}{2}$, then $N - p$ has precisely *one* prime divisor p_1 from $[N^{1/10}, N^{1/3})$ and, of course, none from $[2, N^{1/10})$. Hence $N - p = p_1 m$, $(m, P(N^{1/3})) = 1$, so that m has at most *two* prime factors. If m has exactly two prime factors, so that

$$N - p = p_1 p_2 p_3, \qquad N^{1/3} \leqslant p_2 < p_3,$$

we must have $p_2 < (N/p_1)^{\frac{1}{2}}$; otherwise $p_1 p_2 p_3 > p_1 (N/p_1) = N$, which is impossible. But then the weight of $N - p$ receives a contribution $-\frac{1}{2}$ from the double sum in parentheses, and so is reduced to 0. Hence m has at most *one* prime factor if the weight is to be $\frac{1}{2}$, and then $N - p$ is again at worst a P_2.

Since removing the restrictions implied by Σ' adds at most

$$O\Big(\sum_{p|N} 1\Big) + O\Big(\sum_{N^{1/10} \leqslant p_1 < N^{1/3}} \Big(\frac{N}{p_1{}^2} + 1\Big)\Big)$$
$$+ O\Big(\sum_{N^{1/10} \leqslant p_1 < N^{1/3}} \ \sum_{N^{1/3} \leqslant p_2 < (N/p_1)^{1/2}} \Big(\frac{N}{p_1 p_2{}^2} + 1\Big)\Big) = O(N^{9/10})$$

terms to the sum, we deduce from (2.1) that

$$\begin{aligned} |\{N - p : p \leqslant N, N - p = P_2\}| &\geqslant S(\mathscr{G}; \mathfrak{P}_N, N^{1/10}) \\ &- \tfrac{1}{2} \sum_{\substack{N^{1/10} \leqslant p_1 < N^{1/3} \\ p_1 \in \mathfrak{P}_N}} S(\mathscr{G}_{p_1}; \mathfrak{P}_N, N^{1/10}) \\ &- \tfrac{1}{2} \sum_{\substack{N^{1/10} \leqslant p_1 < N^{1/3} \\ p_1 \in \mathfrak{P}_N}} \ \sum_{\substack{N^{1/3} \leqslant p_2 < (N/p_1)^{1/2} \\ p_2 \in \mathfrak{P}_N}} \left|\left\{p_3 : p_3 \leqslant \frac{N}{p_1 p_2}, N - p_1 p_2 p_3 = p\right\}\right| \\ &+ O(N^{9/10}). \end{aligned} \tag{2.2}$$

Of the three leading terms on the right, the first two are estimated on the basis of Theorem 8.4, (5.2) (for the first of the two) and Theorem 8.3, (4.1) (for the second), both with $\alpha = \frac{1}{2}$ and the second in conjunction with Lemma 3.5 (Bombieri's theorem). The estimation follows precisely the steps described in detail in Chapter 9 and we leave it to the reader as an exercise; together they contribute at least

$$(li\, N)\, W(N^{1/10}) \left\{f(5) - \frac{1}{2}\int_3^{10} F\left(5 - \frac{10}{t}\right)\frac{dt}{t}\right\} \{1 - B_{27} \log^{-1/15} N\}$$
$$\geqslant \prod_{p>2}\left(1 - \frac{1}{(p-1)^2}\right) \prod_{2<p|N} \frac{p-1}{p-2} \cdot \frac{N}{\log^2 N}$$
$$\times\, 20e^{-\gamma} \left\{f(5) - \frac{1}{2}\int_3^{10} F\left(5 - \frac{10}{t}\right)\frac{dt}{t}\right\} (1 - B_{28} \log^{-1/15} N)$$

(cf. the argument leading to (9.2.4)). By repeated application of (8.2.7), (2.8) and (2.9) we obtain

$$5f(5) = 2e^{\gamma}\left(\log 4 + \int_3^4 \frac{du}{u} \int_2^{u-1} \frac{\log(t-1)}{t}\, dt\right)$$

and

$$5\int_3^{10} F\left(5 - \frac{10}{t}\right)\frac{dt}{t} = 2e^{\gamma}\left(\log 8 + 5\int_3^4 \frac{du}{u(5-u)} \int_2^{u-1} \frac{\log(t-1)}{t}\, dt\right),$$

so that

$$20e^{-\gamma}\left\{f(5) - \frac{1}{2}\int_3^{10} F\left(5 - \frac{10}{t}\right)\frac{dt}{t}\right\}$$

$$= 8\left(\frac{1}{2}\log 2 - \int_3^4 \frac{u - \frac{5}{2}}{5 - u}\cdot\frac{du}{u}\int_2^{u-1}\frac{\log(t-1)}{t}\,dt\right).$$

Since

$$\log x \leqslant \frac{x-1}{2} + \frac{x-1}{x+1}, \qquad 1 \leqslant x \leqslant 2,$$

we have

$$\int_2^{u-1}\frac{\log(t-1)}{t}\,dt \leqslant \int_2^{u-1}\left(\frac{t-2}{2t} + \frac{t-2}{t^2}\right)dt, \qquad 3 \leqslant u \leqslant 4,$$

$$= \frac{1}{2}(u-3) + \frac{2}{u-1} - 1,$$

and therefore†

$$\int_3^4 \frac{u-\frac{5}{2}}{5-u}\cdot\frac{du}{u}\int_2^{u-1}\frac{\log(t-1)}{t}\,dt \leqslant \frac{1}{2}\int_3^4 \frac{(u-\frac{5}{2})(u-3)}{5-u}\frac{du}{u}$$

$$+ 2\int_3^4 \frac{(u-\frac{5}{2})}{(5-u)(u-1)}\frac{du}{u} - \int_3^4 \frac{u-\frac{5}{2}}{5-u}\frac{du}{u}$$

$$= 5\log 2 - 3\log 3 + \tfrac{1}{2}\log 2 - \tfrac{1}{2} < 0{\cdot}0165.$$

Hence

$$20e^{-\gamma}\left\{f(5) - \frac{1}{2}\int_3^{10} F\left(5 - \frac{10}{t}\right)\frac{dt}{t}\right\} \geqslant 8\left(\tfrac{1}{2}\log 2 - 0{\cdot}0165\right)$$

$$> 8 \times 0{\cdot}33007 = 2{\cdot}64056$$

and we deduce from (2.2), if N is large enough, that

$$|\{N - p : p \leqslant N,\, N - p = P_2\}|$$

$$> 2{\cdot}64 \prod_{p>2}\left(1 - \frac{1}{(p-1)^2}\right)\prod_{2<p|N}\frac{p-1}{p-2}\cdot\frac{N}{\log^2 N} - \tfrac{1}{2}S_0, \qquad (2.3)$$

where

$$S_0 = \sum_{\substack{N^{1/10} \leqslant p_1 < N^{1/3} \leqslant p_2 < (N/p_1)^{1/2} \\ (p_1p_2, N) = 1}} \left|\left\{p : p \leqslant \frac{N}{p_1p_2},\, N - (p_1p_2)p = p'\right\}\right|. \qquad (2.4)$$

† Numerical integration gives the sharper upper bound 0·01489.

It suffices to prove that

$$S_0 < 5{\cdot}28 \prod_{p>2}\left(1-\frac{1}{(p-1)^2}\right)\prod_{2<p\mid N}\frac{p-1}{p-2}\cdot\frac{N}{\log^2 N} \qquad (N>N_0), \qquad (2.5)$$

and from now on we shall concentrate on S_0; in fact, Chen's argument will lead us to an appreciably sharper upper bound than (2.5).

The counting number

$$\left|\left\{p: p \leqslant \frac{N}{p_1p_2}, N-(p_1p_2)p = p'\right\}\right| \qquad (2.6)$$

is precisely that studied in Theorem 3.12 (with $k=1$, $x=N/(p_1p_2)$, $a=-p_1p_2$ and $b=N$); if we applied the theorem directly, we should get†

$$\prod_{p>2}\left(1-\frac{1}{(p-1)^2}\right)\prod_{2<p\mid N}\frac{p-1}{p-2}.N.8(1+\varepsilon)$$

$$\times \sum_{\substack{N^{1/10}\leqslant p_1<N^{1/3}\leqslant p_2<(N/p_1)^{1/2}\\ (p_1p_2,N)=1}} \frac{1}{(p_1-1)(p_2-1)\log^2(N/p_1p_2)}$$

as an upper estimate of S_0, and this turns out to be too large, by a numerical factor. We have therefore to proceed more carefully. From now on, let

$$\mathscr{P} = \mathscr{P}_N = \{p_1p_2 : N^{1/10} \leqslant p_1 < N^{1/3} \leqslant p_2 < (N/p_1)^{1/2}\},$$

so that

$$|\mathscr{P}_N| \ll N^{2/3};$$

and, for each element p_1p_2 of $\mathscr{P}_N$, let

$$\mathscr{A} = \mathscr{A}(p_1,p_2) = \{N-(p_1p_2)p : p \leqslant N/(p_1p_2)\}.$$

Then, as in the proof of Theorem 3.12, with $K = Np_1p_2$,

$$\begin{aligned} S_0 &\leqslant \sum_{\substack{p_1p_2\in\mathscr{P}_N\\(p_1p_2,N)=1}} (S(\mathscr{A}(p_1,p_2); \mathfrak{P}_{Np_1p_2}, z)+z) \\ &\leqslant \sum_{\substack{p_1p_2\in\mathscr{P}_N\\(p_1p_2,N)=1}} S(\mathscr{A}(p_1,p_2); \mathfrak{P}_{Np_1p_2}, z) + O(N^{11/12}) \end{aligned} \qquad (2.7)$$

† Throughout we write $\log(N/p_1p_2)$ for $\log\left(\dfrac{N}{p_1p_2}\right)$

provided that $z \leqslant N^{1/4}$. If now we were to apply (3.8.1) (with x, a, b interpreted as above) we should arrive, taking $z^2 = N^{\frac{1}{2}-\varepsilon}$, at

$$S_0 \leqslant \prod_{p>2}\left(1 - \frac{1}{(p-1)^2}\right) \prod_{2<p|N} \frac{p-1}{p-2}$$

$$\times \frac{N}{\log N} 8(1+\varepsilon) \sum_{p_1p_2 \in \mathscr{P}_N} \frac{1}{(p_1 - 1)(p_2 - 1)(\log (N/p_1p_2)}$$

$$+ \sum_{d<N^{1/2-\varepsilon}} \mu^2(d)\, 3^{\nu(d)} \sum_{p_1p_2 \in \mathscr{P}_N} E\left(\frac{N}{p_1p_2}\,;\, d\right).$$

If the leading term were, in fact, the dominant term, then (2.5) would follow at once; for, according to Lemma 11.1 below,

$$8 \sum_{p_1p_2 \in \mathscr{P}_N} \frac{1}{(p_1 - 1)(p_2 - 1)\log (N/p_1p_2)} < 4.$$

However, the remainder term

$$\sum_{d<N^{1/2-\varepsilon}} \mu^2(d) 3^{\nu(d)} \sum_{p_1p_2 \in \mathscr{P}_N} E\left(\frac{N}{p_1p_2}\,;\, d\right)$$

cannot be estimated simply by appealing directly to Lemma 3.5. Indeed, we shall have to go back to (2.6) and start from first principles. Since our argument will ultimately be analytic it is convenient to introduce the classical device of "weighting" the primes p with von Mangoldt's function $\Lambda(n)$. But first we place on record the following result.

LEMMA 11.1. *If* $N > N_0$,

$$\sum_{p_1p_2 \in \mathscr{P}_N} \frac{1}{p_1p_2 \log (N/p_1p_2)} < \frac{0{\cdot}493}{\log N}.$$

Proof. Using

$$\sum_{N^{1/3} \leqslant p_2 < x} \frac{1}{p_2} = \log\left(\frac{\log x}{\log N^{1/3}}\right) + O\left(\frac{1}{\log N}\right),$$

we have by Stieltjes integration that

$$\sum_{N^{1/3} \leqslant p_2 < (N/p_1)^{1/2}} \frac{1}{p_2 \log (N/p_1p_2)} = \int_{N^{1/3}}^{(N/p_1)^{1/2}} \frac{dx}{x(\log x)\log (N/p_1x)} + O\left(\frac{1}{\log^2 N}\right)$$

for $N^{1/10} \leqslant p_1 < N^{1/3}$; and using Stieltjes integration a second time,

$$\sum_{p_1 p_2 \in \mathscr{P}_N} \frac{1}{p_1 p_2 \log (N/p_1 p_2)}$$

$$= \int_{N^{1/10}}^{N^{1/3}} \frac{dy}{y \log y} \int_{N^{1/3}}^{(N/y)^{1/2}} \frac{dx}{x (\log x) \log (N/xy)} + O\left(\frac{1}{\log^2 N}\right)$$

$$= \int_{1/10}^{1/3} \frac{d\alpha}{\alpha} \int_{1/3}^{(1-\alpha)/2} \frac{d\beta}{\beta(1-\alpha-\beta)} \cdot \frac{1}{\log N} + O\left(\frac{1}{\log^2 N}\right).$$

The double integral on the right is equal to

$$\int_{1/10}^{1/3} \frac{d\alpha}{\alpha(1-\alpha)} \log (2 - 3\alpha).$$

If, as a preliminary check, we were now to apply again the inequality

$$\log x \leqslant \tfrac{1}{2}(x-1) + (x-1)/(x+1) \qquad (1 \leqslant x \leqslant 2),$$

we should obtain as an upper estimate of this double integral,

$$\int_{1/10}^{1/3} \frac{d\alpha}{\alpha(1-\alpha)} \left(\tfrac{1}{2}(1-3\alpha) + \frac{1-3\alpha}{3-3\alpha}\right)$$

$$= \int_{1/10}^{1/3} \left(\frac{5}{6}\frac{1}{\alpha} - \frac{2}{3}\cdot\frac{1}{1-\alpha} - \frac{2}{3}\frac{1}{(1-\alpha)^2}\right) d\alpha$$

$$= \tfrac{5}{6}\log \tfrac{10}{3} - \tfrac{2}{3}\log \tfrac{27}{20} - \tfrac{7}{27} < 0{\cdot}538.$$

This alone would suffice to prove the qualitative part of Theorem 11.1. The better estimate given here follows by a straightforward numerical integration.†

From now on, let

$$z^2 = N^{\frac{1}{2}-\varepsilon} \tag{2.8}$$

where

$$0 < \varepsilon < \tfrac{1}{100}, \tag{2.9}$$

and write

$$P(z) = \prod_{\substack{p<z \\ (p,N)=1}} p. \tag{2.10}$$

Then the counting number (2.6) is at most

$$z + \left|\left\{p : p \leqslant \frac{N}{p_1 p_2}, (N - p_1 p_2 p, P(z)) = 1\right\}\right|$$

† Actually, numerical integration yields the estimate $0{\cdot}490995\,(1+\varepsilon)\,|\varepsilon| \leqslant 10^{-5}$.

so that, by (2.4)

$$S_0 \leqslant \sum_{p_1p_2 \in \mathscr{P}_N} \left| \left\{ p : p \leqslant \frac{N}{p_1p_2}, (N - p_1p_2p, P(z)) = 1 \right\} \right| + O(N^{11/12}).$$

Next, writing $r = p_1p_2$ and letting $\varepsilon_0 = (\log N)^{-\frac{1}{3}}$,

$$\sum_{\substack{n \leqslant N/r \\ (N-rn, P(z)) = 1}} \Lambda(n) \geqslant \sum_{\substack{(N/r)^{1-\varepsilon_0} < p \leqslant N/r \\ (N-rp, P(z)) = 1}} \log p > (1 - \varepsilon_0) \log \frac{N}{r}$$

$$\times \left(\left| \left\{ p : p \leqslant \frac{N}{r}, (N - rp, P(z)) = 1 \right\} \right| - (N/r)^{1-\varepsilon_0} \right)$$

and it is easy to check that

$$\sum_{p_1p_2 \in \mathscr{P}_N} \left(\frac{N}{p_1p_2} \right)^{1-\varepsilon_0} \ll N^{1-\varepsilon_0/3}.$$

Hence

$$S_0 < \frac{1}{1 - \varepsilon_0} \sum_{p_1p_2 \in \mathscr{P}_N} \frac{1}{\log (N/p_1p_2)} \sum_{\substack{n \leqslant N/(p_1p_2) \\ (N - p_1p_2n, P(z)) = 1}} \Lambda(n) + O(N^{1-\varepsilon_0/3})$$

$$\leqslant (1 + 2\varepsilon_0) S_1 + O(N^{1-\varepsilon_0/3}), \qquad \varepsilon_0 = (\log N)^{-\frac{1}{3}}, \tag{2.11}$$

where

$$S_1 = \sum_{\substack{m \leqslant N \\ (N-m, P(z)) = 1}} \Lambda_0(m), \qquad \Lambda_0(m) = \sum_{\substack{p_1p_2n = m \\ p_1p_2 \in \mathscr{P}_N}} \frac{\Lambda(n)}{\log (N/p_1p_2)}. \tag{2.12}$$

We observe here, for future reference, that since $p_1p_2 \leqslant N^{2/3}$ for $p_1p_2 \in \mathscr{P}_N$,

$$\Lambda_0(m) \leqslant \frac{3}{\log N} \sum_{n|m} \Lambda(n) = \frac{3 \log m}{\log N} \leqslant 3 \quad \text{if} \quad m \leqslant N. \tag{2.13}$$

We now proceed to the estimation of S_1 from above.

3. Application of Selberg's Upper Sieve

We begin as if we were estimating $S(\mathscr{A}(p_1,p_2); \mathfrak{P}_N, z)$ from above: with z given by (2.8), we choose

$$\lambda_d = \frac{\mu(d)}{\prod_{p|d} \left(1 - \frac{1}{p-1}\right)} \frac{G_d(z/d)}{G(z)} \quad (\mu(d) \neq 0, \; (d, N) = 1) \tag{3.1}$$

where, with $g(p) = 1/(p-2)$ if $(p, N) = 1$ and 0 otherwise,

$$G_d(x) = \sum_{\substack{m<x \\ (m,d)=1}} \mu^2(m)\, g(m) \quad \text{and} \quad G(x) = G_1(x); \tag{3.2}$$

it follows that, as usual, $\lambda_1 = 1$, $\lambda_d = 0$ for $d \geqslant z$ and $|\lambda_d| \leqslant 1$ for all the d's for which λ_d has been defined, and also that (cf. Chapter 3)

$$\sum_{d_1|P(z)} \sum_{d_2|P(z)} \frac{\lambda_{d_1}\lambda_{d_2}}{\phi(D)} = \frac{1}{G(z)}, \qquad D = [d_1, d_2]. \tag{3.3}$$

Accordingly, by (2.12) we have

$$S_1 \leqslant \sum_{m\leqslant N} \Lambda_0(m) \Bigg(\sum_{\substack{d|P(z) \\ d|N-m}} \lambda_d\Bigg)^2 = \sum_{d_1, d_2|P(z)} \lambda_{d_1}\lambda_{d_2} \sum_{\substack{m\leqslant N \\ m\equiv N \bmod D}} \Lambda_0(m)$$

$$= \sum_{d_1,d_2|P(z)} \frac{\lambda_{d_1}\lambda_{d_2}}{\phi(D)} \sum_{\chi \bmod D} \bar{\chi}(N) \sum_{m\leqslant N} \chi(m)\Lambda_0(m)$$

where the inner summation is over all Dirichlet characters mod D. We put

$$\psi_0(N, \chi) = \sum_{m\leqslant N} \chi(m)\Lambda_0(m), \tag{3.4}$$

so that if χ_0 denotes the principal character,

$$\psi_0(N, \chi_0) = \sum_{\substack{m\leqslant N \\ (m,D)=1}} \Lambda_0(m) = \sum_{m\leqslant N} \Lambda_0(m) - \sum_{\substack{m\leqslant N \\ (m,D)>1}} \Lambda_0(m).$$

Hence, by (3.3),

$$S_1 \leqslant \Bigg(\sum_{m\leqslant N} \Lambda_0(m)\Bigg)\Bigg/G(z) + S_2 + S_3 \tag{3.5}$$

where

$$S_2 = \sum_{\substack{D<z^2 \\ D|P(z)}} \frac{\mu^2(D)\,3^{\nu(D)}}{\phi(D)} \sum_{\substack{m\leqslant N \\ (m,D)>1}} \Lambda_0(m) \tag{3.6}$$

and

$$S_3 = \sum_{\substack{D<z^2 \\ D|P(z)}} \frac{\mu^2(D)\,3^{\nu(D)}}{\phi(D)} \Bigg| \sum_{\substack{\chi \bmod D \\ \chi\neq\chi_0}} \bar{\chi}(N)\,\psi_0(N, \chi)\Bigg|. \tag{3.7}$$

We have, by (3.6.3), that

$$\frac{1}{G(z)} \leqslant 2\prod_{p>2}\left(1 - \frac{1}{(p-1)^2}\right) \prod_{2<p|N} \frac{p-1}{p-2} \cdot \frac{1}{\log z}\left(1 + O\left(\frac{1}{\log z}\right)\right) \tag{3.8}$$

and from this we deduce

LEMMA 11.2. *If $N > N_0$, and $z = N^{1/4-\varepsilon/2}$, then*

$$\frac{1}{G(z)} \sum_{m \leqslant N} \Lambda_0(m)$$

$$\leqslant \prod_{p>2} \left(1 - \frac{1}{(p-1)^2}\right) \prod_{2<p|N} \frac{p-1}{p-2} \cdot \frac{3{\cdot}944\, N}{\log^2 N} \cdot \left(1 + O\left(\frac{1}{\log N}\right)\right).$$

Proof. We have

$$\sum_{m \leqslant N} \Lambda_0(m) = \sum_{p_2 p_1 \in \mathscr{P}_N} \frac{1}{\log (N/p_1 p_2)} \sum_{n \leqslant N/(p_1 p_2)} \Lambda(n),$$

and, by the prime number theorem, we know that

$$\sum_{n \leqslant x} \Lambda(n) = x \left(1 + O\left(\frac{1}{\log x}\right)\right), \qquad x \to \infty.$$

Since $p_1 p_2 < N^{2/3}$ for $(p_1 p_2) \in \mathscr{P}_N$, we deduce that

$$\sum_{m \leqslant n} \Lambda_0(m) = N \left(\sum_{p_1 p_2 \in \mathscr{P}_N} \frac{1}{p_1 p_2 \log (N/p_1 p_2)}\right) \left(1 + O\left(\frac{1}{\log N}\right)\right),$$

and the result of the lemma now follows from (3.8) and Lemma 11.1.

LEMMA 11.3. *If $N > N_0$*

$$S_2 \ll N^{9/10} \log^4 N.$$

Proof. By (2.12), and recalling that if $p_1 p_2 \in \mathscr{P}$, then $p_2 \geqslant N^{1/3} > z$,

$$\sum_{\substack{m \leqslant N \\ (m,D)>1}} \Lambda_0(m) \leqslant \sum_{p_1 p_2 \in \mathscr{P}} \frac{1}{\log (N/p_1 p_2)} \sum_{\substack{n \leqslant N/(p_1 p_2) \\ (n,D)>1}} \Lambda(n)$$

$$+ \sum_{n \leqslant N} \Lambda(n) \sum_{\substack{N^{1/10} \leqslant p_1 < N/n \\ p_1 | D}} \sum_{p_2 \leqslant N/(p_1 n)} \frac{1}{\log (N/p_1 p_2)}.$$

But

$$\sum_{\substack{n \leqslant x \\ (n,D)>1}} \Lambda(n) = \sum_{p|D} (\log p) \sum_{p^m \leqslant x} 1 = \sum_{p|D} (\log p) \left[\frac{\log x}{\log p}\right] \leqslant (\log x)\, \nu(D),$$

whence

$$\sum_{\substack{m \leqslant N \\ (m,D)>1}} \Lambda_0(m) \ll |\mathscr{P}|\, \nu(D) + \frac{N}{\log N} \sum_{n \leqslant N} \frac{\Lambda(n)}{n} \sum_{\substack{N^{1/10} \leqslant p_1 \\ p_1 | D}} \cdot \frac{1}{p_1}$$

$$\ll (|\mathscr{P}| + N^{9/10})\, \nu(D).$$

Substituting in (3.6) we obtain

$$S_2 \ll N^{9/10} \sum_{D<z^2} \frac{\mu^2(D)\, 3^{\nu(D)}\, \nu(D)}{\phi(D)} \ll N^{9/10} \sum_{D<z^2} \frac{\mu^2(D)\, 3^{\nu(D)}}{D} \log D$$

since it is well known that $\nu(D) \ll (\log D)/(\log\log D)$ and that $\phi(D) \gg D/\log\log D$. The result follows at once by reference to Lemma 3.4.

If now we combine (3.5) with Lemmas 11.2 and 11.3 we arrive at

$$S_1 \leqslant 3{\cdot}944 \prod_{p>2} \left(1 - \frac{1}{(p-1)^2}\right) \prod_{2<p|N} \frac{p-1}{p-2} \cdot \frac{N}{\log^2 N} \left(1 + O\left(\frac{1}{\log N}\right)\right) + S_3 \tag{3.9}$$

and it remains to estimate S_3, given by (3.7) (with (3.4)). We shall see that $S_3 = o(N \log^{-2} N)$, (cf. (6.1) below), and then Theorem 11.1 will follow at once from this, (3.9), (2.11) and (2.3); note that (3.9) will have lead to an appreciably better upper estimate than (2.5).

The estimation of S_3 will occupy the next three sections.

4. Transition to Primitive Characters

We require the following elementary result.

LEMMA 11.4. *Let $m \neq 1$ be an integer and d an odd, square-free integer. Then*

$$\left| \sum_{\chi \bmod d}^{*} \chi(m) \right| \leqslant |(m-1,\ d)|,$$

where $\sum_\chi^*$ denotes (here and later) summation over all primitive characters mod d.

Proof. The sum on the left is multiplicative, and therefore the expression on the left is equal to

$$\prod_{p|d} \left| \sum_{\chi \bmod p}^{*} \chi(m) \right| = \prod_{p|d} \left| \sum_{\substack{\chi \bmod p \\ \chi \neq \chi_0}} \chi(m) \right|.$$

But

$$\sum_{\substack{\chi \bmod p \\ \chi \neq \chi_0}} \chi(m) = \begin{cases} p-2 & \text{if} \quad m \equiv 1 \bmod p \\ -1 & \text{otherwise;} \end{cases}$$

hence the expression is equal to

$$\prod_{\substack{p|d\\ p|m-1}} (p-2) \leqslant \prod_{p|(d,\, m-1)} p = |(d, m-1)|.$$

We now turn to S_3, as defined by (3.7). Each non-principal character χ mod D is induced by a unique primitive character χ^* mod d, where $1 < d|D$, and $\chi(m) = \chi^*(m)$ whenever $(m, D) = 1$. Hence, by (3.4),

$$\psi_0(N, \chi) = \psi_0(N, \chi^*) - \sum_{\substack{m \leqslant N\\ (m, D/d) > 1}} \Lambda_0(m)\, \chi^*(m),$$

and accordingly

$$S_3 \leqslant \sum_{\substack{D<z^2\\ D|P(z)}} \frac{\mu^2(D)\, 3^{\nu(D)}}{\phi(D)} \sum_{1<d|D} \left| \sum_{\chi \bmod d}^{*} \bar{\chi}(N)\, \psi_0(N, \chi) \right|$$

$$+ \sum_{\substack{D<z^2\\ D|P(z)}} \frac{\mu^2(D)\, 3^{\nu(D)}}{\phi(D)} \left| \sum_{1<d|D} \sum_{\chi \bmod d}^{*} \bar{\chi}(N) \sum_{\substack{m \leqslant N\\ (m, D/d)>1}} \Lambda_0(m)\, \chi(m) \right|.$$

In each summation over D we write $D = d\delta$ and use the fact that then $\phi(D) = \phi(d)\phi(\delta)$ (since the factor $\mu^2(D)$ ensures that $(d, \delta) = 1$). It follows that

$$S_3 \leqslant \left\{ \sum_{\substack{1<d<z^2\\ d|P(z)}} \frac{\mu^2(d)\, 3^{\nu(d)}}{\phi(d)} \left| \sum_{\chi \bmod d}^{*} \bar{\chi}(N)\, \psi_0(N, \chi) \right| \right\} \sum_{\delta<z^2} \frac{\mu^2(\delta)\, 3^{\nu(\delta)}}{\phi(\delta)}$$

$$+ \sum_{\substack{d\delta<z^2\\ d\delta|P(z)}} \frac{\mu^2(d\delta)\, 3^{\nu(d\delta)}}{\phi(d\delta)} \sum_{\substack{m \leqslant N\\ (m,\delta)>1}} \Lambda_0(m) \left| \sum_{\chi \bmod d}^{*} \bar{\chi}(N)\, \chi(m) \right|$$

$$\ll (\log N)^3 \sum_{\substack{1<d<z^2\\ d|P(z)}} \frac{\mu^2(d)\, 3^{\nu(d)}}{\phi(d)} \left| \sum_{\chi \bmod d}^{*} \bar{\chi}(N)\, \psi_0(N, \chi) \right|$$

$$+ \sum_{\substack{d<z^2\\ d|P(z)}} \frac{\mu^2(d)\, 3^{\nu(d)}}{\phi(d)} \sum_{\substack{\delta<z^2\\ \delta|P(z)\\ (\delta,d)=1}} \frac{\mu^2(\delta)\, 3^{\nu(\delta)}}{\phi(\delta)} \sum_{\substack{m \leqslant N\\ (m,\delta)>1}} \Lambda_0(m)\, (N-m, d)$$

by Lemma 11.4. We shall write this inequality briefly as

$$S_3 \ll S_4 \log^3 N + S_5,$$

and we shall deal first with S_5. We have, for $m \leqslant N$, that

$$\sum_{d<z^2} \frac{\mu^2(d)}{\phi(d)} 3^{\nu(d)} (N-m, d) = \sum_{d<z^2} \frac{\mu^2(d)}{\phi(d)} \sum_{\substack{k|N-m \\ k|d}} \phi(k)$$

$$= \sum_{k|N-m} \mu^2(k)\phi(k) \sum_{\substack{d<z^2 \\ k|d}} \frac{\mu^2(d)3^{\nu(d)}}{\phi(d)} \ll (\log N)^3 \sum_{k|N-m} \mu^2(k) 3^{\nu(k)} \ll N^{\varepsilon},$$

so that

$$S_5 \ll N^{\varepsilon} \sum_{\delta<z^2} \frac{\mu^2(\delta)3^{\nu(\delta)}}{\phi(\delta)} \sum_{\substack{m\leqslant N \\ (m,\delta)>1}} \Lambda_0(m)$$

$$\ll N^{\varepsilon} \sum_{\delta<z^2} \frac{\mu^2(\delta)3^{\nu(\delta)}}{\phi(d)} (|\mathscr{P}| + N^{9/10})\,\nu(\delta)$$

by the argument that was used in the proof of Lemma 11.3. Hence

$$S_5 \ll N^{9/10+\varepsilon} \log^4 N$$

and it follows at once that

$$S_3 \leqslant S_4 \log^3 N + O(N^{19/20}), \tag{4.1}$$

where

$$S_4 = \sum_{\substack{1<d<z^2 \\ d|P(z)}} \frac{\mu^2(d)3^{\nu(d)}}{\phi(d)} \left|\sum_{\chi}{}^{*} \bar{\chi}(N)\,\psi_0(N,\chi)\right|. \tag{4.2}$$

The factor $3^{\nu(d)}$ could make subsequent calculations unnecessarily complicated, and we therefore remove it now, by applying Cauchy's inequality (cf. Lemma 3.5). This intermediate step is also psychologically reassuring, for we shall find in the process that S_4 (or a sum rather like it which, if anything, is larger) is at any rate not too large. We have at once

$$S_4 \leqslant S_5^{1/2} S_6^{1/2}$$

where

$$S_5 = \sum_{\substack{1<d<z^2 \\ d|P(z)}} \frac{\mu^2(d)}{\phi(d)} \sum_{\chi}{}^{*} |\psi_0(N,\chi)| \tag{4.3}$$

and

$$S_6 = \sum_{\substack{1<d<z^2 \\ d|P(z)}} \frac{\mu^2(d)9^{\nu(d)}}{\phi(d)} \left|\sum_{\chi}{}^{*} \bar{\chi}(N)\,\psi_0(N,\chi))\right|.$$

Then, by appealing again to Lemma 11.4 (see also the argument leading to (4.1)),

$$\begin{aligned}
S_6 &< \sum_{\substack{1<d<z^2\\ d|P(z)}} \frac{\mu^2(d)9^{\nu(d)}}{\phi(d)} \sum_{m\leqslant N} \Lambda_0(m)\,(N-m,d)\\
&\leqslant \sum_{m\leqslant N} \Lambda_0(m) \sum_{k|N-m} \phi(k) \sum_{\substack{1<d<z^2\\ k|d}} \frac{\mu^2(d)9^{\nu(d)}}{\phi(d)}\\
&\ll (\log N)^9 \sum_{m\leqslant N} \Lambda_0(m) \sum_{\substack{k|N-m\\ k<z^2}} \mu^2(k)\,9^{\nu(k)}\\
&\ll (\log N)^9 \sum_{k<z^2} \mu^2(k)\,9^{\nu(k)} \sum_{\substack{m\leqslant N\\ m\equiv N \bmod k}} \Lambda_0(m)\\
&\ll N\,(\log N)^9 \sum_{k<z^2} \frac{\mu^2(k)\,9^{\nu(k)}}{k}
\end{aligned}$$

in view of (2.13), Lemma 3.4 and because $N/k \gg 1$ for $k < z^2$. Hence

$$S_6 \ll N \log^{18} N,$$

so that

$$S_4 \ll (N^{1/2} \log^9 N)\, S_5^{1/2}, \tag{4.4}$$

where S_5 is given by (4.3) above; and it remains to deal with S_5.

We begin by disposing of the small d's. From a classical result in prime number theory, sometimes referred to as the Siegel–Walfisz theorem, we know, if $x \geqslant 3$ and χ is a non-principal character mod d, that

$$\psi(x,\chi) = \sum_{n\leqslant x} \Lambda(n)\,\chi(n) \ll x \exp\left(-\frac{1}{200}\sqrt{\log x}\right)$$

uniformly in d for $1 < d < \log^{200} x$. But

$$\psi_0(N,\chi) = \sum_{p_1p_2\in\mathscr{P}} \frac{\chi(p_1p_2)}{\log(N/p_1p_2)}\,\psi\left(\frac{N}{p_1p_2},\chi\right),$$

so that, since $N/(p_1p_2) \geqslant N^{1/3}$ for all elements p_1p_2 of $\mathscr{P}$, we may infer at once (in view of Lemma 11.1) that

$$\psi_0(N,\chi) \ll N \exp\left(-\frac{1}{400}\sqrt{\log N}\right)$$

uniformly in d for $1 < d \leqslant \log^{100} N$. Hence, by (4.3) and writing

$$Q = \log^{100} N, \tag{4.5}$$

we have

$$S_5 \leqslant \sum_{\substack{Q<d<z^2\\(d,N)=1}} \frac{\mu^2(d)}{\phi(d)} \sum_{\chi}{}^{*} |\psi_0(N,\chi)| + O\left(N\exp\left(-\frac{1}{500}\sqrt{\log N}\right)\right). \tag{4.6}$$

5. Application of Contour Integration

It is well-known that if

$$E(x) = \begin{cases} 1, & x > 1, \\ 0, & 0 < x < 1, \end{cases}$$

then

$$\left| \int_{a-iT}^{a+iT} \frac{x^s}{s}\, ds - 2\pi i\, E(x) \right| \leqslant \frac{2x^a}{T|\log x|} \qquad (a > 0, T > 0, x > 0, x \neq 1).$$

Hence, if $a > 1$,

$$\left| \frac{1}{2\pi i} \int_{a-iT}^{a+iT} \frac{(N+\frac{1}{2})^s}{s} \left\{ \sum_{m=1}^{\infty} \frac{\Lambda_0(m)\chi(m)}{m^s} \right\} ds - \psi_0(N,\chi) \right|$$

$$\leqslant (\pi T)^{-1} \sum_{m=1}^{\infty} \left(\frac{N+\frac{1}{2}}{m}\right)^a \Lambda_0(m) \left| \log \frac{N+\frac{1}{2}}{m} \right|^{-1}$$

$$\leqslant \frac{3}{\pi T} \sum_{m=1}^{\infty} \left(\frac{N+\frac{1}{2}}{m}\right)^a \left| \log \frac{N+\frac{1}{2}}{m} \right|^{-1} \ll \frac{N\log N}{T}$$

by a standard calculation on taking

$$a = 1 + \frac{1}{\log N}. \tag{5.1}$$

Since Λ_0 is plainly the convolution of Λ with the arithmetic function equal to $1/\log(N/r)$ for $r \in \mathscr{P}$ and 0 otherwise, we deduce that

$$\psi_0(N,\chi) = \frac{-1}{2\pi i} \int_{a-iT}^{a+T} \frac{(N+\frac{1}{2})^s}{s} \frac{L'(s,\chi)}{L(s,\chi)} P(s,\chi)\, ds + O\left(\frac{N\log N}{T}\right) \tag{5.2}$$

where

$$L(s,\chi) = \sum_{n=1}^{\infty} \frac{\chi(n)}{n^s}, \quad \frac{L'}{L}(s,\chi) = -\sum_{n=1}^{\infty} \frac{\chi(n)\Lambda(n)}{n^s}$$

and

$$P(s, \chi) = \sum_{r \in \mathscr{P}} \frac{\chi(r)}{r^s} \frac{1}{\log(N/r)} \qquad (\mathscr{R}s > 1).$$

We write

$$\frac{L'}{L}(s, \chi) = -\left(\sum_{n > u^2} + \sum_{1 \leqslant n \leqslant u^2}\right) \frac{\chi(n)\Lambda(n)}{n^s} = M_1(s, \chi) + M_2(s, \chi),$$

say, where u is a number satisfying $1 \leqslant u \leqslant z^2$. Since P and M_2 are both finite sums we may move the line of integration from $\sigma = a$ to $\sigma = \frac{1}{2}$ so far as that portion of the integrand which corresponds to PM_2 is concerned. Hence we deduce from (5.2) that

$$\psi_0(N, \chi)$$

$$\ll N \int_{-T}^{T} |(PM_1)(a + it, \chi)| \frac{dt}{1 + |t|} + N^{\frac{1}{2}} \int_{-T}^{T} |(PM_2)(\tfrac{1}{2} + it, \chi)| \frac{dt}{1 + |t|}$$

$$+ \frac{N}{T} \int_{\frac{1}{2}}^{a} |(PM_2)(\sigma + iT, \chi)| \, d\sigma + \frac{N}{T} \log N. \qquad (5.3)$$

It is convenient at this point to introduce, for any function $f(s, \chi)$, $s = \sigma + it$, the notations

$$\mathscr{B}_0(\sigma, f) = \sum_{\substack{d < u \\ (d, N) = 1}} \frac{d}{\phi(d)} \sum_{\chi}{}^{*} \int_{-T}^{T} |f(\sigma + it, \chi)| \frac{dt}{1 + |t|}$$

and

$$\mathscr{B}_1(\sigma, f) = \sum_{\substack{d < u \\ (d, N) = 1}} \frac{d}{\phi(d)} \sum_{\chi}{}^{*} |f(\sigma + it, \chi)|.$$

In this terminology (5.3), when combined with (4.6), yields that

$$\sum_{d < u} \frac{d}{\phi(d)} \sum_{\chi}{}^{*} |\psi_0(N, \chi)| \ll N\mathscr{B}_0(a, PM_1) + N^{\frac{1}{2}} \mathscr{B}_0(\tfrac{1}{2}, PM_2)$$

$$+ \frac{N}{T}\left(\max_{\frac{1}{2} \leqslant \sigma \leqslant a} \mathscr{B}_1(\sigma, PM_2) + \log N\right). \qquad (5.4)$$

By the inequalities of Schwarz and Cauchy we have that

$$\mathscr{B}_i(\sigma, f_1 f_2) \leqslant \mathscr{B}_i^{\frac{1}{2}}(\sigma, f_1^2)\, \mathscr{B}_i^{\frac{1}{2}}(\sigma, f_2^2) \qquad (i = 1, 2) \qquad (5.5)$$

and we shall use this remark to estimate each of the first three expressions on the right of (5.4) by means of the large sieve.

6. Application of the Large Sieve

We begin by quoting from the literature the following large sieve estimates.

Define, for any complex numbers a_n,

$$\mathfrak{G}(s, \chi) = \sum_{y<n\leqslant x+y} \frac{a_n\,\chi(n)}{n^s}, \qquad s = \sigma + it, \qquad \sigma \geqslant \tfrac{1}{2},$$

and suppose that $1 \leqslant u < z^2$. Then

LEMMA 11.5. *We have*

$$\sum_{d<u} \frac{d}{\phi(d)} \sum_{\chi}{}^{*} |\mathfrak{G}(s, \chi)|^2 \ll (u^2 + x) \sum_{y<n\leqslant x+y} \frac{|a_n|^2}{n^{2\sigma}}.$$

LEMMA 11.6. *We have*

$$\sum_{d<u} \frac{d}{\phi(d)} \sum_{\chi}{}^{*} \int_{-T}^{T} \frac{|\mathfrak{G}(\sigma+it, \chi)|^2}{\sigma + |t|}\,dt \ll \sum_{y<n\leqslant x+y} (u^2 \log T + n) \frac{|a_n|^2}{n^{2\sigma}},$$

which holds also with x infinite when $\sigma = a$.

From these we deduce readily

COROLLARY 11.5.1. *If $\frac{1}{2} \leqslant \sigma \leqslant a$, then*

$$\mathscr{B}_1(\sigma, P^2) \ll (u^2 + N^{2/3}) \log^{-2} N$$

and

$$\mathscr{B}_1(\sigma, M_2{}^2) \ll u^2 \log^2 N.$$

COROLLARY 11.6.1. *If $T \ll N^2$, then*

$$\mathscr{B}_0(a, P^2) \ll \left(\frac{u^2}{N^{13/30}} + \log^3 N\right) (\log N)^{-1},$$

$$\mathscr{B}_0(\tfrac{1}{2}, P^2) \ll (u^2 + N^{2/3}) (\log N)^{-1},$$

$$\mathscr{B}_0(a, M_1{}^2) \ll \log^4 N,$$

and

$$\mathscr{B}_0(\tfrac{1}{2}, M_2{}^2) \ll u^2 \log^3 N.$$

If now we apply (5.5) we obtain from Corollaries 11.5.1 and 11.6.1 that

$$\mathscr{B}_0(a, PM_1) \ll (uN^{-13/60} + \log^{3/2} N) \log^{-3/2} N,$$

$$\mathscr{B}_0(\tfrac{1}{2}, PM_2) \ll u(u + N^{1/3}) \log N$$

and

$$\max_{\frac{1}{2} \leqslant \sigma \leqslant a} \mathscr{B}_1 (\sigma, PM_2) \ll u(u + N^{1/3}).$$

It follows from (5.4), on taking $T = N^2$, *that*

$$\begin{aligned}\sum_{d<u} \frac{d}{\phi(d)} \sum_{\chi}{}^{*} |\psi_0 (N, \chi)| &\ll (uN^{47/60} + N \log^{3/2} N) \log^{-3/2} N \\ &\quad + (u^2 N^{\frac{1}{2}} + uN^{5/6}) \log N \\ &\quad + \frac{1}{T} (u^2 N + uN^{4/3} + N \log N) \\ &\ll N + (uN^{5/6} + u^2 N^{\frac{1}{2}}) \log N + u^2 N^{-2/3}.\end{aligned}$$

If we now apply Abel summation we derive readily that

$$\sum_{Q<d<z^2} \frac{1}{\phi(d)} \sum_{\chi}{}^{*} |\psi_0 (N, \chi)| \ll \frac{N}{Q};$$

hence, by (4.5) and (4.6),

$$S_5 \ll N/Q.$$

Retracing our steps now, we deduce from (4.4) that

$$S_4 \ll \frac{N}{\log^{41} N},$$

from (4.1) that

$$S_3 \ll \frac{N}{\log^{38} N}, \tag{6.1}$$

from (3.9) that

$$S_1 \leqslant 3{\cdot}944 \prod_{p>2} \left(1 - \frac{1}{(p-1)^2}\right) \prod_{2<p|N} \frac{p-1}{p-2} \cdot \frac{N}{\log^2 N} \left(1 + O\left(\frac{1}{\log N}\right)\right)$$

and hence, finally, from (2.11), that

$$S_0 \leqslant 3{\cdot}944 \prod_{p>2} \left(1 - \frac{1}{(p-1)^2}\right) \prod_{2<p|N} \frac{p-1}{p-2} \cdot \frac{N}{\log^2 N} \left(1 + O\left(\frac{1}{\sqrt{\log N}}\right)\right).$$

This is a better result than (2.5) and, indeed, completes the proof of Chen's Theorem, in the form stated.

NOTES

11.1. Theorem 11.1: Chen [3]. The numerical integrations mentioned in the text allow us to replace 0·67 by 0·689.

Theorem 11.1. Remark 1: It would be interesting to know whether the more elaborate weighting procedure of Chapter 9 could be adapted to the purpose of Theorem 11.1. This might lead to numerical improvements and could be important.

Remark 2: Vaughan's observation from Chapter 9 (see Notes p. 268) may readily be transferred here; *either* $2p + 1$ is prime infinitely often, *or* $2p + 1 = p_1 p_2$ infinitely often, in which case the equation $\tau(n + 1) = \tau(n)$ has infinitely many solutions.

11.2. (2.2), (2.4) *et seq.*: Observe Chen's important innovation here—the first two terms on the right of (2.2) correspond to a straightforward Kuhn weighting but the third expression is attacked in terms of sifting, not $\mathcal{G}_{p_1p_2}$, but the different sequences $\{N - (p_1 p_2)p \colon p \leqslant N/(p_1 p_2)\}$.

11.4. The theory of Dirichlet characters is discussed in e.g. Prachar [6], Chapter IV.6.

We make frequent use here, and later, of the inequality

$$\sum_{d<y} \frac{\mu^2(d)k^{\nu(d)}}{\phi(d)} \ll \log^k y, \qquad y \geqslant 2.$$

The expression on the left is at most

$$\prod_{p<y}\left(1 + \frac{k}{p-1}\right) < \prod_{p<y}\left(1 + \frac{1}{p-1}\right)^k = \prod_{p<y}\left(1 - \frac{1}{p}\right)^{-k} \ll \log^k y.$$

The Siegel–Walfisz inequality: see e.g. T. Estermann, "Introduction to Modern Prime Number Theory", Cambridge Tract No. 41 (1952), Theorem 52.

11.5. For the calculations involving $E(x)$ see Estermann *loc. cit.*, Theorems 15, 17.

11.6. Lemma 11.5: See Montgomery [2], Theorem 2.5.

Lemma 11.6 (Gallagher): see Montgomery [2], Lemma 1.10 and the preceding Lemma 11.5, from which our result follows by integration by parts.

The argument after (5.2), shown to us by Dr. Vaughan, is due to his student, Mr. P. M. Ross, and represents a substantial simplification of the analytic part of Chen's own proof, and we are indebted to him for permitting us to use it. Further simplification is possible and we refer the reader for this to a forthcoming paper by Ross.

Bibliography†

I.

A. Theoretical Contributions

Merlin [1, 2]; Brun [1, 3, 5, 7, 8]; Chang [1]; Buchstab [2]; Kuhn [1]; Rényi [1, 2]; Le Veque [1]; A. Selberg [1, 2, 3, 4]; Kuhn [2, 3]; Hooley [1]; Fluch [1]; Barban [4, 7]; Levin [5, 6]; Ankeny [1]; Ankeny–Onishi [1]; Levin [8, 10]; Miech [1, 2, 3]; Jurkat–Richert [1]; Buchstab [9]; Vinogradov [7]; Levin–Maksudov [1]; Halberstam–Jurkat–Richert [1]; Buchstab [10]; Hooley [5]; Montgomery [1]; Bombieri [2]; Richert [1]; A. Selberg [5]; Montgomery [2]; Iwaniec [1, 2]; Halberstam–Richert [2]; Huxley [2]; Kobayashi [1]; Chen [3]; Montgomery–Vaughan [1, 2]; Halberstam–Richert [5].

B. Surveys

a) Brun's sieve: Brun [9, 10]; Landau [1, 2, 3]; James [3, 4]; Trost [1]; Odlyzko [1].

b) Selberg's sieve: Prachar [6]; Hua [2]; Linnik [3]; Gelfond–Linnik [1]; Halberstam–Roth [1]; Richert [3]; A. Selberg [5]; Odlyzko [1].

II. Methodological Developments

Ankeny–Onishi [1]; Barban [1, 4]; Brun [8]; Buchstab [2, 10]; Čulanovskiĭ [1]; Estermann [1]; Fluch [1]; Halberstam [1]; Halberstam–Jurkat–Richert [1]; Halberstam–Richert [4, 5]; Hooley [2, 5]; James [1, 3]; Jurkat–Richert [1]; Klimov [2, 3]; Landau [1]; Lenskoĭ [4]; Levin [1, 3, 5, 9, 10]; van Lint–Richert [2]; Mientka [1]; Ožigova [1]; Pan [1]; Prachar [6]; Rademacher [1]; Ricci [8]; Rieger [10, 12, 13]; Schwarz [2]; A. Selberg [4]; Shapiro–Warga [1]; Tartakovskiĭ [2]; Uchiyama [1, 2, 3, 4]; Vinogradov [1, 2]; Wang [1, 2, 8, 11].

III. Applications

a) Elementary sieve applications: Chang [1]; Ducci [1, 2]; Erdös [3, 7, 14, 17]; Hooley [3]; Hua [2]; Norton [2]; Prachar [6, 9]; Rankin [1]; Schönhage [1]; Vinogradov–Linnik [1].

b) Fundamental Lemma type applications: Barban [3, 5, 9]; Barban–Vinogradov [1]; Brun [7]; Buchstab [7, 8]; Chowla–Briggs [1]; Eda–Yamano [1]; Erdös–Kac [1];

† The references listed under I, A are in chronological order, all the others are in alphabetic order.

Halberstam–Richert [3, 4, 5]; Hooley [1]; Jurkat–Richert [1]; Kátai [6]; Kubilius [1, 2]; Lavrik [3]; Le Veque [1]; Levin [9]; Rademacher [1]; A. Selberg [1]; Uždavinis [1, 2, 3]; Warlimont [1].

c) Brun's upper sieve: Barner [1]; Bays [1]; Bredihin [1]; de Bruijn [2]; Brun [2, 3, 4, 5, 6, 7]; Buchstab [1, 2, 3, 7]; Chung [1]; Cugiani [3, 4]; Erdös [1, 2, 4, 5, 6, 7, 8, 9, 10, 12, 13, 15, 16, 19]; Erdös–Prachar [1]; Erdös–Rényi [1]; Gelfond–Linnik [1]; Goldfeld [2]; Hardy–Littlewood [2]; Heilbronn [1]; Heilbronn–Landau–Scherk [1]; Hooley [1, 8, 9]; Hua [2]; Klimov [4]; Knödel [1, 2, 3, 4]; Kuhn [1]; Kuzjašev–Čečuro [1]; Landau [1]; Motohashi [2, 3]; Odlyzko [1]; Porter [1]; Prachar [1, 2, 3, 4, 8, 10]; Rademacher [3]; Rankin [2, 3, 4]; Ricci [1, 3, 5, 6, 7, 9, 10, 11, 12]; Rieger [6, 7, 8, 9, 10, 14, 18]; Romanoff [1]; Roux [1]; Schnirelmann [1]; Segal [1]; S. Selberg [2]; Tartakovskiĭ [1]; Titchmarsh [1]; Vaughan [1]; Viola [1].

d) Selberg's upper sieve: Ankeny [1]; Ankeny–Onishi [1]; Barban [3, 6, 7, 8, 10]; Barban–Vinogradov–Levin [1]; Barban–Levin [1]; Bateman–Horn [1, 2]; Bateman–Stemmler [1]; Bombieri–Davenport [1, 2]; Chen [2, 3]; Chowla–Briggs [1]; Čulanovskiĭ [1]; Dodson [1]; Elliott [1]; Elliott–Halberstam [1]; Erdös [4, 12, 16]; Faĭziev [1]; Gelfond–Linnik [1]; Goldfeld [1]; Halberstam [1, 2]; Halberstam–Roth [1]; Hall [1]; Haneke [1]; Haselgrove [1]; Hooley [1, 4]; Hua [2]; Huxley [1]; Indlekofer–Schwarz [1]; Karšiev [1]; Kátai [1, 2, 3, 4, 5]; Klimov [1, 2, 3]; Klimov–Pil'tjaĭ–Šeptickaja [1]; Kondakova–Klimov [1]; Lavrik [1, 2]; Lenskoĭ [1, 2, 3, 5]; Levin [5]; Levin–Maksudov [1]; Linnik [3]; van Lint–Richert [2]; Maksudov [1]; Montgomery [2]; Montgomery–Vaughan [1]; Motohashi [1, 4, 5]; Ožigova [1]; Pan [3]; Prachar [1, 2, 3, 5, 6, 8, 10]; Ramachandra [1, 2, 3]; Rieger [5, 6, 7, 9, 10, 11, 12, 13, 14, 15, 16]; Rodriquez [1]; Russell [1]; Schwarz [1, 3, 4]; A. Selberg [1, 3, 4]; S. Selberg [3]; Shapiro–Warga [1]; Siebert [1]; Uchiyama [1]; Vinogradov [4, 7]; Wang [1, 2, 3, 5, 8, 11]; Wang–Hsieh–Yu [1]; Webb [1]; Yin [1, 2].

e) Brun's lower sieve: de Bruijn [2]; Brun [4, 5, 6, 7]; Buchstab [1, 2, 3, 7, 8]; Chowla–Erdös–Straus [1]; Erdös [6, 14, 16]; Estermann [1]; Hooley [3]; Hua [2]; Iwaniec [1, 2]; James [2, 3, 4]; Kuhn [1, 2, 3]; Linnik [1]; Odlyzko [1]; Prachar [7]; Rademacher [1]; Rényi [1, 2]; Ricci [1, 2, 3, 5, 6, 7, 9]; Rogers [1]; Schinzel [3]; Schinzel–Wang [1]; Tartakovskiĭ [1, 2]; Wang [2, 7, 10, 11, 12, 13].

f) Selberg's lower sieve: Ankeny [1]; Ankeny–Onishi [1]; Barban [6, 7, 8, 10]; Buchstab [9, 10]; Chen [1, 2, 3]; Elliott [2]; Fluch [1, 2]; Gelfond–Linnik [1]; Greaves [1, 2]; Halberstam–Jurkat–Richert [1]; Halberstam–Richert [2]; Halberstam–Roth [1]; Halberstam–Rotkiewicz [1]; Hooley [1]; Hsieh [1, 2]; Hua [2]; Jurkat–Richert [1]; Jutila [1]; Kuhn [1, 2, 3]; Levin [1, 2, 4, 5, 8, 10]; Levin–Maksudov [1]; Maksudov [1]; Miech [1, 2, 3]; Mientka [1]; Pan [1, 2]; Porter [2]; Richert [1, 2]; Rieger [9, 17]; Schinzel [1]; A. Selberg [2, 3, 4]; Togashi–Uchiyama [1]; Uchiyama–Uchiyama [1]; Uchiyama [3, 4, 5]; Vaughan [2]; Vinogradov [1, 2, 3, 7]; Wang [1, 2, 3, 4, 5, 6, 8, 9, 11, 12, 13].

IV. Extension of Sieve Methods to Algebraic Domains

Andruhaev [1]; Fogels [1, 2, 3, 4, 5, 6]; Lenskoĭ [2, 4]; Levin–Maksudov [1]; Levin–Tuljaganova [1]; Rademacher [2]; Rieger [1, 2, 3, 4]; Sarges [1]; Schaal [1]; Tatuzawa [1]; Tuljaganova [1]; Vinogradov [5, 6]; Wilson [1]; Zaikina [1].

V. Variants of Sieve Methods

a) Sieving by squares and higher powers of primes: Cugiani [1, 2]; Erdös [11]; Hooley [6, 7]; Ricci [2, 4, 6, 7].

b) Others: Chang [1]; Prachar [6, p. 157]; Gallagher [3]; A. Selberg [3]; Vinogradov–Linnik [1].

VI. Aids

a) The G-functions: Ankeny–Onishi [1]; Barban [5, 8, 10]; Čulanovskiĭ [1]; Halberstam–Richert [1]; Hsieh [1, 2]; Klimov [1, 2]; Kubilius [2]; Levin [1, 5, 6, 7, 8, 9]; Levin–Faĭnleĭb [1, 2]; Levin–Tuljaganova [1]; Levinson [1, 2]; van Lint–Richert [1]; Onishi [1]; Pan [1]; Prachar [6]; Rankin [1]; Schwarz [1]; Shapiro–Warga [1]; Vinogradov [2]; Wang [2, 8, 11]; Wirsing [1, 2].

b) Combinatorial identities: Barban [5]; de Bruijn [2]; Brun [5, 7, 11]; Buchstab [1, 2, 4, 5, 6, 10]; Hua [2]; Jurkat–Richert [1]; Rademacher [1, 3]; Tartakovskiĭ [1]; Uždavinis [3]; Vinogradov [2].

c) $\Psi(x, y)$: de Bruijn [4, 5]; Norton [1].

d) The estimating functions: Ankeny [1]; Ankeny–Onishi [1]; de Bruijn [1, 2, 3, 4, 5]; Buchstab [1, 6, 7]; Hagedorn [1]; Halberstam–Richert [2]; Hua [1, 2]; Jurkat–Richert [1]; Norton [1]; Porter [2]; Uchiyama [3, 4, 6].

e) Hypothesis H and other conjectures: Barban [8]; Bateman–Horn [1, 2]; Bateman–Stemmler [1]; Buchstab [10]; Davenport–Schinzel [1]; Dickson [1]; Elliott–Halberstam [1, 2]; Erdös [12, 16]; Hardy–Littlewood [1, 2]; Hensley–Richards [1]; Porter [1]; Schinzel [1, 2]; Schinzel–Sierpiński [1]; Shah–Wilson [1]; Siebert [3]; Stein–Stein [1]; Turán [1]; Vinogradov [3].

f) Bombieri's prime number theorem and similar results: Ankeny–Onishi [1]; Barban [2, 6, 7, 8, 10]; Bombieri [1]; Chen [3]; Davenport–Halberstam [1]; Gallagher [1, 2]; Jutila [1]; Levin [10]; Montgomery [2]; Pan [1, 2, 3]; Rényi [1, 2]; Siebert [2]; Siebert–Wolke [1]; Vinogradov [7]; Wolke [1, 2].

g) Special results from number theory: Nagel [1]; Rosser–Schoenfeld [1].

h) Numerical results: Ankeny–Onishi [1]; Buchstab [2, 3, 9, 10]; Halberstam–Richert [2]; Hsieh [1]; Jurkat–Richert [1]; Levin [8]; Porter [2]; A. Selberg [5]; Wang [1, 2, 9].

References

ANDRUHAEV, H. M.

1. The addition problem for prime and near-prime numbers in algebraic number fields. (Russian)
Dokl. Akad. Nauk SSSR **159** (1964), 1207–1209 = *Soviet Math. Dokl.* **5** (1964), 1666–1668. MR **30,** 1116.

ANKENY, N. C.

1. Applications of the sieve.
Proc. Sympos. Pure Math. **8** (1965), 113–118. MR **31,** 141.

ANKENY, N. C. and ONISHI, H.

1. The general sieve.
Acta Arith. **10** (1964/65), 31–62. MR **29,** 4740.

BARBAN, M. B.

1. Selberg's sieve applied to a certain estimate from below. (Russian)
Dokl. Akad. Nauk UzSSR **1959,** no. 3, 7–8.
2. New applications of the "great sieve" of Yu. V. Linnik. (Russian)
Akad. Nauk UzSSR Trudy Inst. Mat. No. 22 (1961), 1–20. MR **30,** 1990.
3. The normal order of additive arithmetic functions on sets of "shifted" primes. (Russian. English summary)
Acta Math. Acad. Sci. Hungar. **12** (1961), 409–415. MR **24,** A3145.
4. An analogue of the law of large numbers for additive number-theoretic functions defined on a set of "shifted" primes. (Russian)
Dokl. Akad. Nauk UzSSR **1961,** no. 12, 8–12.
5. On a theorem of I. P. Kubiljus. (Russian. Uzbek summary)
Izv. Akad. Nauk UzSSR Ser. Fiz.-Mat. Nauk **1961,** no. 5, 3–9. MR **25,** 2051; Corrigendum; *ibid.* **1963,** no. 1, 82–83.
6. The density of zeros of Dirichlet L-series and the problem of the addition of primes and almost primes. (Russian)
Dokl. Akad. Nauk UzSSR **1963,** no. 1, 9–10.
7. Analogues of the divisor problem of Titchmarsh. (Russian. English summary)
Vestnik Leningrad. Univ. **18** (1963), no. 4, 5–13. MR **28,** 57.
8. The "density" of the zeros of Dirichlet L-series and the problem of the sum of primes and "near primes". (Russian)
Mat. Sbornik (N.S.) **61 (103)** (1963), 418–425. MR **30,** 1992.
9. On the number of divisors of "translations" of the prime number-twins. (Russian. English summary)
Acta Math. Acad. Sci. Hungar. **15** (1964), 285–288. MR **30,** 1102.
10. The "large sieve" method and its application to number theory. (Russian)
Uspehi Mat. Nauk **21** (1966), no. 1 (127), 51–102 = *Russian Math. Surveys* **21** (1966), no. 1, 49–103. MR **33,** 7320.

BARBAN, M. B. and VINOGRADOV, A. I.
1. On the number-theoretic basis of probabilistic number theory. (Russian) *Dokl. Akad. Nauk SSSR* **154** (1964), 495–496 = *Soviet Math. Dokl.* **5** (1964), 96–98. MR **29,** 1194.

BARBAN, M. B., VINOGRADOV, A. I. and LEVIN, B. V.
1. Limit laws for functions of the class H of I. P. Kubilius which are defined on a set of "shifted" primes. (Russian. Lithuanian and English summaries) *Litovsk. Mat. Sb.* **5** (1965), 5–8. MR **34,** 5794.

BARBAN, M. B. and LEVIN, B. V.
1. Multiplicative functions of "shifted" primes. (Russian) *Dokl. Akad. Nauk SSSR* **181** (1968), 778–780 = *Soviet Math. Dokl.* **9** (1968), 912–914. MR **40,** 99.

BARNER, K.
1. Zur Abschätzung von Reihen, deren Glieder von rationalen Funktionen einer festen Anzahl sukzessiver Primzahlen gebildet werden. *Monatsh. Math.* **68** (1964), 1–16. MR **31,** 4772.

BATEMAN, P. T. and HORN, R. A.
1. A heuristic asymptotic formula concerning the distribution of prime numbers. *Math. Comp.* **16** (1962), 363–367. MR **26,** 6139.
2. Primes represented by irreducible polynomials in one variable. *Proc. Sympos. Pure Math.* **8** (1965), 119–135. MR **31,** 1234.

BATEMAN, P. T. and STEMMLER, R. M.
1. Waring's problem for algebraic number fields and primes of the form $(p^r-1)/(p^d-1)$. *Illinois J. Math.* **6** (1962), 142–156. MR **25,** 2059.

BAYS, S.
1. Sur un théorème de Viggo Brun et intervalle entre deux nombres premiers consécutifs. *Enseignement Math.* **28** (1929), 287–288.

BOMBIERI, E.
1. On the large sieve. *Mathematika* **12** (1965), 201–225. MR **33,** 5590.
2. On a theorem of van Lint and Richert. *Symposia Mathematica* **4** (INDAM, Rome, 1968/69), 175–180. MR **43,** 4791.

BOMBIERI, E. and DAVENPORT, H.
1. Small differences between prime numbers. *Proc. Roy. Soc. Ser. A* **293** (1966), 1–18. MR **33,** 7314.
2. On the large sieve method. Number Theory and Analysis (Papers in Honor of Edmund Landau), pp. 9–22. New York, 1969. MR **41,** 5327.

BREDIHIN, B. M.
1. The dispersion method and binary additive problems of definite type. (Russian) *Uspehi Mat. Nauk* **20** (1965), no. 2 (122), 89–130 = *Russian Math. Surveys* **20** (1965), no. 2, 85–125. MR **32,** 5618.

DE BRUIJN, N. G.
1. On some linear functional equations. *Publ. Math. Debrecen* **1** (1950), 129–134. MR **12,** p. 106.

2. On the number of uncancelled elements in the sieve of Eratosthenes.
Nederl. Akad. Wetensch. Proc. **53** (1950), 803–812=*Indag. Math.* **12** (1950), 247–256. MR **12,** p. 11.
3. The asymptotic behaviour of a function occurring in the theory of primes.
J. Indian Math. Soc. (*N.S.*) **15** (1951), 25–32. MR **13,** p. 326.
4. On the number of positive integers $\leqslant x$ and free of prime factors $>y$.
Nederl. Akad. Wetensch. Proc. Ser. A **54** (1951), 50–60 = *Indag. Math.* **13** (1951), 2–12. MR **13,** p. 724.
5. On the number of positive integers $\leqslant x$ and free of prime factors $>y$. II.
Nederl. Akad. Wetensch. Proc. Ser. A **69** = *Indag. Math.* **28** (1966), 239–247. MR **34,** 5770.

BRUN, V.
1. Über das Goldbachsche Gesetz und die Anzahl der Primzahlpaare.
Archiv for Math. og Naturvid. B **34** (1915), no. 8, 19pp.
2. Om fordelingen av primtallene i forskjellige talklasser. En øvre begrænsning.
Nyt Tidsskr. f. Math. **27** B (1916), 45–58.
3. Sur les nombres premiers de la forme $ap + b$.
Archiv for Math. og Naturvid. B **34** (1917), no. 14, 9pp.
4. Le crible d'Eratosthène et le théorème de Goldbach.
C. R. Acad. Sci. Paris **168** (1919), 544–546.
5. La série 1/5 +1/7 +1/11 +1/13 +1/17 +1/19 +1/29 +1/31 +1/41 +1/43 +1/59 +1/61 +. . . où les dénominateurs sont "nombres premiers jumeaux" est convergente ou finie.
Bull. Sci. Math. (2) **43** (1919), 100–104.
6. La série 1/5 +1/7 +1/11 +1/13 +1/17 +1/19 +1/29 +1/31 +1/41 +1/43 +1/59 +1/61 +. . . où les dénominateurs sont "nombres premiers jumeaux" est convergente ou finie.
Bull. Sci. Math. (2) **43** (1919), 124–128.
7. Le crible d'Eratosthène et le théorème de Goldbach.
Skr. Norske Vid.-Akad. Kristiania I. **1920,** no. 3, 36pp.
8. Eratosthenes's sold.
Norsk Mat. Tidsskr. **4** (1922), 65–79.
9. Das Sieb des Eratosthenes.
5. *Skand. Mat. Kongr., Helsingfors,* **1922,** 197–203.
10. Untersuchungen über das Siebverfahren des Eratosthenes.
Jber. Deutsch. Math.-Verein. **33** (1924), 81–96.
11. Reflections on the sieve of Eratosthenes.
Norske Vid. Selsk. Skr. (*Trondheim*) **1967,** no. 1, 9pp. MR **36,** 2548.

BUCHSTAB, A. A.
1. Asymptotic estimates of a general number-theoretic function. (Russian. German summary)
Mat. Sbornik (*N.S.*) **2 (44)** (1937), 1239–1246.
2. New improvements in the method of the sieve of Eratosthenes. (Russian. German summary)
Mat. Sbornik (*N.S.*) **4 (46)** (1938), 375–387.
3. Sur la décomposition des nombres pairs en somme de deux composantes dont chacune est formée d'un nombre borné de facteurs premiers.
Dokl. Akad. Nauk SSSR **29** (1940), 544–548. MR **2,** p. 348.
4. On an additive representation of integers. (Russian. English summary)
Mat. Sbornik (*N.S.*) **10 (52)** (1942), 87–91. MR **4,** p. 190.

5. On a relation for the function $\pi(x)$ expressing the number of primes that do not exceed x. (Russian. English summary)
Mat. Sbornik (*N.S.*) **12 (54)** (1943), 152–160. MR **5,** p.35.
6. On those numbers in an arithmetic progression all prime factors of which are small in order of magnitude. (Russian)
Dokl. Akad. Nauk SSSR **67** (1949), 5–8. MR **11,** pp.84, 871.
7. On an asymptotic estimate of the number of numbers of an arithmetic progression which are not divisible by "relatively" small prime numbers. (Russian)
Mat. Sbornik (*N.S.*) **28 (70)** (1951), 165–184. MR **13,** p.626.
8. On an additive representation of integers. (Russian)
Moskov. Gos. Ped. Inst. Učen. Zap. **71** (1953), 45–62. MR **18,** p.17.
9. New results in the investigation of the Goldbach–Euler problem and the problem of prime pairs. (Russian)
Dokl. Akad. Nauk SSSR **162** (1965), 735–738 = *Soviet Math. Dokl.* **6** (1965), 729–732. MR **31,** 2226.
10. Combinatorial intensification of the sieve method of Eratosthenes. (Russian)
Uspehi Mat. Nauk **22** (1967), no. 3 (135), 199–226 = *Russian Math. Surveys* **22** (1967), no. 3, 205–233. MR **36,** 1413.

CHANG, T.-H.
1. Über aufeinanderfolgende Zahlen, von denen jede mindestens einer von n linearen Kongruenzen genügt, deren Moduln die ersten n Primzahlen sind.
Schr. Math. Sem. Inst. Angew. Math. Univ. Berlin **4** (1938), 35–55.

CHEN, J.
1. On large odd numbers as sum of three almost equal primes.
Sci. Sinica **14** (1965), 1113–1117. MR **32,** 5619.
2. On the representation of a large even integer as the sum of a prime and the product of at most two primes. (Chinese)
Kexue Tongbao **17** (1966), 385–386. MR **34,** 7483.
3. On the representation of a larger even integer as the sum of a prime and the product of at most two primes.
Sci. Sinica **16** (1973), 157–176.

CHOWLA, S. and BRIGGS, W. E.
1. On the number of positive integers $\leqslant x$ all of whose prime factors are $\leqslant y$.
Proc. Amer. Math. Soc. **6** (1955), 558–562. MR **17,** p.127.

CHOWLA, S., ERDÖS, P. and STRAUS, E. G.
1. On the maximal number of pairwise orthogonal Latin squares of a given order.
Canad. J. Math. **12** (1960), 204–208. MR **23,** A70.

CHUNG, K.-L.
1. Two remarks on Viggo Brun's method.
Sci. Rep. Nat. Tsing Hua Univ. (*A*) **4** (1940), 249–255. MR **3,** p.68.

CUGIANI, M.
1. Sulla rappresentazione degli interi come somme di una potenza e di un numero libero da potenze.
Ann. Mat. Pura Appl. (4) **33** (1952), 135–143. MR **14,** p.356.
2. Sui valori di un polinomio che risultano liberi da potenze.
Ann. Mat. Pura Appl. (4) **35** (1953), 291–298. MR **15,** p.603.
3. Sulle "catene" di numeri primi consecutivi a differenze limitata.
Ann. Mat. Pura Appl. (4) **36** (1954), 121–132. MR **15,** p.935.

4. Nuovi risultati sulle "catene" di numeri primi.
Ann. Mat. Pura Appl. (4) **38** (1955), 309–320. MR **17**, p.127.

ČULANOVSKIĬ, I. V.

1. Certain estimates connected with a new method of Selberg in elementary number theory. (Russian)
Dokl. Akad. Nauk SSSR **63** (1948), 491–494. MR **10**, p.355.

DAVENPORT, H. and HALBERSTAM, H.

1. Primes in arithmetic progressions.
Michigan Math. J. **13** (1966), 485–489. MR **34**, 156; Corrigendum: *ibid.* **15** (1968), 505. MR **38**, 2099.

DAVENPORT, H. and SCHINZEL, A.

1. A note on certain arithmetical constants.
Illinois J. Math. **10** (1966), 181–185. MR **32**, 5632.

DICKSON, L. E.

1. A new extension of Dirichlet's theorem on prime numbers.
Messenger of Math. **33** (1903/04), 155–161.

DODSON, M. M.

1. The average order of two arithmetical functions.
Acta. Arith. **16** (1969/70), 71–84. MR **40**, 1353.

DUCCI, E.

1. Sulla totalità dei numeri primi.
Giorn. Mat. Battaglini **74** (1936), 21–28.
2. Sulla totalità dei numeri primi. II.
Giorn. Mat. Battaglini **74** (1936), 139–148.

EDA, Y. and YAMANO, G.

1. On the number of prime factors of integers.
Sci. Rep. Kanazawa Univ. **14** (1969), 13–20. MR **41**, 1671.

ELLIOTT, P. D. T. A.

1. On sequences of integers.
Quart. J. Oxford (2) **16** (1965), 35–45. MR **40**, 2632.
2. The distribution of primitive roots.
Canad. J. Math. **21** (1969), 822–841. MR **40**, 104.

ELLIOTT, P. D. T. A. and HALBERSTAM, H.

1. Some applications of Bombieri's theorem.
Mathematika **13** (1966), 196–203. MR **34**, 5788.
2. A conjecture in prime number theory.
Symposia Mathematica **4** (INDAM, Rome, 1968/69), 59–72. MR **43**, 1943.

EL'NATANOV, B. A.

1. Asymptotic estimate of special arithmetic sums taken over numbers with "large" and "small" prime divisors. (Russian. Tajiki summary)
Dokl. Akad. Nauk Tadžik. SSR **9** (1966), no. 8, 8–13. MR **34**, 1283.

ERDÖS, P.

1. On the difference of consecutive primes.
Quart. J. Oxford **6** (1935), 124–128.
2. On the normal number of prime factors of $p-1$ and some related problems concerning Euler's ϕ-function.
Quart. J. Oxford **6** (1935), 205–213.

3. Note on sequences of integers no one of which is divisible by any other.
J. London Math. Soc. **10** (1935), 126–128.
4. On the easier Waring problem for powers of primes. I.
Proc. Cambridge Philos. Soc. **33** (1937), 6–12.
5. The difference of consecutive primes.
Duke Math. J. **6** (1940), 438–441. MR **1,** p.292.
6. On some applications of Brun's method.
Acta Univ. Szeged. Sect. Sci. Math. **13** (1949), 57–63. MR **10,** p.684.
7. Problems and results on the differences of consecutive primes.
Publ. Math. Debrecen **1** (1949), 33–37. MR **11,** p.84.
8. Some problems and results in elementary number theory.
Publ. Math. Debrecen **2** (1951), 103–109. MR **13,** p.627.
9. On the sum $\sum_{k=1}^{x} d(f(k))$.
J. London Math. Soc. **27** (1952), 7–15. MR **13,** p.438.
10. Some unsolved problems.
Michigan Math. J. **4** (1957), 291–300. MR **20,** 5157.
11. Über die kleinste quadratfreie Zahl einer arithmetischen Reihe.
Monatsh. Math. **64** (1960), 314–316. MR **22,** 9476.
12. Some unsolved problems.
Magyar Tud. Akad. Mat. Kutató Int. Közl. **6** (1961), 221–254. MR **31,** 2106.
13. Some remarks on number theory. II. (Hungarian. Russian and English summaries)
Mat. Lapok **12** (1961), 161–169. MR **26,** 2411.
14. On the integers relatively prime to n and on a number-theoretic function considered by Jacobsthal.
Math. Scand. **10** (1962), 163–170. MR **26,** 3651.
15. Quelques problèmes de théorie des nombres.
Monographies de L'Enseignement Mathématique, No. 6, pp. 81–135.
L'Enseignement Mathématique, Université, Geneva, 1963. MR **28,** 2070.
16. Some recent advances and current problems in number theory.
Lectures on Modern Mathematics **3** (New York, 1965), 196–244. MR **31,** 2191.
17. On the distribution of divisors of integers in the residue classes (mod d).
Bull. Soc. Math. Grèce (N.S.) **61** (1965), fasc. 1, 27–36. MR **34,** 7474.
18. On some properties of prime factors of integers.
Nagoya Math. J. **27** (1966), 617–623. MR **34,** 4220.
19. Some remarks on the iterates of the ϕ and σ functions.
Colloq. Math. **17** (1967), 195–202. MR **36,** 2573.

ERDÖS, P. and KAC, M.
1. The Gaussian law of errors in the theory of additive number theoretic functions.
Amer. J. Math. **62** (1940), 738–742. MR **2,** p.42.

ERDÖS, P. and PRACHAR, K.
1. Sätze und Probleme über p_k/k.
Abh. Math. Sem. Univ. Hamburg **25** (1961/62), 251–256. MR **25,** 3901.

ERDÖS, P. and RÉNYI, A.
1. Some problems and results on consecutive primes.
Simon Stevin **27** (1950), 115–125. MR **11,** p.644.

ERDÖS, P. and TURÁN, P.
1. On some new questions on the distribution of prime numbers.
Bull. Amer. Math. Soc. **54** (1948), 371–378. MR **9,** p.498.

ESTERMANN, T.

1. Eine neue Darstellung und neue Anwendungen der Viggo Brunschen Methode. *J. Reine Angew. Math.* **168** (1932), 106–116.

FAĬZIEV, R. F.

1. The number of integers, expressible in the form of a sum of two primes, and the number of k-twin pairs. (Russian. Tajiki summary) *Dokl. Akad. Nauk Tadžik. SSR* **12** (1969), no. 2, 12–16. MR **40,** 5566.

FLUCH, W.

1. Verwendung der Zeta-Funktion beim Sieb von Selberg. *Acta Arith.* **5** (1959), 381–405. MR **23,** A1614.
2. Bemerkung über quadratfreie Zahlen in arithmetischen Progressionen. *Monatsh. Math.* **72** (1968), 427–430. MR **39,** 6840.

FOGELS, E. K.

1. Analogue of the Brun–Titchmarsh theorem. (Russian. Latvian summary) *Latvijas PSR Zinātņu Akad. Fiz. Mat. Inst. Raksti.* **2** (1950), 46–58. MR **13,** p. 725.
2. On the distribution of prime ideals. (Russian) *Dokl. Akad. Nauk SSSR* **140** (1961), 1029–1032 = *Soviet Math. Dokl.* **2** (1961), 1322–1326. MR **24,** A1906.
3. On the zeros of Hecke's L-functions. III. *Acta Arith.* **7** (1961/62), 225–240.
4. On the abstract theory of primes. I. *Acta Arith.* **10** (1964/65), 137–182. MR **29,** 5802.
5. On the abstract theory of primes. II. *Acta Arith.* **10** (1964/65), 333–358. MR **30,** 4738.
6. On the zeros of L-functions. *Acta Arith.* **11** (1965), 67–96. MR **31,** 1230; Corrigendum: *ibid.* **14** (1967/68), 435. MR **37,** 4033.

GALLAGHER, P. X.

1. The large sieve. *Mathematika* **14** (1967), 14–20. MR **35,** 5411.
2. Bombieri's mean value theorem. *Mathematika* **15** (1968), 1–6. MR **38,** 5724.
3. A larger sieve. *Acta Arith.* **18** (1971), 77–81. MR **45,** 214.

GELFOND, A. O. and LINNIK, YU. V.

1. Elementary methods in analytic number theory. Moscow, 1962, 272pp. = Chicago, 1965, xii + 242pp. MR **32,** 5575. = Oxford–New York–Toronto, 1966, xi + 232pp. MR **34,** 1252.

GOLDFELD, M.

1. On the number of primes p for which $p + a$ has a large prime factor. *Mathematika* **16** (1969), 23–27. MR **39,** 5493.
2. A further improvement of the Brun–Titchmarsh theorem. *J. London Math. Soc.* (to appear)

GREAVES, G.

1. Large prime factors of binary forms. *J. Number Theory* **3** (1971), 35–59. MR **42,** 5909.
2. An application of a theorem of Barban–Davenport and Halberstam. *Bull. London Math. Soc.* **6** (1974), 1–9.

HAGEDORN, H. W.
1. Sieve methods and polynomial sequences.
Acta Arith. **28,** no. 3. (to appear).

HALBERSTAM, H.
1. On the distribution of additive number-theoretic functions. III.
J. London Math. Soc. **31** (1956), 14–27. MR **17,** p.461.
2. Footnote to the Titchmarsh–Linnik divisor problem.
Proc. Amer. Math. Soc. **18** (1967), 187–188. MR **34,** 4221.

HALBERSTAM, H., JURKAT, W. and RICHERT, H.-E.
1. Un nouveau résultat de la méthode du crible.
C. R. Acad. Sci. Paris Sér. A-B **264** (1967), A920-A923. MR **36,** 6374.

HALBERSTAM, H. and RICHERT, H.-E.
1. Mean value theorems for a class of arithmetic functions.
Acta Arith. **18** (1971), 243–256. MR **44,** 6626.
2. The distribution of polynomial sequences.
Mathematika **19** (1972), 25–50.
3. Brun's method and the Fundamental Lemma.
Proc. Sympos. Pure Math. **24** (1973), 247–249.
4. Brun's method and the Fundamental Lemma.
Acta Arith. **24** (1973), 113–133.
5. Brun's method and the Fundamental Lemma. II.
Acta Arith. (to appear).

HALBERSTAM, H. and ROTH, K. F.
1. Sequences.
Oxford, 1966, xx + 291pp. MR **35,** 1565.

HALBERSTAM, H. and ROTKIEWICZ, A.
1. A gap theorem for pseudoprimes in arithmetic progressions.
Acta Arith. **13** (1968), 395–404. MR **37,** 1329.

HALL, R. R.
1. The divisors of $p - 1$.
Mathematika **20** (1973), 87–97.

HANEKE, W.
1. Über eine Turánsche Ungleichung mit reellen Charakteren.
Acta Arith. **16** (1969/70), 315–326. MR **43,** 4776.

HARDY, G. H. and LITTLEWOOD, J. E.
1. Note on Messrs. Shah and Wilson's paper entitled: "On an empirical formula connected with Goldbach's Theorem".
Proc. Cambridge Philos. Soc. **19** (1919), 245–254.
2. Some problems of "partitio numerorum"; III: On the expression of a number as a sum of primes.
Acta Math. **44** (1923), 1–70.

HASELGROVE, C. B.
1. Some theorems in the analytic theory of numbers.
J. London Math. Soc. **26** (1951), 273–277. MR **13,** p.438.

HEILBRONN, H.
1. Über die Verteilung der Primzahlen in Polynomen.
Math. Ann. **104** (1931), 794–799.

HEILBRONN, H., LANDAU, E. and SCHERK, P.
1. Alle grossen ganzen Zahlen lassen sich als Summe von höchstens 71 Primzahlen darstellen.
Časopis Pěst. Mat. **65** (1936), 117–141.

HENSLEY, D. and RICHARDS, I.
1. On the incompatibility of two conjectures concerning primes.
Proc. Sympos. Pure Math. **24** (1973), 123–127.

HOOLEY, C.
1. On the representation of a number as the sum of two squares and a prime.
Acta Math. **97** (1957), 189–210. MR **19,** p.532.
2. On the representation of a number as the sum of a square and a product.
Math. Z. **69** (1958), 211–227. MR **20,** 3107.
3. On the difference of consecutive numbers prime to n.
Acta Arith. **8** (1962/63), 343–347. MR **27,** 5741.
4. On the representations of a number as the sum of two cubes.
Math. Z. **82** (1963), 259–266. MR **27,** 5742.
5. On the greatest prime factor of a quadratic polynomial.
Acta Math. **117** (1967), 281–299. MR **34,** 4225.
6. On the square-free values of cubic polynomials.
J. Reine Angew. Math. **229** (1968), 147–154. MR **36**, 3738.
7. On the power free values of cubic polynomials.
Mathematika **14** (1967), 21–26. MR **35,** 5405.
8. On the Brun–Titchmarsh theorem.
J. Reine Angew. Math. **255** (1972), 60–79. MR **46,** 3463.
9. On the Brun–Titchmarsh theorem. II.
Proc. London Math. Soc. (3) (to appear).
10. On the greatest prime factor of $p + a$.
Mathematika (to appear).

HSIEH, S.
1. Distribution of the triplet of almost prime numbers. (Chinese)
Shuxue Jinzhan **8** (1965), 71–77. MR **37,** 5170.
2. On the representation of a large even number as a sum of a prime and the product of at most three primes. (Chinese)
Shuxue Jinzhan **8** (1965), 209–216. MR **37,** 5171.

HUA, L.-K.
1. Estimation of an integral. (Chinese)
Chungkow Kao Hiao **2** (1951), 393–402.
2. Die Abschätzung von Exponentialsummen und ihre Anwendung in der Zahlentheorie.
Enzykl. math. Wiss., I, 2, Heft 13. Teil I. Leipzig, 1959. 123pp. MR **24,** A94.

HUXLEY, M. N.
1. On the differences of primes in arithmetical progressions.
Acta Arith. **15** (1968/69), 367–392. MR **39,** 5494.
2. The distribution of prime numbers.
Oxford, 1972, x + 128pp.

INDLEKOFER, K.-H. and SCHWARZ, W.
1. Über B-Zwillinge.
Arch. Math. **23** (1972), 251–256.

IWANIEC, H.

1. On the error term in the linear sieve.
Acta Arith. **19** (1971), 1–30. MR **45,** 5104.
2. Primes of the type $\phi(x, y) + A$ where ϕ is a quadratic form.
Acta Arith. **21** (1972), 203–234. MR **46,** 3466.

JAMES, R. D.

1. A problem in additive number theory.
Trans. Amer. Math. Soc. **43** (1938), 296–302.
2. Integers which are not represented by certain ternary quadratic forms.
Duke Math. J. **5** (1939), 948–962. MR **1,** p.200.
3. On the sieve method of Viggo Brun.
Bull. Amer. Math. Soc. **49** (1943), 422–432. MR **4,** p.265.
4. Recent progress in the Goldbach problem.
Bull. Amer. Math. Soc. **55** (1949), 246–260. MR **10,** pp.515, 856.

JURKAT, W. B. and RICHERT, H.-E.

1. An improvement of Selberg's sieve method. I.
Acta Arith. **11** (1965), 217–240. MR **34,** 2540.

JUTILA, M.

1. A statistical density theorem for *L*-functions with applications.
Acta Arith. **16** (1969), 207–216. MR **40,** 5557.

KARŠIEV, A. K.

1. The generalized problem of divisors of Titchmarsh. (Russian. Uzbek summary)
Izv. Akad. Nauk UzSSR Ser. Fiz.-Mat. Nauk **11** (1967), no. 6, 21–28. MR **36,** 6364.

KÁTAI, I.

1. On sets characterizing number-theoretical functions.
Acta Arith. **13** (1968), 315–320. MR **36,** 3742.
2. On the sum $\sum d(f(n))$.
Acta. Sci. Math. (*Szeged*) **29** (1968), 199–206. MR **39,** 5496.
3. On a classification of primes.
Acta Sci. Math. (*Szeged*) **29** (1968), 207–212. MR **39,** 5497.
4. On sets characterizing number-theoretical functions. II. The set of "prime plus one" 's is a set of quasi-uniqueness.
Acta Arith. **16** (1969/70), 1–4. MR **40,** 106.
5. On an algorithm for additive representation of integers by prime numbers.
Ann. Univ. Sci. Budapest Eötvös Sect. Math. **12** (1969), 23–27. MR **41,** 6806.
6. Some remarks on additive arithmetical functions. (Lithuanian and Russian summaries)
Litovsk. Mat. Sb. **9** (1969), 515–518. MR **42,** 1785.

KLIMOV, N. I.

1. Upper estimates of some number theoretical functions. (Russian)
Dokl. Akad. Nauk SSSR **111** (1956), 16–18. MR **19,** p.251.
2. Combination of elementary and analytic methods in the theory of numbers. (Russian)
Uspehi Mat. Nauk **13** (1958), no. 3 (81), 145–164. MR **20,** 3841.
3. Almost prime numbers. (Russian)
Uspehi Mat. Nauk **16** (1961), no. 3 (99), 181–188 = *Amer. Math. Soc. Transl.* (2) **46** (1965), 48–56. MR **23,** A2398.

4. Apropos the computations of Šnirel' man's constant. (Russian)
Volž. Mat. Sb. Vyp. **7** (1969), 32–40. MR **44,** 6633.

KLIMOV, N. I., PIL'TJAĬ, G. Z. and ŠEPTICKAJA, T. A.

1. The representation of natural numbers as sums of a bounded number of prime numbers. (Russian)
Kuĭbyšev Gos. Ped. Inst. Učen. Zap. Vyp. **1** (1971), 44–47.

KNÖDEL, W.

1. Ein Satz über Primzahlen.
Anz. Öster. Akad. Wiss. Math.-Nat. Kl. **86** (1949), 112–116. MR **12,** p.676.
2. Sätze über Primzahlen.
Monatsh. Math. **55** (1951), 62–75. MR **12,** p.676; MR **14,** p.1277.
3. Sätze über Primzahlen. II.
Monatsh. Math. **56** (1952), 137–143. MR **14,** p.355.
4. Primzahldifferenzen.
J. Reine Angew. Math. **195** (1955), 202–209. MR **17,** 1057; MR **18,** 1118.

KOBAYASHI, I.

1. A note on the Selberg sieve and the large sieve.
Proc. Japan Acad. **49** (1973), 1–5.

KONDAKOVA, L. F. and KLIMOV, N. I.

1. Certain additive problems. (Russian)
Volž. Mat. Sb. Vyp. **7** (1969), 41–44. (loose errata) MR **44,** 6634.

KUBILIUS, I. P.

1. Probabilistic methods in the theory of numbers. (Russian)
Uspehi Mat. Nauk **11** (1956), no. 2 (68), 31–66. MR **18,** p.17. = *Amer. Math. Soc. Transl.* (2) **19** (1962), 47–85. MR **24,** A1266.
2. Probabilistic methods in the theory of numbers.
Vilna, 1959, 164pp. MR **23,** A134; 2nd ed. Vilna, 1962, 221pp. MR **26,** 3691. = Providence, R.I., 1964, xviii + 182pp. MR **28,** 3956.

KUHN, P.

1. Zur Viggo Brun'schen Siebmethode. I.
Norske Vid. Selsk. Forh., Trondhjem **14** (1941), no. 39, 145–148. MR **8,** p.503.
2. Neue Abschätzungen auf Grund der Viggo Brunschen Siebmethode.
12. *Skand. Mat. Kongr., Lund,* **1953,** 160–168. MR **16,** p.676.
3. Über die Primteiler eines Polynoms.
Proc. Internat. Congress Math., Amsterdam, 1954, **2,** 35–37.

KUZJAŠEV, A. A. and ČEČURO, E. F.

1. The representation of large integers by sums of primes. (Russian)
Studies in Number Theory, No. 3 (Russian), pp.46–50. Izdat. Saratov. Univ., Saratov, 1969. MR **42,** 4517.

LANDAU, E.

1. Die Goldbachsche Vermutung und der Schnirelmannsche Satz.
Nachr. Akad. Wiss. Göttingen Math.-Phys. Kl. **1930,** 255–276.
2. Über einige neuere Fortschritte der additiven Zahlentheorie.
Cambridge Tract No. 35, London, 1937, 94pp.
3. Elementary number theory.
New York, 1958, 256pp. MR **19,** p.1159.

LAVRIK, A. F.

1. On the distribution of k-twin primes. (Russian)
Dokl. Akad. Nauk SSSR **132** (1960), 1258–1260 = *Soviet Math. Dokl.* **1** (1960), 764–766. MR **28,** 3017.
2. Binary case of an additive problem with squares of primes. (Russian)
Dokl. Akad. Nauk SSSR **140** (1961), 529–532 = *Soviet Math. Dokl.* **2**(1961), 1232–1236.
3. The theory of quasi-prime numbers. (Russian)
Dokl. Akad. Nauk SSSR **152** (1963), 544–547 = *Soviet Math. Dokl.* **4**(1963), 1355–1359. MR **27,** 4805.

LENSKOĬ, D. N.

1. On the representation of prime numbers by polynomials of two variables. (Russian. English summary)
Vestnik Leningrad. Univ. **18** (1963), no. 4, 150–154. MR **28,** 1163.
2. An upper bound for certain number-theoretic functions in algebraic number fields. (Russian)
Dokl. Akad. Nauk SSSR **150** (1963), 251–253 = *Soviet Math. Dokl.* **4** (1963), 641–643. MR **27,** 138.
3. On a certain generalization of twins. (Russian)
Volž. Mat. Sb. Vyp. **1** (1963), 114–118. MR **35,** 4178.
4. On the arithmetic properties of the values of polynomials. (Russian. English summary)
Vestnik Leningrad. Univ. **19** (1964), no. 2, 19–28. MR **30,** 75.
5. On the theory of prime numbers. (Russian)
Dokl. Akad. Nauk SSSR **169** (1966), 266–268 = *Soviet Math. Dokl.* **7** (1966), 904–906. MR **35,** 2847.

LE VEQUE, W. J.

1. On the size of certain number-theoretic functions.
Trans. Amer. Math. Soc. **66** (1949), 440–463. MR **11,** p.83.

LEVIN, B. V.

1. Estimates from below for the number of nearly-prime integers belonging to some general sequences. (Russian. English summary)
Vestnik Leningrad. Univ. **15** (1960), no. 7, 48–65. MR **22,** A7985.
2. The weak Landau problem and its generalization. (Russian)
Uspehi Mat. Nauk **16** (1961), no. 2 (98), 123–125. MR **26,** 3677.
3. On the sieve method. (Russian)
Taškent. Gos. Univ. Naučn. Trudy, Vyp. **189** (1961), 31–36.
4. Distribution of "near primes" in polynomial sequences. (Russian)
Dokl. Akad. Nauk UzSSR **1962,** no. 11, 7–9.
5. Distribution of "near primes" in polynomial sequences. (Russian)
Mat. Sbornik (N.S.) **61 (103)** (1963), 389–407. MR **30,** 1991.
6. On a class of problems in number theory reducing to differential equations with a retarded argument. (Russian)
Taškent. Gos. Univ. Naunč. Trudy, Vyp. **228** (1963), 56–68.
7. The estimation of special sums and products related to the sieve method. (Russian)
Taškent. Gos. Univ. Naučn. Trudy, Vyp. **228** (1963), 69–79.
8. On the least almost prime number in an arithmetic progression and the sequence $k^2x^2 + 1$. (Russian)
Uspehi Mat. Nauk **20** (1965), no. 4 (124), 158–162. MR **32,** 5612.

9. Comparison of A. Selberg's and V. Brun's sieves. (Russian)
Uspehi Mat. Nauk **20** (1965), no. 5 (125), 214–220. MR **32,** 7534.
10. A one-dimensional sieve. (Russian)
Acta Arith. **10** (1964/65), 387–397. MR **31,** 4774.

LEVIN, B. V. and FAĬNLEĬB, A. S.
1. The asymptotic behavior of sums of multiplicative functions. (Russian)
Dokl. Akad. Nauk UzSSR **1965,** no. 11, 5–8. MR **37,** 5172.
2. Application of certain integral equations to questions of the theory of numbers. (Russian)
Uspehi Mat. Nauk **22** (1967), no. 3 (135), 119–197 = *Russian Math. Surveys* **22** (1967), no. 3, 119–204. MR **37,** 5174.

LEVIN, B. V. and MAKSUDOV, I. G.
1. Distribution of almost prime numbers in polynomials in *n* variables. (Russian. Uzbek summary)
Izv. Akad. Nauk UzSSR Ser. Fiz.-Mat. Nauk **10** (1966), no. 3, 15–23. MR **33,** 7316.

LEVIN, B. V. and TULJAGANOVA, M. I.
1. The sieve of A. Selberg in algebraic number fields. (Russian. Lithuanian and English summaries)
Litovsk. Mat. Sb. **6** (1966), 59–73. MR **34,** 7492.

LEVINSON, N.
1. Summing certain number theoretic series arising in the sieve.
J. Math. Anal. Appl. **22** (1968), 631–645. MR **37,** 1338.
2. Correction and addendum to: "Summing certain number theoretic series arising in the sieve".
J. Math. Anal. Appl. **25** (1969), 710–716. MR **38,** 5739.

LINNIK, YU. V.
1. On certain results relating to positive ternary quadratic forms. (English. Russian summary)
Mat. Sbornik (N.S.) **5 (47)** (1939), 453–471. MR **2,** p.36.
2. A remark on products of three primes. (Russian)
Dokl. Akad. Nauk SSSR **72** (1950), 9–10. MR **11,** p.644.
3. The dispersion method in binary additive problems.
Leningrad, 1961, 208pp. MR **25,** 3920. = Providence, R.I., 1963, x + 186pp. MR **29,** 5804.

VAN LINT, J. H. and RICHERT, H.-E.
1. Über die Summe $\Sigma_{n \leqslant x,\, p(n) < y}\, \mu^2(n)/\phi(n)$.
Nederl. Akad. Wetensch. Proc. Ser. A **67** = *Indag. Math.* **26** (1964), 582–587. MR **30,** 1994.
2. On primes in arithmetic progressions.
Acta Arith. **11** (1965), 209–216. MR **32,** 5613.

MAKSUDOV, I. G.
1. Distribution of almost primes in certain sequences. (Russian. Uzbek summary)
Dokl. Akad. Nauk UzSSR **1966,** no. 7, 3–6. MR **41,** 8356.

MERLIN, J.
1. Sur quelques théorèmes d'Arithmétique et un énoncé qui les contient.
C. R. Acad. Sci. Paris **153** (1911), 516–518.

2. Un travail de Jean Merlin sur les nombres premiers.
Bull. Sci. Math. (2) **39** (1915), 121–136.

MIECH, R. J.

1. Almost primes generated by a polynomial.
Acta Arith. **10** (1964/65), 9–30. MR **29,** 1174.
2. Primes, polynomials and almost primes.
Acta Arith. **11** (1965), 35–56. MR **31,** 3390.
3. A uniform result on almost primes.
Acta Arith. **11** (1966), 371–391. MR **34,** 1287.

MIENTKA, W. E.

1. An application of the Selberg sieve method.
J. Indian Math. Soc. (N.S.) **25** (1961), 129–138. MR **27,** 2490.

MONTGOMERY, H. L.

1. A note on the large sieve.
J. London Math. Soc. **43** (1968), 93–98. MR **37,** 184.
2. Topics in multiplicative number theory.
Lecture Notes in Math. **227** (1971), Berlin and New York.

MONTGOMERY, H. L. and VAUGHAN, R. C.

1. On the large sieve.
Mathematika **20** (1973), 119–134.
2. Hilbert's inequality.
J. London Math. Soc. (2) **8** (1974), 73–82.

MOTOHASHI, Y.

1. On the distribution of prime numbers which are of the form $x^2 + y^2 + 1$.
Acta Arith. **16** (1969/70), 351–363. MR **44,** 5284.
2. A note on the least prime in an arithmetic progression with a prime difference.
Acta Arith. **17** (1970), 283–285. MR **42,** 3030.
3. On the distribution of prime numbers which are of the form "$x^2 + y^2 + 1$". II.
Acta Math. Acad. Sci. Hungar. **22** (1971), 207–210. MR **44,** 5285.
4. An improvement of the Brun–Titchmarsh theorem.
Nagoya Math. J. (to appear)
5. On some improvements of the Brun–Titchmarsh theorem.
J. Math. Soc. Japan **26** (1974), 306–323.

NAGEL, T.

1. Généralisation d'un théorème de Tchebycheff.
J. Math. Pures Appl. (8) **4** (1921), 343–356.

NORTON, K. K.

1. Numbers with small prime factors, and the least kth power non-residue.
Memoirs of the American Mathematical Society, No. 106, 1971, ii + 106pp.
MR **44,** 3948.
2. Bounds for sequences of consecutive power residues. I.
Proc. Sympos. Pure Math. **24** (1973), 213–220.

ODLYZKO, A. M.

1. Sieve methods.
Senior thesis, California Institute of Technology, Pasadena, 1971, ii + 146pp.

ONISHI, H.

1. A Tauberian theorem on Dirichlet series.
J. Number Theory **5** (1973), 55–57. MR **47,** 4956.

OŽIGOVA, E. P.

1. Modification of the method of the "sieve of Eratosthenes" given by A. Selberg. (Russian)
   Uspehi Mat. Nauk **8** (1953), no. 3 (55), 119–124. MR **15,** p.202.

PAN, C.

1. On the representation of an even integer as the sum of a prime and an almost prime. (Chinese)
   Acta Math. Sinica **12** (1962), 95–106 = *Chinese Math.-Acta* **3** (1963), 101–112. MR **29,** 4727. = *Sci. Sinica* **11** (1962), 873–888. (Russian) MR **27,** 1427.
2. On the representation of even numbers as the sum of a prime and a product of not more than 4 primes. (Chinese)
   Acta Sci. Natur. Univ. Shangtung **1962,** no. 2, 40–62 = *Sci. Sinica* **12** (1963), 455–473. (Russian) MR **28,** 73.
3. A new application of the Yu. V. Linnik large sieve method. (Chinese)
   Acta Math. Sinica **14** (1964), 597–606 = *Chinese Math.-Acta* **5** (1964), 642–652. MR **30,** 3871. = *Sci. Sinica* **13** (1964), 1045–1053. (Russian) MR **30,** 3872.

PORTER, J. W.

1. The generalized Titchmarsh–Linnik divisor problem.
   Proc. London Math. Soc. (3) **24** (1972), 15–26. MR **45,** 5098.
2. Some numerical results in the Selberg sieve method.
   Acta Arith. **20** (1972), 417–421. MR **46,** 5268.

PRACHAR, K.

1. Über Primzahldifferenzen.
   Monatsh. Math. **56** (1952), 304–306. MR **14,** p.727.
2. Über Primzahldifferenzen. II.
   Monatsh. Math. **56** (1952), 307–312. MR **14,** p.727.
3. Über ein Resultat von A. Walfisz.
   Monatsh. Math. **58** (1954), 114–116. MR **16,** p.114.
4. Bemerkung zu einer Arbeit von Erdös und Rényi und Berichtigung.
   Monatsh. Math. **58** (1954), 117. MR **16,** p.221.
5. Über die Lösungszahl eines Systems von Gleichungen in Primzahlen.
   Monatsh. Math. **59** (1955), 98–103. MR **17,** p.14.
6. Primzahlverteilung.
   Berlin–Göttingen–Heidelberg, 1957, x + 415pp. MR **19,** p.393.
7. Über die kleinste Primzahl einer arithmetischen Reihe.
   J. Reine Angew. Math. **206** (1961), 3–4. MR **23,** A2399.
8. Über die Differenzen aufeinanderfolgender Primzahlen.
   Bull. Soc. Math. Phys. Serbie **14** (1962), 165–168. MR **32,** 1177.
9. Sätze über quadratfreie Zahlen.
   Monatsh. Math. **66** (1962), 306–312. MR **25,** 3887.
10. Über Zahlen, die sich als Summe einer Primzahl und einer "kleinen" Potenz darstellen lassen.
    Monatsh. Math. **68** (1964), 409–420. MR **30,** 1097.

RADEMACHER, H.

1. Beiträge zur Viggo Brunschen Methode in der Zahlentheorie.
   Abh. Math. Sem. Univ. Hamburg **3** (1924), 12–30.
2. Über die Anwendung der Viggo Brunschen Methode auf die Theorie der algebraischen Zahlkörper.
   S.-B. Preuss. Akad. Wiss. **1923,** 211–218.

3. Lectures on elementary number theory.
New York–Toronto–London, 1964, ix + 146pp. MR **30,** 1079.

RAMACHANDRA, K.
1. A note on numbers with a large prime factor.
J. London Math. Soc. (2) **1** (1969), 303–306. MR **40,** 118.
2. A note on numbers with a large prime factor. II.
J. Indian Math. Soc. (*N.S.*) **34** (1970), 39–48. MR **45,** 8616.
3. A note on numbers with a large prime factor. III.
Acta Arith. **19** (1971), 49–62. MR **45,** 8617.

RANKIN, R. A.
1. The difference between consecutive prime numbers.
J. London Math. Soc. **13** (1938), 242–247.
2. The difference between consecutive prime numbers. II.
Proc. Cambridge Philos. Soc. **36** (1940), 255–266. MR **1,** p.292.
3. The difference between consecutive prime numbers. III.
J. London Math. Soc. **22** (1947), 226–230. MR **9,** p.498.
4. The difference between consecutive prime numbers. IV.
Proc. Amer. Math. Soc. **1** (1950), 143–150. MR **11,** p.644.

RÉNYI, A.
1. On the representation of an even number as the sum of a single prime and a single almost-prime number. (Russian)
Dokl. Akad. Nauk SSSR **56** (1947), 455–458. MR **9,** p.136.
2. On the representation of an even number as the sum of a single prime and a single almost-prime number. (Russian)
Izv. Akad. Nauk SSSR. Ser. Mat. **12** (1948), 57–78. MR **9,** p.413. = *Amer. Math. Soc. Transl.* (2) **19** (1962), 299–321. MR **24,** A1264.

RICCI, G.
1. Sui grandi divisori primi delle coppie di interi in posti corrispondenti di due progressioni aritmetiche. Applicazione del metodo di Brun.
Ann. Mat. Pura Appl. (4) **11** (1933), 91–110.
2. Ricerche aritmetiche sui polinomi.
Rend. Circ. Mat. Palermo **57** (1933), 433–475.
3. Ricerche aritmetiche sui polinomi. II.
Rend. Circ. Mat. Palermo **58** (1934), 190–208.
4. Sull'aritmetica additiva degl'interi liberi da potenze.
Tôhoku Math. J. **41** (1935), 20–26.
5. Su la congettura di Goldbach e la costante di Schnirelmann.
Boll. Un. Mat. Ital. **15** (1936), 183–187.
6. Su la congettura di Goldbach e la costante di Schnirelmann. I.
Ann. Scuola Norm. Sup. Pisa (2) **6** (1937), 71–90.
7. Su la congettura di Goldbach e la costante di Schnirelmann. II.
Ann. Scuola Norm. Sup. Pisa (2) **6** (1937), 91–116.
8. Recenti risultati nel campo dell'Aritmetica. Il problema di Goldbach.
Rend. Sem. Mat. Fis. Milano **13** (1939), 204–226.
9. La differenza di numeri primi consecutivi.
Univ. e Politec. Torino. Rend. Sem. Mat. **11** (1952), 149–200. MR **14,** p.727.
10. Sul coefficiente di Viggo Brun.
Ann. Scuola Norm. Sup. Pisa (3) **7** (1953), 133–151. MR **15,** p.202.

11. Sull'andamento della differenza di numeri primi consecutivi.
Riv. Mat. Univ. Parma **5** (1954), 3–54. MR **16,** p.675.
12. Recherches sur l'allure de la suite $(p_{n+1} - p_n)/\log p_n$.
Colloque sur la Théorie des Nombres, Bruxelles, **1955,** 93–106. MR **18,** p.112.

RICHERT, H.-E.
1. Selberg's sieve with weights.
Mathematatika **16** (1969), 1–22. MR **40,** 119.
2. Selberg's sieve with weights.
Symposia Mathematica **4** (INDAM, Rome, 1968/69), 73–80. MR **45,** 1873.
3. Selberg's sieve with weights.
Proc. Sympos. Pure Math. **20** (1971), 287–310. MR **47,** 3286.

RIEGER, G. J.
1. Verallgemeinerung der Siebmethode von A. Selberg auf algebraische Zahlkörper. I.
J. Reine Angew. Math. **199** (1958), 208–214. MR **20,** 3115.
2. Verallgemeinerung der Siebmethode von A. Selberg auf algebraische Zahlkörper. II.
J. Reine Angew. Math. **201** (1959), 157–171. MR **24,** A1903.
3. Verallgemeinerung der Siebmethode von A. Selberg auf algebraische Zahlkörper. III.
J. Reine Angew. Math. **208** (1961), 79–90. MR **28,** 3024.
4. On the prime ideals of smallest norm in an ideal class mod f of an algebraic number field.
Bull. Amer. Math. Soc. **67** (1961), 314–315. MR **23,** A2412.
5. Über ein lineares Gleichungssystem von Prachar mit Primzahlen.
J. Reine Angew. Math. **213** (1963), 103–107. MR **29,** 85.
6. Über die Differenzen von drei aufeinanderfolgenden Primzahlen.
Math. Z. **82** (1963), 59–62. MR **27,** 3594.
7. On linked binary representations of pairs of integers: some theorems of the Romanov type.
Bull. Amer. Math. Soc. **69** (1963), 558–563. MR **27,** 3593.
8. Über die Summe beliebiger und die Differenz aufeinanderfolgender Primzahlen.
Elem. Math. **18** (1963), 104–105. MR **27,** 5739.
9. Über die Folge der Zahlen der Gestalt $p_1 + p_2$.
Arch. Math. **15** (1964), 33–41. MR **28,** 3023.
10. Anwendung der Siebmethode auf einige Fragen der additiven Zahlentheorie. I.
J. Reine Angew. Math. **214/215** (1964), 373–385. MR **29,** 86.
11. Anwendung der Siebmethode auf einige Fragen der additiven Zahlentheorie. II.
Math. Nachr. **28** (1964/65), 207–217. MR **31,** 131.
12. Über die Anzahl der als Summe von zwei Quadraten darstellbaren und in einer primen Restklasse gelegenen Zahlen unterhalb einer positiven Schranke. I.
Arch. Math. **15** (1964), 430–434. MR **30,** 1111.
13. Aufeinanderfolgende Zahlen als Summen von zwei Quadraten.
Nederl. Akad. Wetensch. Proc. Ser. A **68** = *Indag. Math.* **27** (1965), 208–220. MR **31,** 147.
14. Über p_k/k und verwandte Folgen.
J. Reine Angew. Math. **221** (1966), 14–19. MR **32,** 1178.
15. Über die natürlichen und primen Zahlen der Gestalt $[n^c]$ in arithmetischer Progression.
Arch. Math. **18** (1967), 35–44. MR **34,** 5793.

16. Über die Summe aus einem Quadrat und einem Primzahlquadrat.
J. Reine Angew. Math. **231** (1968), 89–100. MR **37,** 5177.
17. On polynomials and almost-primes.
Bull. Amer. Math. Soc. **75** (1969), 100–103. MR **38,** 2104.
18. Über ein additives Problem mit Primzahlen.
Arch. Math. **21** (1970), 54–58. MR **41,** 3421.

RODRIQUEZ, G.
1. Sul problema dei divisori di Titchmarsh. (English summary)
Boll. Un. Mat. Ital. (3) **20** (1965), 358–366. MR **33,** 5574.

ROGERS, K.
1. A note on orthogonal Latin squares.
Pacific J. Math. **14** (1964), 1395–1397. MR **33,** 5501.

ROMANOFF, N. P.
1. Über einige Sätze der additiven Zahlentheorie.
Math. Ann. **109** (1934), 668–678.

ROSSER, J. B. and SCHOENFELD, L.
1. Approximate formulas for some functions of prime numbers.
Illinois J. Math. **6** (1962), 64–94. MR **25,** 1139.

ROUX, D.
1. Sulla distribuzione degli interi rappresentabili come somma di due quadrati.
Ist. Lombardo Sci. Lett. Rend. Cl. Sci. Mat. Nat. (3) **21 (90)** (1956), 137–140. MR **18,** p.18.

RUSSELL, P. B.
1. Primes generated by polynomials in one variable.
Office of Naval Research, Project No. 043–194, 1964, iv + 161pp.

SARGES, H.
1. Eine Anwendung des Selbergschen Siebes auf algebraische Zahlkörper.
Dissertation, Marburg, 1973, 40pp.

SCHAAL, W.
1. Obere und untere Abschätzungen in algebraischen Zahlkörpern mit Hilfe des linearen Selbergschen Siebes.
Acta Arith. **13** (1968), 267–313. MR **36,** 5099.

SCHINZEL, A.
1. Remarks on the paper "Sur certaines hypothèses concernant les nombres premiers".
Acta Arith. **7** (1961/62), 1–8. MR **24,** A70.
2. A remark on a paper of Bateman and Horn.
Math. Comp. **17** (1963), 445–447. MR **27,** 3609.
3. On sums of roots of unity. Solution of two problems of R. M. Robinson.
Acta Arith. **11** (1966), 419–432. MR **34,** 1302.

SCHINZEL, A. and SIERPIŃSKI, W.
1. Sur certaines hypothèses concernant les nombres premiers.
Acta Arith. **4** (1958), 185–208; Corrigendum: *ibid.* **5** (1959), 259. MR **21,** 4936.

SCHINZEL, A. and WANG, Y.
1. A note on some properties of the functions $\phi(n)$, $\sigma(n)$ and $\theta(n)$.
Ann. Polon. Math. **4** (1958), 201–213. MR **20,** 1655; Corrigendum: *ibid.* **19** (1967), 115. MR **35,** 142.

SCHNIRELMANN, L.
1. Über additive Eigenschaften von Zahlen.
Math. Ann. **107** (1933), 649–690.

SCHÖNHAGE, A.
1. Eine Bemerkung zur Konstruktion grosser Primzahllücken.
Arch. Math. **14** (1963), 29–30. MR **26,** 3680.

SCHWARZ, W.
1. Weitere, mit einer Methode von Erdös–Prachar erzielte Ergebnisse.
Math. Nachr. **23** (1961), 327–348. MR **25,** 3004.
2. Über die Summe $\Sigma_{n \leqslant x} \phi(f(n))$ und verwandte Probleme.
Monatsh. Math. **66** (1962), 43–54. MR **25,** 2052.
3. Weitere, mit einer Methode von Erdös–Prachar erzielte Ergebnisse. II.
Monatsh. Math. **68** (1964), 75–80. MR **28,** 1182.
4. Berichtigung zu der Arbeit "Weitere, mit einer Methode von Erdös–Prachar erzielte Ergebnisse".
Math. Nachr. **52** (1972), 385. MR **47,** 165.

SEGAL, B.
1. Généralisation du théorème de Brun.
Dokl. Akad. Nauk SSSR **1930,** 501–507.

SELBERG, A.
1. On an elementary method in the theory of primes.
Norske Vid. Selsk. Forh., Trondhjem **19** (1947), no. 18, 64–67. MR **9,** p.271.
2. Twin prime problem.
Manuscript 16pp. + Appendix 7pp. (unpublished)
3. On elementary methods in prime number-theory and their limitations.
11. *Skand. Mat. Kongr., Trondheim* **1949,** 13–22. MR **14,** p.726.
4. The general sieve method and its place in prime number theory.
Proc. Internat. Congress Math., Cambridge, Mass. **1** (1950), 286–292. MR **13,** p.438.
5. Sieve methods.
Proc. Sympos. Pure Math. **20** (1971), 311–351. MR **47,** 3286.

SELBERG, S.
1. The number of cancelled elements in the sieve of Eratosthenes.
Norsk Mat. Tidsskr. **26** (1944), 79–84. MR **8,** p.317.
2. An upper bound for the number of cancelled numbers in the sieve of Eratosthenes. (Norwegian)
Norske Vid. Selsk. Forh., Trondhjem **19** (1946), no. 2, 3–6. MR **9,** p.332.
3. A generalisation of a theorem of Romanoff.
Norske Vid. Selsk. Forh., Trondhjem **35** (1962), 91–95. MR **26,** 2427.

SHAH, N. M. and WILSON, B. M.
1. On an empirical formula connected with Goldbach's Theorem.
Proc. Cambridge Philos. Soc. **19** (1919), 238–244.

SHAPIRO, H. N. and WARGA, J.
1. On the representation of large integers as sums of primes. I.
Comm. Pure Appl. Math. **3** (1950), 153–176. MR **12,** p.244.

SIEBERT, H.
1. Darstellung als Summe von Primzahlen.
Diplomarbeit, Marburg, 1968, 55pp.

2. Über die Verteilung der Werte von $\mu(n)$ und $\lambda(n)$ in arithmetischen Progressionen mod p.
Dissertation, Marburg, 1969, 40pp.
3. On a question of P. Turán.
Acta Arith. (to appear)

SIEBERT, H. and WOLKE, D.
1. Über einige Analoga zum Bombierischen Primzahlsatz.
Math. Z. **122** (1971), 327–341.

STEIN, M. L. and STEIN, P. R.
1. New experimental results on the Goldbach conjecture.
Math. Mag. **38** (1965), 72–80. MR **32,** 4109.

TARTAKOVSKIĬ, V. A.
1. Sur quelques sommes du type de Viggo Brun.
Dokl. Akad. Nauk SSSR **23** (1939), 121–125.
2. La méthode du crible approximatif "électif".
Dokl. Akad. Nauk SSSR **23** (1939), 126–129.

TATUZAWA, T.
1. Additive prime number theory in an algebraic number field.
J. Math. Soc. Japan **7** (1955), 409–423. MR **18,** p.113.

TITCHMARSH, E. C.
1. A divisor problem.
Rend. Circ. Mat. Palermo **54** (1930), 414–429.

TOGASHI, A. and UCHIYAMA, S.
1. On the representation of large even integers as sums of two almost primes. I.
J. Fac. Sci. Hokkaido Univ. Ser. I **18** (1964), 60–68. MR **29,** 3420.

TROST, E.
1. Primzahlen.
Basel–Stuttgart, 1953, 95pp. MR **15,** p.401; 2nd ed. Basel–Stuttgart, 1968, 100pp. MR **41,** 160.

TULJAGANOVA, M. I.
1. The Euler–Goldbach problem for an imaginary quadratic field. (Russian. Uzbek summary)
Izv. Akad. Nauk UzSSR Ser. Fiz.-Mat. Nauk **1963,** no. 1, 11–17. MR **31,** 2229.

TURÁN, P.
1. Über die Siegel–Nullstelle der Dirichletschen Funktionen.
Acta Arith. **24** (1973), 135–141.

UCHIYAMA, M. and UCHIYAMA, S.
1. On the representation of large even integers as sums of a prime and an almost prime.
Proc. Japan Acad. **40** (1964), 150–154. MR **29,** 2234.

UCHIYAMA, S.
1. A note on the sieve method of A. Selberg.
J. Fac. Sci. Hokkaido Univ. Ser. I **16** (1962), 189–192. MR **27,** 1432.
2. A further note on the sieve method of A. Selberg.
J. Fac. Sci. Hokkaido Univ. Ser. I **17** (1963), 79–83. MR **29,** 1196.
3. On a theorem concerning the distribution of almost primes.
J. Fac. Sci. Hokkaido Univ. Ser. I **17** (1963), 152–159. MR **28,** 2101.

4. On the distribution of almost primes in an arithmetic progression.
J. Fac. Sci. Hokkaido Univ. Ser. I **18** (1964), 1–22. MR **30,** 1108.
5. On the representation of large even integers as sums of two almost primes. II.
J. Fac. Sci. Hokkaido Univ. Ser. I **18** (1964), 69–77. MR **29,** 3421.
6. On a differential-difference equation.
J. Fac. Sci. Hokkaido Univ. Ser. I **19** (1966), 59–65. MR **34,** 4736.
7. On the representation of large even integers as sums of a prime and an almost prime. II.
Proc. Japan Acad. **43** (1967), 567–571. MR **37,** 179.

UŽDAVINIS, R. V.
1. On the distribution of the values of additive number-theoretic functions of integral polynomials. (Russian)
Trudy Akad. Nauk Litov. SSR Ser. B **1959,** no. 2 (18), 9–29.
2. Some limit theorems for additive arithmetic functions. (Russian. Lithuanian and German summaries)
Litovsk. Mat. Sb. **1** (1961), no. 1–2, 355–364. MR **26,** 2390.
3. Arithmetic functions on the sets of values of integer-valued polynomials. (Russian. Lithuanian and German summaries)
Litovsk. Mat. Sb. **2** (1962), no. 2, 253–280. MR **28,** 1164.

VAUGHAN, R. C.
1. On a problem of Erdös, Straus and Schinzel.
Mathematika **17** (1970), 193–198. MR **44,** 6600.
2. A remark on the divisor function $d(n)$.
Glasgow Math. J. **14** (1973), 54–55. MR **47,** 169.

VINOGRADOV, A. I.
1. On the connections between the sieve of Eratosthenes and the Riemann ζ-function. (Russian)
Vestnik Leningrad Univ. **11** (1956), no. 13, 142–146. MR **20,** 3835.
2. Application of $\zeta(s)$ to the sieve of Eratosthenes. (Russian)
Mat. Sbornik (N.S.) **41 (83)** (1957), 49–80; Corrigendum: *ibid.* 415–416. MR **20,** 3836. = *Amer. Math. Soc. Transl.* (2) **13** (1960), 29–60. MR **22,** A3720.
3. Estimates for binary problems. (Russian. English summary)
Vestnik Leningrad. Univ. **14** (1959), no. 7, 26–31. MR **21,** 1957.
4. Generalization of a lemma of Erdös. (Russian. English summary)
Vestnik Leningrad. Univ. **15** (1960), no. 19, 124–126. MR **23,** A2404.
5. Estimates from below by the sieve process in algebraic number fields. (Russian)
Dokl. Akad. Nauk SSSR **154** (1964), 13–15 = *Soviet Math. Dokl.* **5** (1964), 6–9. MR **28,** 3025.
6. The sieve method in algebraic fields. Lower bounds. (Russian)
Mat. Sbornik (N.S.) **64 (106)** (1964), 52–78. MR **29,** 1195.
7. The density hypothesis for Dirichlet L-series. (Russian)
Izv. Akad. Nauk SSSR Ser. Mat. **29** (1965), 903–934. MR **33,** 5579; Corrigendum: *ibid.* **30** (1966), 719–720. MR **33,** 2607.

VINOGRADOV, A. I. and LINNIK, YU. V.
1. Hyperelliptic curves and the least prime quadratic residue. (Russian)
Dokl. Akad. Nauk SSSR **168** (1966), 259–261 = *Soviet Math. Dokl.* **7** (1966), 612–614. MR **35,** 125.

VIOLA, C.

1. On the diophantine equations $\prod_0^k x_i - \sum_0^k x_i = n$ and $\sum_0^k \frac{1}{x_i} = \frac{a}{n}$.
   Acta Arith. **22** (1973), 339–352.

WANG, Y.

1. On the representation of large even integer as a sum of a product of at most three primes and a product of at most four primes. (Chinese. English summary)
   Acta Math. Sinica **6** (1956), 500–513.
2. On the representation of large even integer as a sum of a prime and a product of at most 4 primes. (Chinese. English summary)
   Acta Math. Sinica **6** (1956), 565–582. MR **20,** 4530.
3. On some properties of integral valued polynomials. (Chinese. English summary)
   Advancement in Math. **3** (1957), 416–423. MR **20,** 4531.
4. On sieve methods and some of the related problems.
   Sci. Record (*N.S.*) **1** (1957), no. 1, 9–12. MR **23,** A2400.
5. On sieve methods and some of their applications.
   Sci. Record (*N.S.*) **1** (1957), no. 3, 1–5. MR **19,** p.533.
6. On the representation of large even number as a sum of two almost-primes.
   Sci. Record (*N.S.*) **1** (1957), no. 5, 15–19. MR **23,** A874.
7. A note on some properties of the arithmetical functions $\phi(n)$, $\sigma(n)$ and $d(n)$. (Chinese. English summary)
   Acta Math. Sinica **8** (1958), 1–11 = *Chinese Math.-Acta* **8** (1966), 585–598 = *Amer. Math. Soc. Transl.* (2) **37** (1964), 143–156. MR **20,** 4533.
8. On sieve methods and some of their applications. I. (Chinese. English summary)
   Acta Math. Sinica **8** (1958), 413–429. MR **21,** 1958. = *Sci. Sinica* **8** (1959), 357–381. MR **21,** 4944.
9. On sieve methods and some of their applications. (Chinese. English summary)
   Acta Math. Sinica **9** (1959), 87–100. MR **21,** 5615. = *Sci. Sinica* **11** (1962), 1607–1624. MR **26,** 3685.
10. On the least primitive root of a prime. (Chinese)
    Acta Math. Sinica **9** (1959), 432–441. MR **22,** A4659. = *Sci. Sinica* **10** (1961), 1–14. MR **24,** A702.
11. On the representation of large integer as a sum of a prime and an almost prime. (Chinese. English summary)
    Acta Math. Sinica **10** (1960), 168–181 = *Chinese Math.-Acta* **1** (1962), 181–195. MR **27,** 5733. = *Sci. Sinica* **11** (1962), 1033–1054. MR **27,** 1424.
12. A note on the maximal number of pairwise orthogonal Latin squares of a given order.
    Sci. Sinica **13** (1964), 841–843. MR **30,** 3866.
13. On the maximal number of pairwise orthogonal Latin squares of order s; an application of the sieve method. (Chinese)
    Acta Math. Sinica **16** (1966), 400–410 = *Chinese Math.-Acta* **8** (1966), 422–432. MR **34,** 2483.

WANG, Y., HSIEH, S. and YU, K.

1. Remarks on the difference of consecutive primes.
   Sci. Sinica **14** (1965), 786–788. MR **32,** 5617.

WARLIMONT, R.

1. Eine Bemerkung zu einem Ergebnis von N. G. de Bruijn.
Monatsh. Math. **74** (1970), 273–276. MR **42,** 225.

WEBB, W. A.

1. On $4/n = 1/x + 1/y + 1/z$.
Proc. Amer. Math. Soc. **25** (1970), 578–584. MR **41,** 1639.

WILSON, R. J.

1. The Selberg sieve for a lattice.
Combinatorial Theory and its Applications **3** (Proc. Colloq., Balatonfüred 1969), 1141–1149. MR **47,** 167.

WIRSING, E.

1. Über die Zahlen, deren Primteiler einer gegebenen Menge angehören.
Arch. Math. **7** (1956), 263–272. MR **18,** p.642.
2. Das asymptotische Verhalten von Summen über multiplikative Funktionen.
Math. Ann. **143** (1961), 75–102. MR **24,** A1241.

WOLKE, D.

1. Über die mittlere Verteilung der Werte zahlentheoretischer Funktionen auf Restklassen. I.
Math. Ann. **202** (1973), 1–25.
2. Über die mittlere Verteilung der Werte zahlentheoretischer Funktionen auf Restklassen. II.
Math. Ann. **204** (1973), 145–153.

YIN, W.

1. Note of the representation of large integers as sums of primes.
Bull. Acad. Polon. Sci. Cl. III. **4** (1956), 793–795. MR **19,** p.16.
2. Remarks on the representation of large integers as sums of primes. (Chinese. English summary)
Acta Sci. Nat. Univ. Pekinensis **1956,** no. 3, 323–326. MR **22,** 7990.

ZAIKINA, N. G.

1. Distribution of the numbers of an imaginary quadratic field composed of small prime ideals. (Russian)
Moskov. Gos. Ped. Inst. Učen. Zap. **108** (1957), 261–272. MR **22,** A12101.